Watershed Management and Applications of AI

Artificial Intelligence (AI) in Engineering
Series Editors: Kaushik Kumar and J. Paulo Davim

This new series will target and provide a collection of textbooks and research books on Artificial Intelligence (AI) applied in engineering disciplines. The series of books will provide an understanding of AI using common language and incorporate tools and applications to assistant in the learning process. They will focus on areas such as Artificial Intelligence and Philosophy, Applications of AI in Mechatronics, AI in Automation, AI in Manufacturing, AI and Industry 4.0, Cognitive Aspects of AI, Intelligent Robotics, Smart Robots, COBOTS, Machine Learning, Conscious Computers, Intelligent Machines, to name just a few areas that the series will have books in.

Nature-Inspired Optimization in Advanced Manufacturing Processes and Systems
Edited by Ganesh M. Kakandikar and Dinesh G. Thakur

Watershed Management and Applications of AI
Sandeep Samantaray, Abinash Sahoo, and Dillip K. Ghose

Artificial Intelligence in Mechanical and Industrial Engineering
Edited by Kaushik Kumar, Divya Zindani, and J. Paulo Davim

For more information on this series, please visit: www.routledge.com/Artificial-Intelligence-AI-in-Engineering/book-series/CRCAIIE

Watershed Management and Applications of AI

Sandeep Samantaray, Abinash Sahoo, and Dillip K. Ghose

CRC Press
Taylor & Francis Group
Boca Raton London New York

CRC Press is an imprint of the
Taylor & Francis Group, an **informa** business

CRC Press
Boca Raton and London
First edition published 2021
by CRC Press
6000 Broken Sound Parkway NW, Suite 300, Boca Raton, FL 33487-2742
and by CRC Press

2 Park Square, Milton Park, Abingdon, Oxon, OX14 4RN

Library of Congress Cataloging-in-Publication Data

Names: Samantaray, Sandeep, author. | Sahoo, Abinash, author. | Ghose, Dillip K., author. Title: Watershed management and applications of AI / Sandeep Samantaray, Abinash Sahoo, and Dillip K. Ghose. Description: First edition. | Boca Raton, FL : CRC Press, 2021. | Series: Artificial intelligence (AI) in engineering | Includes index. Identifiers: LCCN 2020052949 (print) | LCCN 2020052950 (ebook) | ISBN 9780367763060 (hardback) | ISBN 9781003168041 (ebook) Subjects: LCSH: Watershed management—Data processing. | Artificial intelligence. Classification: LCC TC413 .S26 2021 (print) | LCC TC413 (ebook) | DDC 627/.5—dc23 LC record available at https://lccn.loc.gov/2020052949LC ebook record available at https://lccn.loc.gov/2020052950

ISBN: 978-0-367-76306-0 (hbk)
ISBN: 978-0-367-76675-7 (pbk)
ISBN: 978-1-003-16804-1 (ebk)

Typeset in Times LT Std
by KnowledgeWorks Global Ltd.

Contents

Preface

Land use and water resource are two major components which necessitate conservation, management, and maintenance practices through different techniques of engineering uses. Complete exposure of principles of hydrology and their useful practices are helpful for planning, design, and management of watershed. In this book, an attempt has been made to provide an in-depth study on surface water, soil conservation, flood control, and groundwater management of watersheds.

The topics in the book are presented with a view of maintaining clarity related to watershed and are arranged in sequence permitting proper understanding of the students. Chapter 1 introduces aspects of hydrologic cycle, sociological impact on hydrologic cycle, and prospective data base for hydrology. Chapter 2 describes watershed characteristics. Chapter 3 provides an idea about erosion and soil conservation measurements. Chapters 4 and 5 deal with water quantity and water quality. Chapter 6 constitutes hydrology of groundwater. Chapter 7 focuses on aspects related to estimating flood and drought conditions. Chapter 8 defines idea of sampling and mechanics of sediment transport. Chapter 9 provides concepts of measuring runoff in watershed. Chapters 10–12 include application of machine learning techniques for prediction of groundwater fluctuation, flood, and sediment measurement in river basin. Chapters 13 and 14 schemed with runoff and drought in watershed using neural networks.

Numerous sources have been referred to for providing relevant and comprehensive information in the book and the authors acknowledge them. We are grateful to the reviewers, research scholars, colleagues, and other resource providers for their help and constructive suggestion towards the development of the book. In particular, we are extremely thankful to Miss Ankita Agnihotri, Mr. Rashmi Ranjan Samal, and Miss Geetanjali Nayak for preparation of the illustrations.

Authors

Dr. Sandeep Samantaray is a PhD research scholar in the National Institute of Technology, Silchar, Assam, India. He has to his credit research publications and presentations on topics such as watershed management, hydrological forecasting, and hydrologic modeling and computing in developing sustainable means of managing the environment. He has to his credit 16 research papers published in reputed journals and 5 book chapters. He has attended 25 international conferences (SCOPUS Indexing). He also has one patent granted and one published.

Dr. Abinash Sahoo is a PhD research scholar in the National Institute of Technology, Silchar, Assam, India. He has published 9 research papers in reputed journals and presented 17 papers in international conferences. He also has one patent published.

Dr. Dillip Kumar Ghose is currently working as a faculty member in the National Institute of Technology, Silchar, Assam, India. He has to his credit 14 research articles published in reputed journals and has presented 25 papers in international conferences. He has filed three patents: one granted and two published. Several consultancy and research projects associated with water resources engineering, bridge design, etc. have been accomplished by him.

1 Introduction to Watershed Management

1.1 INTRODUCTION

Watershed comprises whole land area which shelters water and distributes it to the outlet during rainstorm, and aids in bifurcation of water from rivers and reservoirs (Ojha et al., 2017; Black, 1996). Water is circulated to the exit through small basins or watersheds. To demarcate the boundary of watershed, extreme points are to be decided in a region for contribution of water. Watershed boundary is to be delineated by drawing lines perpendicular to elevated contour land to the individual drainage points of the watershed. Drainage area is a significant parameter for hydrologic design of watershed. Linear measurements are to be made for depicting the size of the watershed. Line which divides the surface runoff between two catchments is termed as topographic water divide or watershed divide. Understanding the components, processes, and water uses in a watershed and their need is an important approach for planning and management.

1.2 HYDROLOGIC CYCLE

Hydrologic cycle is an important process of natural world to transport water from oceans to atmosphere, to global land, and back to sea. All of world's water is subjected to be controlled by this procedure, which enables water to alter its form, location, and availability. This cycle is also identified as global water cycle that defines storage and movement of water among the lithosphere biosphere, hydrosphere, and atmosphere. Water is evaporated through atmosphere by sunrays, merged into clouds as water vapour, condensed and drops to the land in the form of precipitation, and then goes back to atmosphere through multiple hydrologic processes. Hydrologic cycle is defined in terms of some main constituents: precipitation (P), infiltration (I), evaporation (E), transpiration (T), surface runoff (R), evapotranspiration (ET), groundwater flow (G), and storage. This process is simple enough; however, there are a few steps of breakdown for hydrologic cycle (Ramanathan et al., 2001; Held and Soden, 2006). The processes are as shown in Figure 1.1.

As water transports through the hydrologic cycle, its quality changes. Evaporation through solar rays drives ultimate purification of seawater, saline water, and still water in lakes and reservoirs. When rainfall occurs, clouds absorb contaminants which transfer through atmosphere, penetrating into ground, and finally moving through underground reservoirs making water more unusable for living entities. During the transportation process, soil acts as filter and absorbs contaminants from water shipping through them towards groundwater reservoirs or outlets. Though hydrologic cycle looks simple in concept, in reality it is a very complex phenomenon in terms of merged-through complex processes, including precipitation, evapotranspiration, infiltration, and groundwater flow.

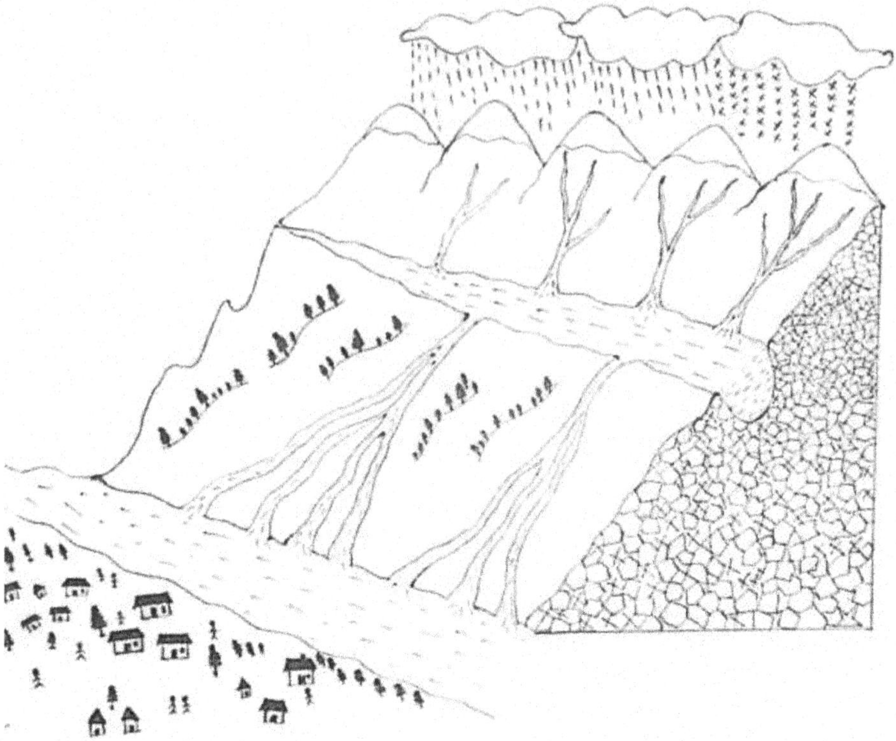

FIGURE 1.1 Hydrologic cycle and its components in a watershed.

1.2.1 Precipitation/Rainfall

Precipitation signifies all forms of water which reaches earth from the atmosphere (Eltahir and Bras, 1996). Rainfall, snowfall, hail, dew, frost, drizzle, glaze, and sleet are the forms of precipitation. Of all these forms, rainfall contributes largely to precipitation causing stream flow or runoff in major rivers. Magnitude of rainfall changes with time and space. Conditions for formation of precipitation are: (i) moisture-bound atmosphere, (ii) sufficient heat and feasible conditions for condensation of water vapour to occur (iii) feasible conditions for condensation of water vapour to occur, (iv) creation of condensation must enrich earth. Net precipitation depends upon meteorological parameters like wind temperature, humidity, and pressure.

1.2.2 Interception

After precipitation occurs over a watershed, all of it does not fall directly on ground. A portion of precipitation may be captured by vegetation before enriching ground following subsequent evaporation. The captured volume of water is known as interception.

Intercepted rainfall may be retained by vegetation as a storage volume on the earth's surface and it returns to atmosphere by evaporation following the process called interception loss (Aston, 1979).

The process which describes how wet leaves shed excess water onto the ground surface is known as *throughfall*. This loss occurs because of evaporation which does not consist of transpiration and *throughfall*.

1.2.3 Evaporation

Evaporation is a process where a liquid state of water is changed to gaseous state at free surface. Molecules of water are in continuous motion in any water body with instantaneous range of velocities. When heat rises, the average speed of molecules increases carrying abundant kinetic energy and these molecules may cross over water surface and water vapour formation occurs (Shuttleworth, 1979). Net emission of water molecules from liquid to gaseous state comprises evaporation. It is a cooling procedure in which latent heat of vaporization is to be given by water body. Rate of evaporation depends on: (i) vapour pressure at water surface and air above the water surface, (ii) temperature of air and water, (iii) wind velocity, (iv) atmospheric pressure, (v) size of water body, and (vi) nature and quality of water.

1.2.4 Transpiration

Water is absorbed by plant root system and escaped through leaves of a living plant through a process called transpiration (Schlesinger and Jasechko, 2014). Transpiration depends upon the following factors: (i) vapour pressure, (ii) temperature, (iii) wind speed, (iv) characteristics of plant, (v) root and leaf system of the plant, and (vi) intensity of light. Transpiration rate also depends upon growth period of a plant.

1.2.5 Evapotranspiration

In any land area, plant loses moisture in the process of transpiration following evaporation of water from soil and water bodies (Allen et al., 2005). Combined effect of evaporation and transpiration processes under one head is termed as evapotranspiration. It is classified as potential evapotranspiration (PET) and actual evapotranspiration (AET):

PET – Sufficient moisture availability for vegetation throughout the area resulting in evapotranspiration is called PET. It depends upon soil and plant factors along with climatic factors.

AET – Real evapotranspiration occurrence in a specific situation is called AET.

Relation between PET and AET – If water supply to the plant is sufficient and soil moisture is at field capacity, then PET = AET, i.e. $\frac{AET}{PET} = 1$. If AET is less than PET, then soil dries out.

1.2.6 Overland Flow

Excess precipitation in the form of runoff moving over land surface to reach channels is called overland flow.

1.2.7 Infiltration

Infiltration is a process by which flow of water into ground occurs through soil surface (Philip, 1969). While water is driven at soil surface, four moisture zones occur:

Zone I: Saturation Zone – The topmost thin layer at the top of the soil approaching to the ground is called saturation zone.
Zone II: Transition Zone – Below the saturation zone, transition occurs which brings transit between transmission zone and saturation zone.
Zone III: Transmission Zone – Below the transition zone, transmission zone occurs where downward movement of moisture takes place and quantity of moisture is above field capacity but below saturation.
Zone IV: Wetting Zone – Soil moisture content will be at or near the field capacity and moisture content decreases with depth.

1.2.8 Groundwater Storage

Below ground, water in soil layer is termed as subsurface water which forms two zones (Andermann et al. 2012): (i) saturated and (ii) aerated.

Saturated or Groundwater Zone – In this zone, all holes of soil are occupied with water. Water table in this zone forms its upper boundary called free surface containing atmospheric pressure.
Zone of Aeration – In this zone, pores of soil are only partly saturated with water. Space amid water table and land surface symbolizes extent of zone of aeration having three subzones known as (i) soil water zone, (ii) capillary fringe, and (iii) intermediate zone.

A part of runoff moves into the ground and gets deposited in the ground reservoir called groundwater storage.

1.2.9 Stream Flow

After ample precipitation has soaked into the soil, absorption of water terminates and subsequently surplus precipitation along with melted snow and ice simply courses off the surface. This water, with the effect of gravity, flows down mountains, hills, and other inclines forming streams and finally joins the rivers (Herschy, Reginald W. 2002). This is called stream flow and is also a standard way in which water passages along the earth's surface. Streams and rivers are pulled by gravitational force up until they merge together to form oceans and lakes.

1.2.10 Hydrologic Budget

Hydrologic budget includes components of the hydrologic cycle and the relationship among them. The relationship for storage water termed as water budget can be given by development of a general equation (Carter et al., 2002):

$$P - R - E - T - G = \Delta S \quad (1.1)$$

considering precipitation (P), surface runoff (R), evaporation (E), transpiration (T), groundwater flow (G), and storage (ΔS). Equation 1.1 is the fundamental equation for all hydrologic modelling. Various applications of this important equation are depicted in later chapters. Calculating the combined effect of evaporation and transpiration, i.e. evapotranspiration (ET), for a region can be reasonably met if precipitation (P), surface runoff (R), and groundwater potential (G) are standardized. For example, in large basins (measured in thousand sq. km), evaluation of groundwater potential can be managed by following surface water hydraulics and boundaries via hydrologic components.

To solve hydrologic budget equation in terms of any one of its parameters, reasonable estimation of other parameters must be made.

1.3 CONCEPTUAL UNDERSTANDING OF SOCIETY'S EFFECT ON THE HYDROLOGIC CYCLE

Population of the world is tremendously increasing, which impacts the land use pattern and has major alterations in characteristics of runoff for any catchment. As a result, land clearance for expansion of agricultural lands increases amount of unprotected soil with decrease in protective cover for natural vegetation. Loss in protective cover reduces potential for percolation and infiltration and hence increases surface runoff potential causing devastating floods. Over the past few decades, urbanization has led to substantial variations in the land use patterns and increased flood issues concentric to urbanized centres like municipal areas, towns, and metro cities. Hence, it has significant effect on the process of hydrologic cycle subjected to urban development. Reduction of storage in terms of volume increases surface runoff and decreases travel time, resulting in excess runoff rate and enhancing soil erosion creating overbank flow. In an effort to recompense for lost natural storage, several areas necessitate replacement of lost storage by development of human made storage. Storm water management is the method for managing efficacy of infiltration on rooftop, and pavement storage. In fact, poor methods of control and management made flood conditions worse in devastating urban and rural regions of watershed.

1.4 CLIMATE DATA IN WATERSHED

Hydrology requires analysis of different types of climatic data in a watershed. Climatic data include precipitation, temperature, relative humidity, pressure, wind movement, solar radiation, and evaporation (Dile and Srinivasan, 2014; Ghose and

TABLE 1.1
Climatic Parameters and Their Measurements

Climatic Data	Measurement Type	Unit	Instrument Used
Precipitation	a. Daily	mm	Rain gauge
	b. Monthly	mm	
	c. Per storm intensity	mm/h	
Temperature	a. Hourly temperature	°C	Thermometer
	b. Maximum daily temperature	°C	
	c. Minimum daily temperature	°C	
Relative humidity	Percentage saturated vapour pressure at some temperature		Hygrometer
Pressure	a. Station level		Barometer
	b. Mean sea level		
Wind movement	a. Wind intensity per day	km/day	Anemometer
	b. Wind intensity per hour		
	c. Direction	km/h	
	d. Wind intensity		
		E, W, N, S (direction)	
Solar radiation	a. Duration	Day time hours	Sunshine recorder
	b. Intensity		
	c. Incoming and outgoing radiation	cal/cm^2/min	
Evaporation		mm/day	Evaporimeter or atmometer

Samantaray, 2019; Eum et al., 2014; Samantaray et al., 2021. Important types of data and instruments used for measurement are enlisted in Table 1.1.

Climate change impact has become a global concern and it is occurring very rapidly at present time (Henderson-Sellers, 1995; Samantaray et al., 2019; Samantaray and Sahoo, 2020a). These effects go well beyond distressing ecologies and societies to upsurge in temperature in the United States and all over the world. Most vital amenities that living beings rely on are energy, transport, wildlife, water, ecosystem and cultivation. However, all these amenities are subject to climate change effects. This phenomenon is happening at present, and it is producing a wide-ranging series of impacts which will affect almost all humans on earth. Severity of every impact depends on our cooperative assortments along with particulars, e.g. specific area and individuals that live in that locality. But altogether, variety of impacts makes this phenomenon as one of the most crucial problems faced by humanity these days.

1.4.1 Water

Variations to water resources can have a huge influence on survival of general people. In few areas, mainly in west part of the United States, drought is a significant issue distressing localities. Accumulation of a smaller amount of snow in foothills is significant in Alaska and West, where snow pack accumulates water for future usage. In northeastern and Midwest states, occurrence of heavy rainstorms has augmented. Hence, in most areas, flood and complications relating to quality of water are expected to be worse due to climate change.

1.4.2 Food

Supply of food is highly dependent on weather and climatic conditions. Even though different irrigation approaches might be malleable, variations like water stress, weather extremes, increased temperatures, and diseases generate complications for ranchers and farmers who place food on our tables.

1.4.3 Health

Well-being of humans is very much susceptible to changing climatic parameters. Change in environmental conditions is projected to create upsurge in waterborne ailments, poor quality of air, more warm stress, and diseases transferred by rodents and insects. Severe weather conditions as mentioned above can cause multiple numerous health issues, which may result in life-threatening consequences.

1.4.4 The Environment

Climate change also has a major influence on the ecosystem. Habitations are being changed, timings of egg-laying and flowering are shifting, and home assortments of different species are fluctuating. Variations are also taking place in ocean because of climate change. Ocean engrosses nearly 30% of carbon dioxide that gets discharged to atmosphere from burning of fossil fuels. Because of this, ocean gradually becomes more acidic which affects aquatic life. Rise in sea levels, owing to melting of land ice sheets and glaciers as well as thermal expansion, has put coastal zones at more threat to storm surge and erosion.

1.5 SOIL EROSION AND SEDIMENTATION

Soil erosion is the removal of soil and rock from the top of the earth's crust due to the movement of external agents of nature like raindrops, running water, wind velocity, and ice with the effect of gravity (Zachar, 2011; Samantaray et al, 2020a, b). Gravitational creeping with external agents promotes soil erosion thus driving away the natural soil from the land surface. Insoluble materials of soil include sand, silt, clay, and organic matter, which are migrated over the surface or through fissures and voids in the soil. Kinetic energy of external agents helps to transport the soil from the origin to other destinations and stabilize through deposition before diminishing of limiting energy. Sedimentation occurs when solid materials are detached from the origin and transported by external agents. Soil loss means net amount of soil moved off from a particular area of a watershed. Sediment yield from any area is the soil loss from slope segments minus the deposition. Deposition occurs at the toe of slopes along the boundaries of watershed in depressions (Lal, 1994). American Society of Civil Engineers (ASCE) defined sedimentation as the process of detachment of soil from original location via transportation and deposition of sediments by erosive agents (impact of rainfall) and transporting agents (runoff)

1.5.1 Causes of Soil Erosion

1.5.1.1 Rainfall and Runoff

Soil erosion is very common particularly due to heavy rainfall (Olabisi, 2012; Samantaray and Ghose, 2020; Samantaray and Sahoo, 2020c, d). Firstly, water causes scattering of large compounds that originate from break down of soil. Usually, rainwater has more impact on lighter ingredients such as finer sand particles, silt, and organic matter. But due to heavy precipitation, larger constituents of soil are also very much affected.

1.5.1.2 Slope of the Land

Physical features of land also contribute to erosion of soil. For instance, land with a tall hill slant will propagate rainwater process or saturation of runoff in a region predominantly because of quicker water movement down the slope.

1.5.1.3 Deficiency of Vegetation

Crops and plants aid in maintaining soil structure, which results in reduction of quantity of soil erosion. Regions with less natural flora might suggest that soil is susceptible to erosion.

1.5.1.4 Wind

Wind plays an important role in promoting erosion and decreasing quality of soil in particular, if structure of soil has already slackened up. But lighter breezes normally do not cause excessive destruction. Lighter or sandy soil is most vulnerable to this type of erosion as it can be easily blown off through air.

1.5.2 Effects of Soil Erosion

An important issue with soil erosion is lack of information regarding how slowly or quickly it will happen. If impact of ongoing climatic events or weather is large, it may be an unhurried evolving procedure that has certainly not been observed (Lal and Moldenhauer, 1987). But an extreme change in weather condition or other effects can add to rapid growth in erosion that could cause countless damage to concerned zone and its occupants. The following are some of the major effects of soil erosion:

Loss of topsoil: Evidently, this is a principal consequence of soil erosion. Since topsoil is very fertile, its removal can create severe damage to harvests done by farmers or effective capability to labour on their land.

Reduced fertile and organic matter: As stated above, removal of topsoil that heavily contains organic matter will decrease potential of land to restore new crops or flora. As new plants or crops cannot be grown fruitfully in that zone, this leads to a cycle of reduced phases of organic nutrients.

Poor drainage: Occasionally, excessive compaction with sand can head to an operational crust which covers the surface stratum, making it more rigid for water to pass through to denser levels. This can assist to stop erosion in some way because of compactly crammed soil. However, if it maintains greater level of runoff resulting from flooding or rainwater, it can have a negative impact on vital topsoil.

Problems with plant reproduction: After erosion of soil in an active crop land, lighter soil properties such as new seeds are buried and destroyed particularly by wind. In return, this affects future production of crop.

Acidity levels of soil: When the soil structure turns out to be compromised and organic nutrients are prominently decreased, there is a greater chance of increase in acidity of soil. This will considerably affect capability of crops and plants to grow.

Pollution of water: A vital problem of runoff is that the soil predominantly utilized for agricultural practices is at greater risk to contamination with fertilizer or pesticide and sedimentation. This can lead to substantial damage to water quality and fish.

1.5.3 Sedimentation

When quantity of soil in the rivulet surpasses capability of water to carry it downstream, rocks and gravel beds, which are significant habitation for insects, fish, and other river lives, can be clogged by the excess soil.Hence, sedimentation is the process where the surplus soil remains in stream after being dropped out of the water (Gregory and Edzwald, 2010)

Naturally occurring aspects also have an impact on deposition into streams and amount and rate of erosion:

1. Most significant physical feature affecting soil erosion is the vegetation. A good vegetation cover infuses the soil with organic matter, binds it together, screens the sediment, shields it from rain, and makes it resilient to runoff. A strong vegetation shield against erosion is one of the best defences.
2. Different climatological parameters having influence on erosion are the frequency, amount, intensity of precipitation, and cold and hot temperatures (Samantaray and Ghose, 2019; Samantaray and Sahoo, 2020b). For example, in the course of frequent rainfall periods, the proportion of runoff is more. And, while quickly thawing soil can lead to increase in rate of erosion, frozen soil is vastly resilient to erosion.
3. Erodibility of soil is also determined by soil characteristics. Among these, one such characteristic is texture, i.e. combination of different size of soil elements. Generally, they fall into three broad classifications: small (clay), medium (silt), and large (sand). Soils most prone to erosion are those with

major amount of medium-size (silt) particles. Sand and clay are less susceptible to erosion.

4. The grade, surface quality (rough or smooth), and combined length of ground slope affect erosion of soil. The lengthier the slope, the sharper the grade, and the smoother the surface, the higher is the possibility of erosion. The quantity of rainfall along with the characteristics of slope regulates the speed of stream. The quick flow of water has a greater potential for sedimentation and erosion.

1.6 SOIL CONSERVATION

The process of preserving the soil is called soil conservation. Basically, it eliminates the upper portion of soil which is useful for microorganisms, nutrients, and organic matter required for plant growth (Hillel, 2007; Morgan, 2009). This is the most commonly applied technique all around the world to protect soil from being eroded away. Gradually the washed away soil ends up in river resources transporting fertilizers and pesticides utilized for agricultural land. For the plants to grow and flourish, healthy soil is important. Therefore, it is very important to adopt various measures for conservation of soil for an environment-friendly way of life.

1.6.1 Methods and Techniques of Soil Conservation

There are various methods designed for preservation of nutrient level in the soil and for prevention of erosion. Some methods are discussed below for soil conservation purposes:

1.6.1.1 Contour Ploughing

It involves ploughing trenches into the desired farmland and planting crop in troughs in the trenches and succeeding contours. It is a very productive way to avoid runoff and thus improve crop yields.

1.6.1.2 Terrace Farming

Terrace farming is a technique of sculpting multiple flat levelled areas into hills. Terrace steps are designed in a way that they are surrounded by a mud wall for preventing runoff and retaining soil nutrients. It is mostly used in underdeveloped countries because of the difficulty faced in the utilization of mechanized equipment in the terraces, e.g. planting rice.

1.6.1.3 Perimeter Runoff Control

It is the practice of planting shrubs, ground cover, and trees around the boundary of farmland that retains essential nutrients in cultivated soil and encumbers surface flow. Application of grass way is a specialized technique of controlling perimeter runoff that utilizes surface friction to channel and disperse the runoff.

1.6.1.4 Windbreaks

Rows of high trees are grown in compact patterns all around the farmland to prevent wind erosion. Evergreen trees could offer year-round protection; however, deciduous trees in particular can be acceptable as long as vegetation is apparent throughout the seasons when the soil is bare.

1.6.1.5 Crop Rotation/Cover Crops

Cover crops such as radishes and turnips are alternated with cash crops for blanketing soil all around the year and harvest green manure that replenishes nitrogen and other serious nutrients (Langdale et al., 1991). Utilization of cover crops can also overpower growth of weeds.

1.6.1.6 No Till Farming

It is the method of cultivating crops throughout the year without varying the soil topography with the help of contouring or tilling. This approach increases the quantity of water that infiltrates the soil and can upsurge organic matter present in soil leading to greater harvests.

1.6.1.7 Salinity Management

After evaporation of water from the soil, its salt is left behind. This leads to soil destruction and loss of nutrients. Saltbush crop cultivation can revitalize the soil quality and substitute the lost nutrients. Humic acids can also be used to prevent this. High salt levels present in soil can often cause changes to the water table by stemming and other sources.

1.6.1.8 River Bank Protection

During flooding events, river banks can often cavern in. Construction of walls along the river banks or planting different valuable types of trees will help in protecting banks for the future and preventing loss of soil down the stream.

1.6.1.9 Impervious Surfaces Reduction

Paved pathways and driveway patios allow rainfall to move freely from them. When the flow of water picks up momentum, it erodes any type of soil in which it flows over after parting from the impervious surfaces. Decrease in amount of such flows around the farmland can avoid erosion.

1.6.1.10 Dry Farming

In low-rainfall areas where amount of precipitation is very less, crops which need very little water must be grown. This will help in preserving the natural levels of nutrients and moisture in the soil.

1.6.1.11 Maintaining pH Levels of Soil

Soil contamination because of acid rains and other contaminants or impurities can cause damage to fertility of soil. Use of a pH indicator to check the acid levels in the soil monthly and treat the soil with ecological and biodegradable chemicals can prevent low yields and loss of crops.

1.7 CONCLUSION

Watershed management is one of the most significant development programmes adopted throughout the world. This is helpful for overall economic development and improvement of socio-economic conditions of the society. This chapter describes in detail the hydrologic processes of a basin and constraints affecting hydrologic processes. Also, it includes a clear study on hydrologic budget and climate data of a watershed or basin. Brief descriptions of sedimentation, soil erosion, and soil conservation are also provided in the chapter.

REFERENCES

Allen, Richard G., Ivan A. Walter, Ronald L. Elliott, Terry A. Howell, Daniel Itenfisu, Marvin E. Jensen, and Richard L. Snyder. 2005. "The ASCE Standardized Reference Evapotranspiration Equation." American Society of Civil Engineers. doi: 10.1061/40499(2000)126

Andermann, Christoff, Laurent Longuevergne, Stéphane Bonnet, Alain Crave, Philippe Davy, and Richard Gloaguen. 2012. "Impact of transient groundwater storage on the discharge of Himalayan Rivers." *Nature Geoscience*, 5 (2), 127–132.

Aston, A.R., 1979. Rainfall interception by eight small trees." *Journal of Hydrology*, 42(3-4), 383–396

Black, Peter E. 1996. Watershed Hydrology, CRC Press.

Carter, Janet M., Daniel G. Driscoll, Joyce E. Williamson, and Van A. Lindquist. 2002. "Atlas of water resources in the Black Hills area, South Dakota." *U.S. Geological Survey, Hydrologic Investigations Atlas HA-747.*

Dile, Yihun Taddele, and Raghavan Srinivasan. 2014. "Evaluation of CFSR climate data for hydrologic prediction in data-scarce watersheds: an application in the Blue Nile River Basin." JAWRA: Journal of the American Water Resources Association, 50 (5), 1226–1241.

Eltahir, Elfatih A. B., and Rafael L. Bras. 1996. "Precipitation recycling." *Reviews of Geophysics*, 34 (3), 367–378.

Eum, Hyung-Il, Yonas Dibike, Terry Prowse, and Barrie Bonsal. 2014. "Inter-comparison of high-resolution gridded climate data sets and their implication on hydrological model simulation over the Athabasca Watershed, Canada." *Hydrological Processes*, 28 (14), 4250–4271.

Ghose, Dillip K. and Sandeep Samantaray. 2019. "Estimating Runoff Using Feed-Forward Neural Networks in Scarce Rainfall Region." In Smart Intelligent Computing and Applications. 53–64, Springer.

Held, Isaac M. and Brian J. Soden. 2006. "Robust responses of the hydrological cycle to global warming." *Journal of Climate*, 19 (21), 5686–5699.

Henderson-Sellers, A. (Ed.). 1995. "Human Effects on Climate Through the Large-Scale Impacts of Land Use Change." In Henderson-Z. Sellers, A Ed., Future Climates of the World: A Modelling Perspective. World Survey of Climatology, vol 16, 433–475, Elsevier Science Ltd.

Herschy, Reginald W. 2002. Streamflow Measurement, CRC Press.

Hillel, Daniel. 2007. Soil in the Environment: Crucible of Terrestrial Life, Elsevier.

Lal, Rattan. 1994. Soil Erosion Research Methods, CRC Press.

Lal, Rattan and William C. Moldenhauer. 1987. "Effects of soil erosion on crop productivity." *Critical Reviews in Plant Sciences*, 5 (4), 303–367.

Langdale, G. W., R. L. Blevins, D. L. Karlen, D. K. McCool, M. A. Nearing, E. L. Skidmore, A. W. Thomas, D. D. Tyler, and J. R. Williams. 1991. "Cover crop effects on soil erosion by wind and water." Cover Crops for Clean Water, 15–22.

Morgan, Royston Philip Charles. 2009. Soil Erosion and Conservation, John Wiley & Sons, Inc.

Ojha, Chandra S. P., Rao Y. Surampalli, András Bárdossy, Tian C. Zhang, and Chih-Ming Kao. 2017. Sustainable Water Resources Management, American Society of Civil Engineers.

Olabisi, Laura Schmitt. 2012. "Uncovering the root causes of soil erosion in the Philippines." *Society and Natural Resources*, 25 (1), 37–51.

Philip, J. Re. 1969. "Theory of Infiltration." *In Advances in Hydroscience*, vol 5, 215–296, Elsevier.

Ramanathan, V. C. P. J., P. J. Crutzen, J. T. Kiehl, and Daniel Rosenfeld. 2001. "Aerosols, climate, and the hydrological cycle." *Science*, 294 (5549), 2119–2124.

Samantaray, Sandeep and Dillip K. Ghose. 2019. "Sediment assessment for a watershed in arid region via neural networks." *Sādhanā*, 44 (10), 219.

Samantaray, Sandeep and Dillip K. Ghose. 2021. "Modelling runoff in a river basin, India: an integration for developing un-gauged catchment." *International Journal of Hydrology Science and Technology*, 10 (3), 248–266.

Samantaray, S. and A. Sahoo. 2020a. "Appraisal of runoff through BPNN, RNN, and RBFN in Tentulikhunti Watershed: a case study." Advances in Intelligent Systems and Computing, 1014. doi:10.1007/978-981-13-9920-6_26.

Samantaray, S. and A. Sahoo. 2020b. "Assessment of sediment concentration through RBNN and SVM-FFA in arid watershed, India." Smart Innovation, Systems and Technologies, 159. doi:10.1007/978-981-13-9282-5_67.

Samantaray, S. and A. Sahoo. 2020c. "Estimation of runoff through BPNN and SVM in Agalpur Watershed." Advances in Intelligent Systems and Computing, 1014. doi:10.1007/978-981-13-9920-6_27.

Samantaray, S. and A. Sahoo. 2020d. "Prediction of runoff using BPNN, FFBPNN, CFBPNN algorithm in arid watershed: a case study." *International Journal of Knowledge-based and Intelligent Engineering Systems*, 24 (3), 243–251.

Samantaray, S., A. Sahoo, and D.K. Ghose. 2019. "Assessment of runoff via precipitation using neural networks: watershed modelling for developing environment in arid region." *Pertanika Journal of Science and Technology*, 27 (4), 2245–2263.

Samantaray, S., A. Sahoo, and D. K. Ghose. 2020a. "Prediction of sedimentation in an arid watershed using BPNN and ANFIS." Lecture Notes in Networks and Systems, 93. doi:10.1007/978-981-15-0630-7_29.

Samantaray, S., A. Sahoo, and D. K. Ghose. 2020b. "Assessment of sediment load concentration using SVM, SVM-FFA and PSR-SVM-FFA in arid watershed, India: a case study." *KSCE Journal of Civil Engineering*, 24, 1–14.

Samantaray, S., A. Sahoo, N. R. Mohanta, P. Biswal, and U. K. Das. 2020. "Runoff prediction using hybrid neural networks in semi-arid watershed, India: a case study." In Communication Software and Networks. 729–736, Springer, Singapore.

Schlesinger, William H. and Scott Jasechko. 2014. "Transpiration in the global water cycle." *Agricultural and Forest Meteorology*, 189, 115–117.

Shuttleworth, W. James. 1979. Evaporation: Widely Cited Institute of Hydrology Report.

Zachar, Dušan. 2011. Soil Erosion, Elsevier.

2 Characteristics of Watershed

2.1 INTRODUCTION

The watershed is an area surrounded by a ridge line which drains its outflow through a common outlet. The other name for the watershed is drainage basin or catchment (Langlein et al., 1947, Miller, 1953). Watershed is a component of hydrological activity. Watershed includes surface water from runoff generated by rainfall which migrates downslope to enrich the outlet. It also includes groundwater beneath the surface of the earth. It is a geographical unit to analyse hydrologic cycle and its constituents. Watershed comprises natural and artificial characteristics, i.e. the characteristics of the surface, subsoil, and climate, geological and topographical features of the environment.

2.2 DELINEATION OF WATERSHED

Delineation of watershed is significant for hydrologic design to demonstrate how a watershed boundary is aligned (Figure 2.1). Watershed boundary is

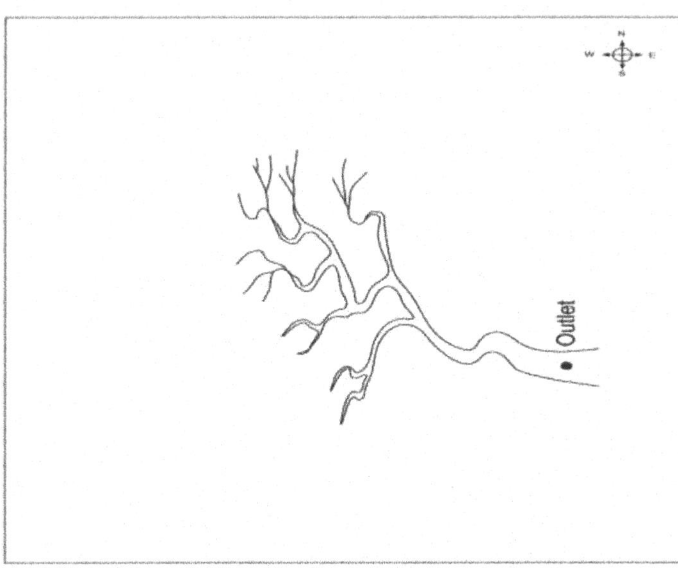

FIGURE 2.1 Watershed boundary.

defined by identifying the points which will contribute water via runoff process to outlet. It is also essential to decide the effective points within a region for contributing water. Extreme points of the catchment signify the boundary of watershed. Consider a topographic schematic map shown in Figure 2.2. If point A is the outlet established at the downstream side of reach AB, then every point circumscribed by elevated CDEF counter will shed water to point A. This will migrate perpendicularly to elevated contours in a path which maximizes the slope. Rainwater dropping at points D and E will move towards upper part

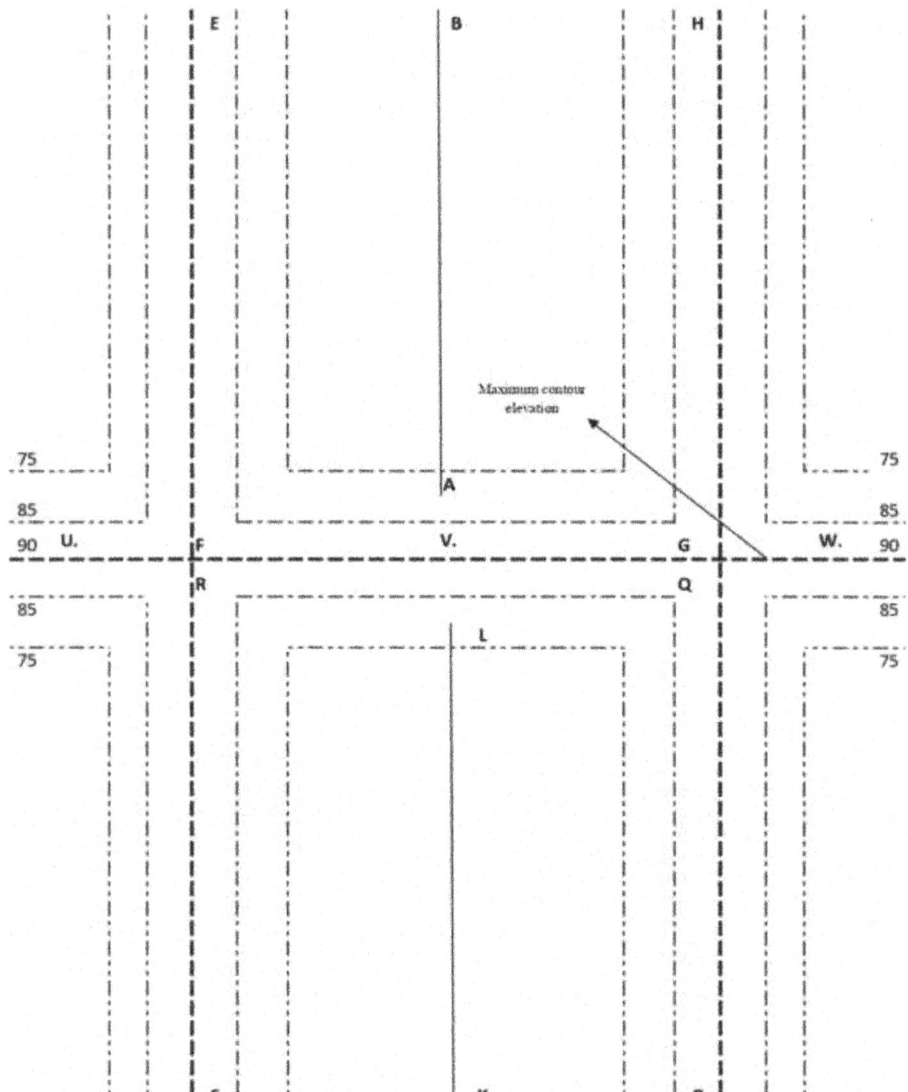

FIGURE 2.2 Delineation of hypothetical watershed.

of channel, i.e. point B. Rainfall dipping at G, H, and K will not travel towards point A as elevations of points G, H, and K are less compared to watershed divide having elevation 80 and hence rainfall at these points will transport to other outlet.

A topographic map for the site is important for understanding the delineation of the watershed. Elevation of all the contours is to be designed from topographic maps to optimize elevated contours in terms of the rainfall runoff contribution to the tributaries via reach lines of watershed.

2.3 DRAINAGE AREA

It is a very significant characteristic of a watershed for hydrologic analysis and design. Drainage area is the measure of volume of water which can be produced from precipitation. In hydrologic design, a common approach is assuming depth of rainfall occurring consistently in the watershed to be constant. With this assumption, available volume of water for runoff is equal to product of depth of precipitation and drainage area. It is necessary as input for development of runoff models, from simple linear models to complex models.

Drainage area can be calculated using an instrument called planimeter. In case of unavailability of planimeter, drainage area will be measured using Stone Age method of counting squares. A transparent sheet showing grid, generally square, is to be placed over the map displaying drainage boundaries. From this, the number of grid blocks inside the boundary is to be computed. Drainage area is equal to the product of number of grid blocks and area of one grid block. Area of one grid block is to be calculated utilizing scale of topographic map from which delineation of watershed boundary is to be done. Accuracy can be maintained by taking care the estimated value of drainage area through counting of grid blocks.

2.4 LINEAR MEASUREMENT

Linear measurement includes the measurement of watershed in one dimension or measurement of channel length. Definition of channel length:

1. The distance measured alongside the central passage from outlet of watershed to basin divide is denoted as L.
2. Distance measured alongside the central passage from outlet of watershed to the channel end shown in figure 2.3 signified as L_c.
3. Distance marked alongside the central passage amid two points situated at 5% and 80% of the distance from outlet is signified as L_{5-80}.

The three lengths L, L_c, L_{5-80} as shown in figure 2.3 possess three parameters of length with respect to time in any watershed. Similarly, any other lengths may also be calculated as per design ideology. Here, the measure of time required for water to travel through the watershed is represented as the time parameter. The velocity of

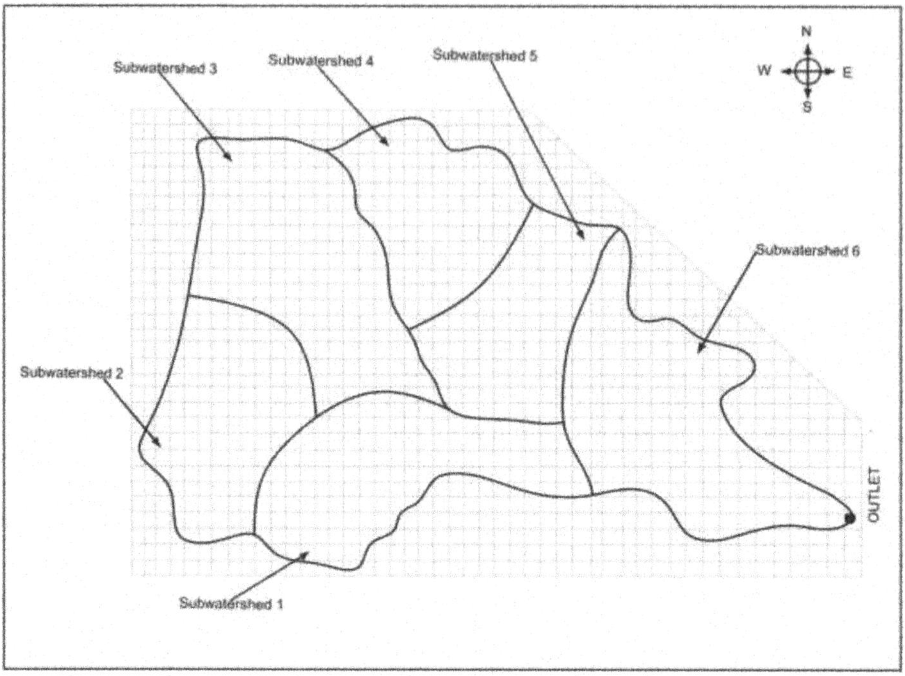

FIGURE 2.3 Delineation of watershed boundary.

water in stream can be measured as the ratio of corresponding length to travel time for covering the length.

The channel system shown in Figure 2.3 is viewed consisting of six reaches: subwatershed 1, subwatershed 2, subwatershed 3, subwatershed 4, subwatershed 5, and subwatershed 6, respectively.

2.5 BASIN SHAPES

Basin shape depends upon the hydrologic design which directly or indirectly depends upon the runoff process. So watershed may have a vast range of shapes, and shape reveals the manner as runoff intermingles with the outlet. Basin shapes may be circular, elliptical, rectangular, or any irregular shape following the flood ways to enrich outlet. Watershed shape will be more apparent when the concept of time area diagram is conversed and introduced. The shape of the basin is replicated by several watershed constraints. The following are the typical parameters which play vital role in designing the basin shape.

Watershed usually circular in shape results in runoff from several portions of basin that will reach the outlet of the basin at equivalent time, whereas in case of an elliptical watershed that has outlet at one end of main axis causes runoff to spread with time, generating a low flood peak than the circular basin. Many parameters of basin have been established for reflecting the shape of basin.

2.5.1 Length to the Centre of Area (L_c)

Length to the centre of area is the distance measured in kilometres alongside the central passage from outlet of basin to observed point of the central passage contrary to centroid of the basin.

2.5.2 Width of Watershed (W)

Width of watershed means average width of watershed in kilometres.

2.5.3 Area of Watershed (A)

It is defined as the product of watershed length along the mainstream and average width of the watershed.

$$\text{Numerically, area of watershed } (A) = L \times W \tag{2.1}$$

2.5.4 Shape Index

Shape index is the ratio of stream length to the width of watershed, i.e. $S_i = \frac{L}{W}$ (2.2)

Also, it can be defined as ratio of square of watershed length to watershed area:

$$S_i = \frac{L^2}{A}$$

2.5.5 Shape Factor (L_s)

It depends upon the combined distance from the outlet of watershed to basin divide (L) and the distance from basin outlet to centroid of area (L_c):

$$L_s = (LL_c)^{0.3} \tag{2.3}$$

where L is the watershed length in metres.

2.5.6 Circularity Ratio (F_c)

It is the basin shape parameter which can be defined as the ratio of perimeter (kilometre) and area (km^2) of watershed:

$$F_c = \frac{P}{4\pi A^{0.5}} \tag{2.4}$$

where P is the perimeter of watershed and A is the area of watershed.

2.5.7 Elongation Ratio (R_e)

Numerically, elongation ratio can be given as

$$R_e = \frac{2}{L_m}\left(\frac{A}{\pi}\right)^{0.5} \tag{2.5}$$

where L_m is the maximum basin length parallel to main drainage lines.

The relationship between R_c and F_c are related by

$$R_c = \frac{1}{F_c^2} \tag{2.6}$$

Length of centre of area is required for deriving synthetic unit hydrograph to understand the basin parameter for calculating cumulative area. Stone Age method may also be utilized for calculating length of reach in any watershed and hence to find out the drainage area outline with the help of a straight pin linear edge and draw a vertical line which approximately rotates the cardboard at 90° and draw a secondary vertical line. The joining of two lines is the centroid of the area. For some cases, approximately three lines are rotated at 60° to provide more accuracy and the centroid of area can be found out accordingly in any watershed.

2.5.8 Basin Length

Theoretically, basin length is the distance travelled by drainage surface, which is more suitably described as hydrologic length (Figure 2.4). This length is utilized to calculate the time taken by water to move from beginning to end of a basin. Therefore, length of the watershed is taken alongside main flow route from exit of watershed to periphery of basin. Because, channel does not lengthen to boundary of basin, it is therefore essential for extending a line from channel end to periphery. The measurement follows a route where largest water volume would normally travel. The length of the basin L_b is the measurement of a basin analogous to its main drainage conduit and width of the basin can be determined in the direction just about right angle to its length. Relationship amid the main length and the area of the drainage basin for small basins is presented as follows:

$$L_b = 1.312\ A^{0.568} \tag{2.7}$$

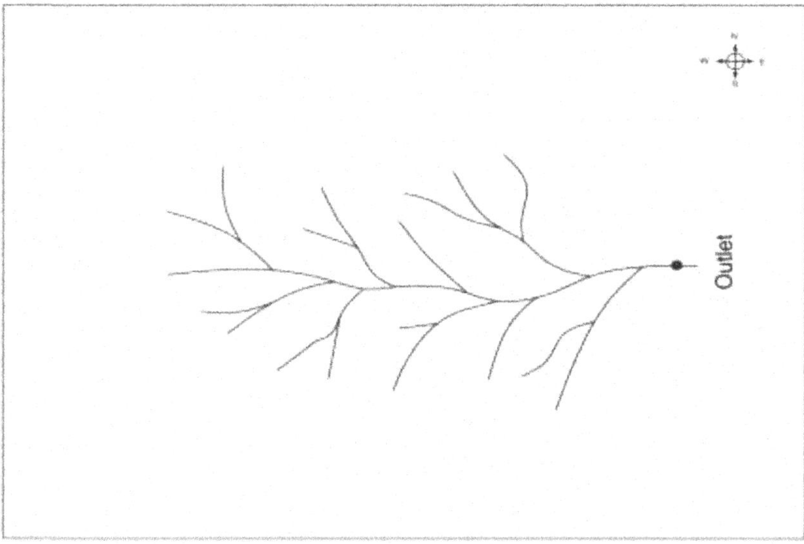

FIGURE 2.4 Delineation of watershed length.

2.6 WATERSHED RELIEF

In a watershed, difference in elevation between uppermost point on perimeter of watershed H_2 and its outlet H_1 is known as watershed relief (H): $H = H_{2\,-}\,H_1$. Few parameters have been established to understand variations in watershed relief. Most common parameters are watershed slope, channel slope, and hypsometric curve.

2.6.1 Channel Slope in Watershed (S_c)

Channel slope in watershed is the ratio of difference in elevation between upper end and lower end of channel to channel length amid two differential points:

$$S_c = \frac{\Delta E}{L_c} \tag{2.8}$$

where ΔE is the elevation difference between two points and L_c is the length of channel between the points. For example, $S_{15-90} = \dfrac{\Delta H_{15-90}}{L_{15-90}}$

ΔE_{15-90} = elevation difference between points describing length of channel L_{15-90}.

When the topographic map of a watershed is available, average slope in percentage can be given as

$$S_p = \frac{C_L X C_i}{A} X100 \tag{2.9}$$

where

S_p = average watershed slope in percentage
C_L = contour length, m
C_i = contour interval, m
A = area of watershed, m^2

In some cases, the slope of the channel is not uniform, where a weighted slope provides an index reflecting influence of slope on hydrologic response of a watershed in better terms. Channel slope index is defined as

$$S_{i=(\frac{n}{k})^2} \tag{2.10}$$

where n is the number of sections into which a channel is distributed, and k can be represented by

$$k = \sum_{i=1}^{n} \frac{1}{\left(\Delta e_i / l_i\right)^{0.5}} \tag{2.11}$$

where Δe_i is the elevation difference between end points of channel section i, and l_i is the length of section i. A channel is classified into sections where each segment shows constant slope.

2.6.1.1 Ruggedness and Geometry Number

The product of basin relief and drainage density, i.e. $H \times D_d$, is called ruggedness number. It varies from 0.05 to 1.0. Strahler (1958) defined geometry number as the ratio of ruggedness number to slope of the ground surface S_g: HD_d / S_g.

2.6.2 Calculation of Slope

Calculate the slope of the channel and watershed slope of subwatershed 1, 2, and 3, as depicted in Figure 2.2. Channel reaches in each subwatershed must be used for calculation of channel slope. Channel slope and watershed slope for subdivision watershed.

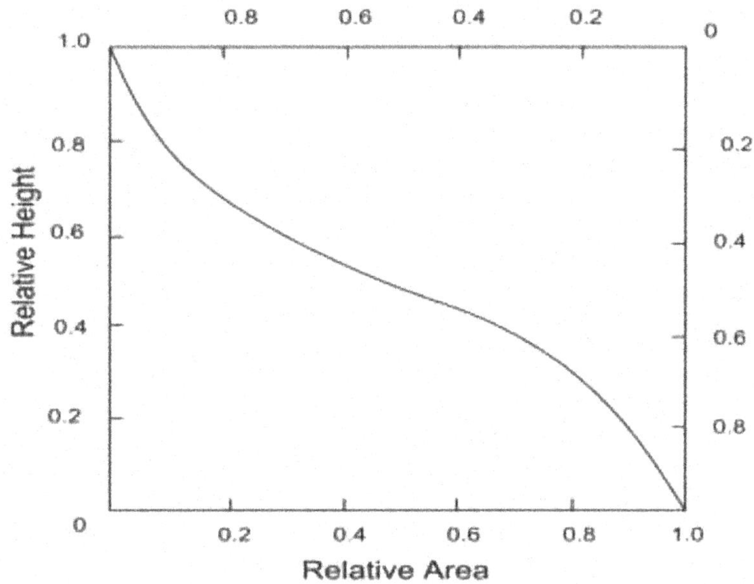

FIGURE 2.5 Hypsometric curve.

2.6.3 Hypsometric Curve

The representation of cumulative relation between the area and elevation in a watershed within elevation intervals is called hypsometric curve. The curve is drawn by plotting elevation (metre) as ordinate and area (m^2) within the watershed as abscissa. It helps in comparing characteristics of area and elevation of a watershed. The comparison may be made by development of a regional hypsometric curve with that of a standard hypsometric curve of the watershed in that region. After standardization of hypsometric curves in every watershed, a mean curve may be developed for region where standardized data are unavailable for the neighbouring watersheds. An index of hypsometric curves for an area-elevation characteristic of a single-valued index of watershed is shown in Figure 2.5.

2.7 DESCRIPTION OF DRAINAGE PATTERN

Drainage pattern is a symbol of flow characteristics of storm runoff. So, to represent the drainage pattern, a number of parameters have been developed depending upon velocity of water flowing over watershed (Schumm, 1956). In general, velocity of water flowing in a channel is greater than the velocity of overland flow. Travel time of runoff for a watershed depends upon overland flow length, which is smaller as compared to travel time for channel flow length. Thus, speed with which runoff enrich the channel can affect the travel time of runoff. To understand the mechanism, Horton's laws play vital role, accordingly a description of drainage pattern is enrolled via Horton's laws.

2.7.1 Horton's Laws

Geomorphology of the basin plays a significant part in conversion of water from terrestrial regions to the channels (streams) and also from channel of one order to another. It can be easily found out by the contour and drainage maps of the basin. The order, the length, the slope and the profile of the channel, and also the density of drainage are very common characteristics of a channel, significant in the estimation of the hydrologic practices of the river basin (Hack, 1957). These features of a watershed depend upon stream order. Stream order is a measure of stream degree channelled through a watershed. Stream length is denoted by its order like first, second, third order etc.

Those channels which do not have tributaries are categorized under first-order sequences. The union of two first-order channels forms a second-order channel. Formation of a third-order channel is due to the union of two second-order channels. Therefore, a sequence of any order comprises two or more tributaries of preceding lower order. This flow management system is known as Horton-Strahler ordering scheme.

$$N_w = R_b^{W-w} \tag{2.12}$$

$$\text{Or, } \log N_w = W \log R_b - w \log R_b = a - b$$

$$(a = W \log R_b,\ b = w \log R_b)$$

where

W = watershed order,
N_w = stream number of order w, and
R_b = bifurcation ratio.

Bifurcation ratio is defined as ratio between numbers of stream channels of a specific order to number of stream channels of one order higher.

$$R_b = {N_w}/{N_{w+1}} \tag{2.13}$$

2.7.1.1 Channel Length

It indicates the channel length of each order. As a geometrical order, mean length of channels of each higher order increases. Hence, the shortest of all channels are the first-order channels. This relationship is called Horton's law of channel lengths and the equation is written as follows:

$$\overline{L_w} = \overline{L_1} R_L^{w-1} \tag{2.14}$$

$$\overline{L_w} = {L_w}/{N_w}$$

where

N_w = number of channels of order w;
L_w = total channel length of order w;
$\overline{L_w}$ = average length of channel of order w;
$\overline{L_1}$ = average first-order stream length;
R_L = ratio of stream and length, usually varying from 1.5 to 3.5.

$$R_L = {L_w}/{L_{w-1}} \tag{2.15}$$

2.7.2 Drainage Density (D_d)

It is the measure of closeness of drain spacing. It is the demonstration of the drainage efficiency of the surface flow and length of the surface flow, in addition to relative proportion index. It is defined as ratio of total stream length to total watershed area. Horton introduced this term by and can be articulated as

$$D_d = L/A \tag{2.16}$$

Or

$$D_d = \frac{\sum_{w=1}^{w} \sum_{i=1}^{N_w} L_{Wi}}{A}$$

where

L = total channel length of all orders
A = area
w = order of basin
N_w = number of different order basins

High drainage density value illustrates a comparatively high density of streams and specifies a quick response to storm.

Horton (1945) suggested utilizing half of reciprocal of drainage density for determining the average length of the surface flow (L_0) for total drainage basin:

$$L_0 = 1/(2D_d) \tag{2.17}$$

where D_d essentially defines the average distance between the streams and L_0 approximates the average length of the overland flow from divides of stream channels.

2.8 TIME PARAMETERS

Hydrologic design involves measure of runoff in a watershed or flood discharge in a river basin. Numerous watershed and hydrologic characteristics are used to reflect the volume of runoff. For calculation, the product of the depth of rainfall and drainage area generally provides volume of water, which is probably accessible for runoff. However, volume prediction is not sufficient enough for several hydrologic design formulations, as dimension of discharge is a measure of time in hydrologic design. Time distribution is important for controlling peak runoff. If a substantial amount of total runoff volume passes a specified site at around same period, then there is a significant possibility of occurrence of flood damage. But, on the contrary, if total runoff volume is distributed over a comparatively longer time period by providing hydraulic structures, then flood damages may be minimized. As the timing of runoff is very significant, most hydrologic models need a characteristic of watershed which will reflect the runoff timing. Thus, time parameters are formulated with rainfall intensity as well as runoff via modelling to understand watershed characteristics. For design of small watershed, time parameter may be an indicator of intensity and rainfall volume is the degree to which there will be attenuation of rainfall for a short phase in hydrologic cycle, which will suggest the attenuation of rainfall for minimal effects. But for design of large watershed, time parameters may be an indication to storage of watershed and impact of storage on distribution of runoff timing. Watershed storage straightaway affects shape, size, and runoff time distribution. Travel time reveals the influence of channel storage on reduction of flood flow. Error in prediction of time parameter will create an error for design period of flood. For estimating time parameters, hydrologic researchers have developed several approaches for defining lag time, peak time, and time of concentration.

Most commonly utilized time parameters are the lag time, time of concentration, time to equilibrium, and time to peak in hydrology (Chow, 1964, Reddi, 2001). These are generally defined in terms of different physical watershed characteristics or distribution of rainfall excess and direct runoff.

2.8.1 Time of Concentration (T_c)

It is a basic parameter of a watershed. The time required for a water particle to flow from most distant point in the watershed to its outlet is called time of concentration. It is computed as the time for runoff to flow from most hydraulically distant point of drainage area to point under study. It is applied for computing peak discharge of a watershed. Peak discharge is based on T_c as it is a function of intensity of rainfall. There can be numerous potential paths taken into consideration to determine longest travel time. Mark and Marek (2011) suggested that designers must detect the flow route alongside which the longest travel time is expected to follow. Sometimes T_c is also computed as the time difference between end of rainfall excess and inflection point.

Calculation of T_c through Equations 2.18–2.20 was developed by Kirpich, FAA (US Federal Aviation Administration), and Kerby in 1940. It is the oldest equation and is undoubtedly the most extensively documented equation. However, with newly developed equations, this is not frequently used in today's era. Equation developed by FAA is most frequently utilized as it utilizes broadly familiar rational coefficient for describing ground cover of watershed. ASCE

(American Society of Civil Engineers) commends use of this equation. Kerby equation is the least used equation among the three and has maximum limitations. There are other equations to determine T_c, but most of them need intensity of rainfall as an input. Hence, utilizing additional equations for determining rainfall intensity to compute peak discharge lead to an iterative procedure as intensity of rainfall is itself a function of T_c.

The following equations are utilized for computation of T_c. All equations given below use variety of units as indicated in Table 2.1. These can be found in Singh (1992), Corbitt (1999), Chow et al. (1988), and Chin (2000).

$$\text{Kirpich equation: } t = 0.0078\ k\ (L/S^{0.5})^{0.77} \tag{2.18}$$

$$\text{FAA equation: } t = 1.8\ (1.1 - c)\ L^{0.5}/(100\ S)^{1/3} \tag{2.19}$$

$$\text{Kerby equation: } t = 0.8268\ (L\ r/S^{0.5})^{0.467} \tag{2.20}$$

where

k = Kirpich adjustment factor
c = rational method runoff coefficient
L = longest watercourse length in the watershed
S = average slope of the watercourse
r = Kerby retardance roughness coefficient
t = time of concentration
V = average velocity in watercourse ($V = L/t$)

Along with the units mentioned above, a variety of units may also be used for computation of time of concentration.

2.9 LAND USE AND LAND COVER

Alterations in land use/land cover (LULC) are affected by natural phenomena and human intrusion such as consumption patterns, agricultural demand and trade, population growth, economic development and urbanization, science and technology, and additional aspects. As a result, information regarding LULC is vital for any sort of natural resources planning and management (Datta, 1986). Precise and timely information regarding detection of change in LULC is exceptionally significant to understand interactions and relationships amid natural phenomena and human interference for enhanced decision-making and management (Lu et al., 2004). There is an enduring demand for up-to-date and accurate information about LULC for all sorts of sustainable development programmes where LULC assists as one of the major input criteria.

Although land cover and land use terms are frequently used interchanging among them, their real meaning are fairly distinctive. Generally, land use is defined as a chain of actions carried out by humans on land, with purpose of obtaining benefits and/or products utilizing land resources, whereas land cover is defined as man-made constructions (buildings, etc.) or vegetation (natural or planted) occurring

TABLE 2.1
Formulae for Calculation of Time of Concentration

Reference	Model	Formula
Izzard (1946)	Izzard	$T_c = 41.025(0.0007i + c)L^{0.33}S^{-0.333}i^{-0.667}$ S = flow path slope (ft/ft) c = retardance coefficient L = flow path length (ft) i = rainfall intensity (in/h)
California Culvert Practice (1955)	California Culvert Practice	$T_c = 60(11.9L^3/H)^{0.385}$ H = difference in elevation between outlet and divide (ft) L = longest water course length (mi)
Kirpich (1940)	Kirpich	$T_c = 0.0078L^{0.77}S^{-0.385}$ S = slope of average watershed (ft/ft) L = channel/ditch length (ft)
Kerby (1959) Hathaway (1945)	Kerby Hathaway	$T_c = 0.8275(LN)^{0.467}S^{-0.233}$ S = overland flow path slope (ft/ft) L = overland flow length (ft) N = flow retardance factor
Williams (1922)	Williams	$T_c = 60LA^{0.4}D^{-1}S^{-0.2}$ S = basin slope (%) L = basin length (mi) D = circular basin diameter (mi) A = basin area (mi^2)
Johnstone and Cross (1949)	Johnstone Cross	$T_c = 300L^{0.5}S^{-0.5}$ S = basin slope (ft/mi) L = basin length (mi)
Morgali and Linsley (1965) Aron and Erborge (1973)	Morgali Linsley	$T_c = 0.94L^{0.6}n^{0.6}S^{-0.3}i^{-0.4}$ i = rainfall intensity (in/h) n = Manning roughness coefficient S = mean overland slope (ft/ft) L = overland flow length (ft)
FAA (1970)	FAA	$T_c = 1.8(1.1 - C)L^{0.5}S^{-0.333}$ S = surface slope (ft/ft) L = length of overland flow (ft) C = rational method runoff coefficient
Chen and Wong (1993) Wong (2005)	Chen-Wong	$T_c = 0.595(3.15)^{0.33k}C^{0.33}L^{0.33(2-k)}S^{-0.33}i^{-0.33(1+k)}$ S = overland plane slope (m/m) i = net rainfall intensity (mm/h) L = overland plane length (m)

TABLE 2.1
(Continued)

Reference	Model	Formula
Yen and Chow (1983)	Yen-Chow	T_c = $Ky(NL/S^{0.5})^{0.6}$ L = length of overland flow (ft) S = average overland slope (ft/ft) N = overland texture factor
Li and Chibber (2008)	TxDOT	T_c = $0.702(1.1 - C)L^{0.5}S^{-0.333}$ S = surface slope (m/m) L = overland flow length (m) C = rational method runoff coefficient
NRCS (1997)	NRCS	T_c = $0.0526[(1000/CN) - 9]L^{0.8}S^{-0.5}$ S = average watershed slope (%) L = flow length (ft) CN = curve number
Papadakis and Kazan (1986)	Papadakis Kazan	T_c = $0.66L^{0.5}n^{0.52}S^{-0.31}i^{-0.38}$ S = average slope of flow path (ft/ft) i = rainfall intensity (in/h) n = roughness coefficient L = flow path length (ft)
U.S. S.C.S. (1975, 1986)	U.S. Soil Conservation service	T_c = $(1/60)\sum A(L/V)$ V = average velocity (ft/s) L = flow path length (ft)
Carter (1961), Nicklow et al. (2006)	Carter	T_c = $100L_m^{0.6}S_m^{-0.3}$ S_m = surface slope (ft/mi) L_m = length of flow (mi)
Eagleson (1962), Nicklow et al. (2006)	Eagleson	T_c = $0.0111(Ln)/R^{-0.666}S^{-0.5}$ R = Hydraulic radius S = average slope of flow path n = roughness coefficient L = flow length
	Passini	T_c = $6.48(AL)^{0.333}S^{-0.5}$ S = average slope of basin (m/m) L = main channel length (km) A = basin area (km^2)
NRCS (2010), Mockus (1961)	Mockus	T_c = 60[L0.8(S ? 1)0.7]/1140Y0.5 Y = average watershed land slope (%) L = flow length (ft) CN = the retardance factor S = maximum potential retention = (1000/CN)–1

on earth surface. Water, ice, bare rock, sand, and similar surfaces also count as land cover. Land cover and land use have certain fundamental variances. Land use directs to the purpose the land serves, such as wildlife habitat, agriculture, or recreation. It doesn't define surface cover on ground. For instance, a recreational land use could occur in shrubland and grasslands, on manicured lawns, or in forests (Brown, 1976). Land cover denotes surface cover on ground, whether urban infrastructure, vegetation, bare soil, water etc (Ram, 1997). It doesn't define usage of land or for land utilization in different ways with similar cover type. For example, a land cover type of forest may be utilized for wildlife management, recreation, or timber production; it might be a popular state park, a protected watershed, or a private land.

Briefly, land use specifies how people are utilizing land, while land cover specifies physical type of land. Land use cannot be determined from satellite imagery, whereas land cover can be determined by studying aerial and satellite imagery. Land cover map gives information regarding current landscape for helping managers to get a better understanding about it. Land cover maps of several previous years are required to observe its changes over time. This information helps managers to assess previous management choices and increase awareness into probable impact of their present decisions before implementing them.

Coastal managers utilize land cover maps and data for better understanding of the effects of different natural phenomena and human utilization of landscape. Maps can provide assistance to managers for modelling water quality issues, assessing urban growth, predicting and evaluating surge and flood impact, tracking losses due to wetland and possible impact from rise in sea level, prioritizing areas for preservation efforts, and comparing land cover variations with environmental effects or connection in socio-economic variations such as increasing population. Understanding both land cover and land use of a track of land offers a wide-ranging picture of a specific area. This data is an essential factor for planning and decision-making procedures for several groups, as it assists them in better understanding about where to plan for different kinds of growth and where to preserve. It also helps in understanding fragmentation or connectivity of different aspects in their community.

2.10 CONCLUSION

Detailed study on watersheds is important for hydrologic scientists, researchers, and field engineers. Without knowing each component/constraints of watershed/catchment, we cannot proceed for further research for sustainable development. Various watershed characteristic like basin shapes, drainage area, width of watershed, area of watershed, shape index, shape factor, basin length, watershed relief in detail are described. For measuring runoff, evaluation of time of concentration is a significant parameter, so detailed study on time of concentration by various researchers/scientists are provided in the above sections. Land cover and land use have a key impact on surface water and groundwater, which affects directly watershed management.

REFERENCES

Aron G, Erborge C.E. 1973. A practical feasibility study of flood peak abatement in urban areas. Report U S Army Corps of Engineers, Sacramento.

Brown, G. W. 1976. Forestry and Water Quality, Oregon State University Book Stores Inc., Corvallis, OR.

California Culvert Practice. 1955. Department of Public Works. Division of Highways, 2th edition Sacramento.

Chen, C.N., Wong, T.S.W. 1993. Critical rainfall duration for maximum discharge from overland plane. *J Hydraul Eng*, 119(9), 1040–1045.

Chin, David A. 2000. Water-Resources Engineering. Prentice-Hall.

Chow, V. T. 1964. Handbook of Applied Hydrology, McGraw-Hill, New York.

Chow, Ven Te, David R. Maidment, and Larry W. Mays. 1988. Applied Hydrology. McGraw-Hill, New York.

Corbitt, Robert A. 1999. Standard Handbook of Environmental Engineering, 2nd ed, McGraw-Hill, New York.

Datta, S. K. 1986. Soil Conservation and Land Management, International Book Distributors, Dehradun.

Hack, J.T. 1957. Studies of Longitudinal Stream Profiles in Virginia and Maryland. U.S. Geological Survey Professional Paper, 294-B.

Horton, R. E. 1945. "Erosion development of streams and their drainage basins: hydrophysical approach to quantitative morphology." *GSA Bulletin*, 56, 275–370.

Izzard, C.F. 1946. Hydraulics of runoff from developed surfaces. In: Proceedings 26th annual meeting highway research board. 26:129–146.

Langlein, W. B., et al. 1947. Topographic Characteristics of Drainage Basins. U.S. Geological Survey Water-supply Paper, 968-C.

Lu, D., Mausel, P., Brondizio, E., Moran, E. 2004. Change detection techniques. Int. J. Remote Sens. 25, 2365–2407.

Mark A, Marek P.E. 2011. Hydraulic design manual. Tex Dep Trans, Design div, Texas.

Miller, V. C. 1953. A Quantitative Geomorphic Study of Drainage Basin Characteristics in the Clinch Mountain Area, Virginia and Tennessee. Project NR 389-042, Technical Repot 3, Columbia University, Department of Geology, ONR, Geography Branch, NY.

Nicklow, J.W. Boulos, P.F., Muleta, M.K. 2006. Surface runoff. Comprehensive urban hydrologic modeling. MWH soft, California.

Papadakis, C.N., Kazan, M.N. 1986. Time of concentration in small rural watersheds. Technical Report 101/08/86/CEE. Civil Engineering Department, University of Cincinnati, Ohio.

Ram, Suresh 1997. Soil and Water Conservation Engineering, Standard Publishers Distributors, Delhi.

Reddi, P. J. 2001. A Textbook of Hydrology, Lanni Publications (Pvt.) Ltd., New Delhi.

Schumm, S. A. 1956. "Evolution of drainage systems and slopes in badlands at Perth Amboy, New Jersey." *GSA Bulletin*, 67, 597–646.

Singh, Vijay P. 1992. Elementary Hydrology. Prentice-Hall. Englewood Cliffs, New Jersey.

Strahler, A.N. 1958. Dimensional analysis applied to fluvially eroded landforms. *Geol Soc Am Bull*, 69, 279–300.

Yen, B.C., Chow V.T. 1983. Local Design Storms. US Dept of Transportation, Fed Highway Administration. Report No FHWA-RD-82-063 to 065, Washington: DC.

3 Soil Erosion and Its Control

3.1 INTRODUCTION

Detachment and removal of soil particle from land surface is the basic cause of soil erosion. It is a natural physical phenomenon which has helped to shape the present form of the earth's surface. Advent of modern civilization has increased the pressure on land, leading to its overexploitation and, subsequently, its degradation. This trigged a very fast pace of soil erosion from land surface because of action of two fluids, water and wind. Soil erosion triggered due to overexploitation of land surface is called accelerated erosion and that caused due to natural phenomena is termed geologic erosion. Geologic erosion of soil is caused mainly by effect of rainfall, atmospheric temperature runoff, wind velocity, topography, and gravitational force. It is a continuous, slow, but constructive process resulting in wearing a way of mountains, building up of coastal plains and flood plains, and development of some of the most fertile valleys of the world such as Indo-Gangetic valley in the Indian subcontinent, Nile Valley in Africa, etc. Accelerated erosion of soil is mainly caused by error in management, such as raising of crops without adopting any soil conservation practices, deforestations, etc. It leads to erosion in excess of threshold value of new soil information, causing severe deterioration of the top surface of the land. Deterioration is sometimes so rapid that it disrupts the equilibrium between relationships of soil-plant environment.

3.2 TYPES OF SOIL EROSION

The process of soil erosion goes through two central phases of activities: detachment of soil particle and their transportation. The major agents responsible for these two phases of activities are wind, water, and gravity. Wind erosion occurs in semi-arid and arid areas where precipitation is scanty and day temperature is very high. Water erosion occurs in regions where precipitation is high and gravity erosion occurs in areas that are near to river, pits, roads, etc. Gravity erosion leads to mass soil movement due to pull of protruding land mass by gravitational forces. As geologic erosion is a part of natural recurring process, it is called as permitted soil erosion. Significant aspects which activate soil erosion process include climatic, hydrologic, geologic, topographical, soil type, and vegetation cover. In addition to these aspects, the socio-economic condition and the level of technical know-how have a substantial impact on rate of exploitation of land surface. Invariably, these factors have a combined effect on soil erosion. Three natural agents – wind, water, and gravity – are primary causes of soil erosion.

- a. Wind
- b. Water
    - 1. Water forms
        - I. Runoff
            - Subsurface runoff
            - Surface runoff
                - Rill erosion
                - Gully erosion
                - Stream-bank erosion
                - Inter-rill erosion
        - II. Reservoir
        - III. Rainfall
    - 2. Glaciers
- c. Mass-soil movement
    - 1. Landslide
    - 2. Debris
    - 3. Creep
    - 4. Landfall

3.2.1 Wind Erosion

It is detachment and transportation of soil particles by the force of moving wind. In areas where precipitation is low, atmospheric temperature during the day is high and velocity of wind is invariably very high. Such climatic conditions generally prevail in semi-arid and arid regions, where velocity of wind is very high. Wind erosion is caused by mismanagement of land resources, such as overgrazing, intensive farming, disforestation, etc. These practices destroy cohesive properties of soil particle making them susceptible to wind erosion. Such erosion slowly leads to formation of sandy tracts, which in turn become more vulnerable to wind erosion.

Wind erosion passes through three phases:

1. Destruction of soil surface by wind forces
2. Transportation of eroded soil to other destinations
3. Deposition of sediment particle when the wind velocity reduces

The soil eroding force is called erosivity of wind. The process of soil erosion caused by wind is very complex in nature because of two opposite forces acting on soil particles: one is the erosivity of wind which tries to blow off soil particles and the other is the erodibility of soil which is the property by which soil particles resist separation from their bond with each other (Chepil 1953). Sandy soils have less cohesion and are eroded easily, whereas clay particles are able to resist erosion to greater extent.

Pasak (1974) developed the following equation to determine the erodibility of soil by wind:

$$S_E = 22.02 - 0.72P - 1.69M_s + 2.64R \tag{3.1}$$

where

S_E is the erodibility of soil by wind $\left({gm}/{m^2} \right)$
P is the content of non-erodible particle in soil (70.8 mm)
M_s is the relative soil moisture, $M_s = {M_0}/{M_n}$
M_0 is the instantaneous moisture and $M_n = {M}/{2.4}$
M is the content of clay particle (0.01 mm) in the soil
R is the wind speed near the soil surface

He found that equations based on clay content are easier for using in the field, and gave the following equation for determining erodibility of soil by wind for different kinds of soil:

$$\text{Sandy soils: } S_E = -269.24 + 24M - 1.65M_0 + 20.92R \tag{3.2}$$

$$\text{Loamy sands: } S_E = 8.95 - 0.63M - 0.5M_0 + 1.22R \tag{3.3}$$

$$\text{Sandy loams: } S_E = 16.09 - 0.58M - 0.18M_0 + 0.42R \tag{3.4}$$

$$\text{Loamy soils: } S_E = 1.55 - 0.05M - 0.01M_0 + 0.08R \tag{3.5}$$

Lal (1990) and Siddoway in 1965 proposed the following parametric model for prediction of erosion by wind:

$$W_E = F(I', K', C', L', V) \tag{3.6}$$

where

W_E = potential average annual erosion
I' = soil erodibility index = ${X_2}/{X_1}$
X_1 = quantity eroded from soil comprising 60% clods (>0.84 mm)
X_2 = quantity eroded under same condition from soil comprising any other proportion of clods more than 0.84 mm
K' = soil ridge roughness factor
C' = wind erosivity, also called the wind erosion climatic factor (Chepil et al., 1963)

$$= 34.483\, {v^3}/{(P - W_E)^2}$$

v = mean annual wind velocity

$$P - W_E = \text{Thornthwaite's } P - W_E \text{ ratio} = 10\left({P}/{W_E} \right)$$

where

P = precipitation >12.7 mm

L' = field length, unsheltered distance across field along prevailing wind erosion direction = $D_f - D_b$

D_f = distance across field

D_b = sheltered distance

V = vegetative cover equivalent quantity = $R_v \times S \times K_o$

R_v = quantity of vegetative cover and mulch = $1.2 \times$ washed oven dry residue

S = kind of vegetative cover = 1.0 for small-grain stubble and solver

0.25 for sorghum stubble and solver

0.20 for maize stubble and solver

2.5 for small grain in seeding and stooling stages)

$$K_o = f(R) = \text{vegitative surface roughness}$$

Gillette et al. (1972) proposed a formula for estimating soil erosion by wind:

$$g_s = \frac{0.295 g_m L I r^{1.5}}{(RK)^{1.26} (\gamma_r)^{1.5}} \tag{3.7}$$

where

g_s = soil erosion (g/cm-s)

g_m = maximum soil flux

L = length of field exposed to wind erosion

I = erodibility index

r = wind drag corrected soil moisture

γ_r = reference wind drag of 3300 kg/ha

R = surface residue kg/ha

K = soil surface roughness, cm

Erosive forces of wind are also called entrainment forces. These forces cause lift, drag, and particle movement through collisions among each other. Chepil 1960 recommended different widths of wind strips for erosion susceptible crops. Soil resisting forces are friction, gravity, and cohesive properties of soil particle. Wind erosion takes place when erosive forces are higher than soil-resisting forces (erodibility properties of soil).

Energy of moving wind causes lift and drag action of soil particles by following Bernaulli's effect. Soil particle gets lifted upward from surface of land due to pressure difference between air mass and land surface, created by reduction in the air pressure due to high velocity of wind and unevenness in pressure created by soil clods and capillaries. Drag is created by the wind velocity, which gives energy for detachment and transportation of soil particles over the land surface. Thus, the velocity of moving wind is proportional to the drag force per unit area, near the ground surface. The drag force is expressed as

$$\tau = Kv^2 \tag{3.8}$$

where

τ = drag force

K = coefficient based on the height of measurement of wind velocity from the ground surface and surface roughness

V = velocity of wind

The profile of wind velocity varies with changes in soil cover and surface roughness. The variation of wind velocity with height is called velocity gradient and in general, an average wind velocity is considered for estimating wind erosion. Chepil and Woodruff (1963) determined average wind velocity as given below:

$$v_z = 5.75 v^* \log \frac{z_0}{k} \tag{3.9}$$

where

v_z = velocity of wind at height z

v^* = friction velocity

K = height above mean aerodynamic surface z_0, where wind velocity is zero

The value of k is zero for small crops and smooth surface.

The wind velocity over solid surface, such as ground surface, is zero at height z_0, but velocity is not zero over vegetation since vegetation is porous and wind is able to flow through it.

Skidmore and Woodruff (1968) determined the magnitude of wind erosion force vector using the following formula:

$$r_j = \sum_{i=1}^{n} \overline{(v_i)}^3 f_i \tag{3.10}$$

where

r_j = wind erosion force vector

$\overline{v_i}$ = mean wind speed with the ith speed

f_i = duration factor in percentage of total observation in the jth direction within the ith speed group

J = subscript indicating the 16 directions of a compass from 0 to 15

The total wind erosion is then given by

$$F_T = \sum_{j=0}^{15} \sum_{i=1}^{n} (v_{ij})^3 f_{ij} \tag{3.11}$$

F_T is a vector quantity. (It is the relative capacity of wind to cause the blowing off of the soil in any direction.)

Bagnold (1941) developed the formula to determine threshold velocity (v_{th}) of wind:

$$v_{th} = \sqrt{(K(\sigma - \rho)gD)} \tag{3.12}$$

- v_{th} is the critical shear velocity of wind above the threshold level, σ and ρ are the particle and air densities, respectively
- G is the acceleration due to gravity
- D is the mean particle diameter
- K is a constant=0.1 (when Reynolds number exceeds 3.5)

The amount of transported soil particle is found by the following formula (Finkel and Noveh, 1986):

$$Q_\delta = C_I v_{th}^3 \tag{3.13}$$

where

- Q_δ amount of soil moved per unit area
- C_I is the constant of proportionality

Hsu (1973) proposed a formula to determine the weight of soil moved per unit area:

$$Q_s = (4.97D - 0.47)(\frac{0.4(\overline{v_z} - 275)}{\sqrt{\text{In}ZgD}})^3 \tag{3.14}$$

where

- Q_s is the tonne per meter width per year:
- Z is the height of measurement of wind speed
- D is the mean diameter of soil particle
- $\overline{v_z}$ is the hourly average mean velocity

3.2.2 Water Erosion

Soil erosion due to water is caused by its two forms: liquid as the flowing water, solid as the glaciers. The impact of rainfall causes splash erosion. Runoff water causes scrapping and transport of soil particle leading to rill, sheet, and gully erosion. Water waves cause erosion of bank sides of reservoir, lakes, and ocean. The subsurface runoff causes soil erosion in the form of pipe erosion, which is also known as tunnel erosion. Glacier erosion causes heavy landslides. In India, glacier erosions are mostly confined to the Himalayan region.

3.3 ESTIMATION OF SOIL EROSION

Several models are developed to estimate the loss of soil from watersheds and outflow of sediment to stream, carried by the runoff from catchment, in the form of suspended load and bed load. Sediment outflow model have been developed based on the historical data of hydrologic parameters and sediment yield. Some of the important model in use are as follows:

1. Soil loss model
   i. Universal soil loss equation (USLE)
   ii. Soil loss equation model for Southern Africa
2. Sediment outflow model
   i. Modified universal soil loss equation (MUSLE)
   ii. Sediment graph model
3. Bed load models
4. Sedimentation model of water storage structures

3.3.1 Soil Loss Model

Evaluation of soil loss from watershed is necessary during assessment of the soil erosion severity and its effect on production of agricultural goods. Soil loss is determined by two methods:

i. Theoretical estimation based on value of watershed parameter
ii. Actual measurements in the field

Soil loss can be estimated as a function of parameters of watershed; sincere attempts have been made for developing soil loss estimation models, starting from 1960s. Most efficient model for estimation of soil loss was given in the year 1965 by Wischmeier and Smith, universally called USLE. This model formed elementary arrangement for most soil loss models, which came after this phase. The notable amongst these are the USLE model for South Africa by Elwell (1978) and Modified USLE (MUSLE) by Williams (1975).

3.3.1.1 Universal Soil Loss Equation

Wischmeier and Smith (1965 and 1978) developed the USLE for prediction of gross soil erosion from agricultural watersheds in the United States. On a close look, the USLE appears to have been based on the Musgrave equation (Musgrave, 1947) that had been developed earlier. The USLE has proved to be very popular, and is being widely used after incorporating the necessary modifications in the values of its parameters for different regions of the world. The equation is especially being applied as a guide for conservation planning of small agricultural watersheds. The equation states that

$$A = RKLSCP \tag{3.16}$$

where

A = estimated gross soil erosion, t/(ha/year)
R = rainfall erosivity factor, J/(ha/year)
K = soil erodibility factor (t/ha)/erosivity factor (R), t/J
L = slope length factor
S = slope gradient factor
C = crop cover or crop management factor
P = supporting conservation practice factor

Magnitude of soil erosion is governed by two forces: detachment of soil particles by effect of rainfall energy, known as erosivity of rain, and capability of soil to resist detachment of its particles by this force, known as erodibility of soil (Foster et al., 1981). This relationship is expressed as follows:

$$\text{Soil erosion} = f[(\text{erosivity of rain}) \times (\text{erodibility of soil})]$$

3.3.1.1.1 *Evaluation of USLE Factor*

3.3.1.1.1.1 Rainfall erosivity factor (R) It is a function of falling rain droplets and intensity of rainfall. Wischmeier and Smith (1958) established that product of raindrop kinetic energy and maximum rainfall intensity over a 30-min duration, in a storm, was best soil loss estimator. This product is called the EI value. It is found that this value provides a very good correlation for soil loss estimation and is the most consistent single estimate of rainfall erosivity potential. EI values are determined from recording rain gauge data of each storm (Smith & Wischmeier 1962). Rainfall mass curve is distributed into small growths and for every increment, the value for rainfall intensity and their raindrop-kinetic energy is computed. From the obtained computed values, maximum rainfall intensity during a 30-min continuous duration (I_{30}) is then determined. The multiple of this value with E gives value for EI_{30}. Rainfall erosivity is computed for each storm, and obtained values are added up for desired period: month, week, year, etc. Kinetic energy is computed by subsequent formula (Wischmeier and Smith, 1978):

$$\text{Kinetic energy of rainfall } (E) = \sum E_i \tag{3.17}$$

where

$$\sum E_i = \sum_{i=1}^{N} (210.3 + 89 \log_{10} I_i)$$

E = total kinetic energy of rainfall
E_i = rainfall kinetic energy of ith increment (m-t/ha-cm)
I_i = average rainfall intensity during ith increment (cm/h)
N = total number of discrete increments

The selection of maximum rainfall intensity during a duration of 30 minutes by Wischmeier and Smith (1978) was based on extensive experimental research. Incidentally, this value has been found to be equally applicable to many parts of India, including Dehradun, by the Central Soil and Water Conservation Research and Training Institute, Dehradun (CS&WCR&TI). In some tropical and subtropical countries of Asia and Africa, it has been reported that the kinetic energies of individual storms, at intensity 25 mm/h, are more appropriate for correlating the soil loss.

3.3.1.1.1.2 Soil erodibility factor (K) The susceptibility rate of soil particles to erosion per unit of rain erosivity factor (R), for a particular soil on a unit plot with a uniform slope of 9% and 22.13 m slope length over a continuous clean fallow land having up and down slope farms, refers to soil erodibility factor. The value of K is found either theoretically or experimentally on the basis of respective clay contents and dust particles (<0.10 mm), organic matter, sandy particle (0.10–2.00 mm), soil permeability, and soil texture. It is computed by the following regression equation:

$$K = 2.8 \times 10^{-7} M^{1.14} (12 - a) + 4.3 \times 10^{-3} (b - 2) + 3.3 \times 10^{-3} (c - 3) \quad (3.18)$$

where

M = particle size parameter (% silt + % very fine sand) (100% of lay)
a = % of organic matter
b = soil structure code (very fine granular = 1; medium or coarse granular = 2; platy, massive, blocky = 4)
c = profile permeability class (rapid = 1; moderate rapid = 2; moderate = 3; slow to moderate = 4; slow = 5; very slow = 6)

Another method used by the scientist at CS & WCRTI, Dehradun (Dhruvanarayan, 1993), comprises direct measurement on unit runoff plots. Formula utilized here is on basis of the USLE and is expressed as follows:

$$K = \text{Total adjusted soil loss } (A) \big/ R \quad (3.18)$$

where

A = expected soil loss from a unit plot having slope 9%; $A = A_0 / S$

A_0 is the observed soil loss in a fallow plot having s% slope and 22.13 m length, S is the slope gradient factor for s% slope. In watersheds having large area, where there is a spatial variation of soil properties, a weighted factor for K is computed by utilizing the subsequent formula:

$$K = \sum_{i=1}^{n} k_i a_i \big/ a \quad (3.19)$$

where

k_i = soil erodibility factor for soil i

a_i = drainage area covered by soil i

a = total drainage area

n is the number of type of soil in the watershed

3.3.1.1.1.3 Slope length factor (L) The slope length is on which overland flow occurs, affecting the rate of soil erosion. On larger slope lengths, there is a higher concentration of overland flow and a higher flow velocity which generates a higher rate of soil erosion. Zingg (1940) found that soil loss has a non-linear relationship with the land slope length, that is, soil loss $\propto (L_p)^m$,

where

L_p = actual slope length

M = ratio of soil loss from field plot length to soil loss from unit plot with a slope length of 22.13 m. The slope length factor is determined by using the following formula:

$$L = \left(L_p / 22.13 \right)^m \tag{3.20}$$

where

L_p = actual unbroken length of slope measured up to the point where the overland flow terminates

M = exponent = 0.5 for slope $\geq 5\%$

0.5 for slope $\geq 5\%$

0.4 for slope $\geq 4\%$

0.3 for slope $\geq 3\%$

0.2 for slope $\geq 1\%$

Dvorak and Novak (1994) have recommended the value of m as 0.5 for 10%. McCool et al. (1989), for rill or inter-rill erosion on a 9% and 22-m-long slope, have recommended that

$$m = \frac{\sin\theta}{\sin\theta + 0.269\,(\sin\theta)^{0.8} + 0.005} \tag{3.21}$$

where θ = field slope in degree $= \tan^{-1}\left(\text{field slope} / 100 \right)$

3.3.1.1.2 Slope gradient factor

On steep slopes the flow velocity is high, which leads to scouring and cutting of soil. Also, soil erosion is high due to splash, because on steep slopes splashed particles are thrown to greater distances on an inclined plane down the slope and the destruction because of impact of raindrop is greater on the soil crust. The slope gradient factor expresses the ratio of soil loss from a plot of known slope to soil loss from a unit plot under identical conditions. Wischmeier and Smith (1965) used subsequent formula for determination of factor *S*:

$$S = \frac{0.43 + 0.30s + 0.043s^2}{6.613} \tag{3.22}$$

where *s* is the slope of the field plot.

In 1978, Wischmeir and Smith modified the *S* factor equation, which is of the following form:

$$S = 65.41\sin^2\theta + 4.56\sin\theta + 0.65 \tag{3.23}$$

McCool et al. (1989) have given subsequent equations to determine value of the factor *S*:

$$S = 3\ (\sin\theta)^{0.8} + 0.56 \text{ for slope 4 m}$$

$$S = 10.8\sin\theta + 0.03 \text{ for slope} > 4 \text{ m and S} < 9\%$$

$$S = 16.8\sin\theta - 0.5 \text{ for slope} > 4 \text{ m and S} \geq 9\%$$

3.3.1.1.2.1 Topographic factor (LS) Wischmeier and Smith (1965) recommended using a combined value of *LS* and called it the topographic factor of slope length and gradient and expressed it by the following formula:

$$LS = \frac{(L_p)^m}{100}(1.36 + 0.97s + 0.1385s^2) \tag{3.24}$$

Also, based on Equations 3.20 and 3.23 for single uniform slope,

$$LS = \left(L_p/22.13\right)^m (0.065 + 4.56\sin\theta + 65.41\sin^2\theta) \tag{3.25}$$

Dvorak and Novak (1994) observed that the USLE model is suitable for straight slopes. In places where the gradient varies, they suggested that the slope be divided into smaller segments of straight gradient, and then the value of *LS* be determined by using the following formula:

$$LS = \sum_{j=1}^{n} \frac{S_j L_{pj}^{1.5} - S_j L_{p(j-1)}^{1.5}}{\left(L_p/22.13\right)^{0.5}} \tag{3.26}$$

where

S_j = factor of slope gradient of the jth segment
L_p = unbroken slope length
L_{pj} = distance from upper end of the slope to the lower boundary of the jth segment
N = number of segment of the slope

3.3.1.1.2.2 Contour-length method for length-slope factor (LS) Williams and Berndt (1976) suggested that contour-length method be used to determine the *LS* values for very large watershed. The formula proposed by him is of the following form:

$$s = \frac{0.25z(LC_{25} + LC_{50} + LC_{75})}{A} \quad (3.27)$$

where

s = average watershed slope
Z = watershed height or relief (km)
$LC_{25}, LC_{50}, LC_{75}$ = contour length at 25%, 50%, 75% of z
A = drainage area (km^2)

To determine average watershed slope length, he proposed the following equation:

$$L_p = \frac{LC \times LB}{2EP\sqrt{LC^2 - LB^2}} \quad (3.28)$$

L_p = average watershed slope length (m)
EP = number of extreme points
LC = total contour length $(LC_{25} + LC_{50} + LC_{75})$ (m)
LB = total contour base length (m)

The following equation was utilized for determination of LS factor:

$$LS = \left(L_p/22.1\right)^m \left(0.065 + 0.454s + 0.0065s^2\right) \quad (3.29)$$

where

LS = length slope factor
m = 0.5 for slope > 5%
m = 0.4 for 1% < slope > 3%
m = 0.2 for slope < 1%

3.3.1.1.2.3 Vegetative cover factor (C) Vegetative cover disperses the impact force of raindrops on soil surface, and shields the soil from splash erosion by adjusting drop size, volume, impact velocity, coefficient of distribution, and kinetic energy of rainfall. Canopy cover is primarily responsible for effectiveness of the vegetative cover. The quality of the cover depends on the foliage characteristics, plant height, and the area covered by the vegetation, whereas the leaf area index depends on the height and density of the canopy, foliage characteristic, and the area covered by different species and affected by the type of vegetation (Babu et al. 1978).

Splash erosion is caused not only by direct impact of raindrop on the bare soil surface, but also by the through fall of raindrop from the canopy cover. The through fall of raindrop depends on the velocity of the fall, which varies with the roughness of the leaf surface, impact velocity of rainfall on the leaf, and the leaf inclination angle. A dense vegetative cover provides a high protective cover to the ground surface, but a higher height of canopy, namely from pines etc., imparts a high terminal velocity to drop of the through fall, which causes heavy soil erosion by splash on the soil surface.

Many types of approaches are used to evaluate the effects of different crop covers on soil erosion. A commonly used method evaluates the intensity of soil erosion under different covers vis-à-vis that from bare soil. In this method, if soil erosion from bare soil is taken to be 100%, then soil erosion from plots with root crops is 60%, cereals 30–40%, fodder crops on agricultural lands 5%, permanent grass land 1%, and woodland 1% (Dvorak and Novak, 1994).

The effectiveness of a crop cover, at different stages of crop growth, varies with its capacity to intercept and dissipate the kinetic energy of rainfall. Wischmeier and Smith (1965) subdivided the crop season into five stages of time periods, which however do not fit under Indian conditions due to shorter crop seasons, namely the rough fallow period from ploughing of sowing, planting (seeding) period, crop establishment period, growth and maturity period, and the residue period spanning from crop harvest to ploughing for the next crop.

The major part of rainfall in India occurs during the monsoon period, which is called the kharif crop season, and during this period the maximum amount of soil erosion takes place. Rao in 1981 and Pratap Narain et al. in 1980, as quoted by Dhruvanarayana (1993), divided the crop growth period into the following three stages:

Stage 1: Germination and seedling. Seedling implies 1 month of crop growth.
Stage 2: Active vegetative and maturity period – 1–2 months of crop growth
Stage 3: Final growth and maturity period – from stage 2 to harvesting of crop

The factor *C* is the ratio of soil loss from land harvested on the basis of specific conditions to corresponding loss from fallow land, continuous, clean-tilled. For calculating the value of factor *C*, the ratio of soil loss for a given crop under specified period (derived from runoff plot data) is given a weightage according to percentage rainfall factor of that specific period (Dhruvanarayana, 1993). Singh et al. (1981) determined the rainfall factors for different crop stages in India.

3.3.1.1.2.4 Conservation practice factor (P) The bare fallow land surface causes maximum soil erosion, especially when cultivation is done up and down the slope, or in others words, cultivated across the contours of the land surface. When a sloping land is put under cultivation, it needs to be protected by practices that will attenuate the runoff velocity, so that much less amounts of soil are carried away by the runoff water. Some of the important practices are strip cropping, contour cultivation, terrace and building systems, and waterways for disposing excess rainfall (i.e. runoff).

The ratio of soil loss from a plot with a particular conservation practice to corresponding loss from a plot having up and down cultivation with equivalent conditions refers to conservation practice factor P. The numerical value of P is always ≤ 1.0. In regions where many practices are in usage, a weighted P value as per area under each practice is taken into consideration.

3.3.1.2 Uses of USLE

The USLE was developed basically to estimate the soil loss caused by soil erosion on an annual basis, but it has also found many other uses for designing soil conservation practices and structures. The equation is most commonly used for determining the values of C and P. The maximum permissible soil loss is fixed as the output, the factors R, K, and LS are kept constant, the values of C and P are varied to get the estimated soil loss equal to the permissible soil loss.

Similarly, the design dimensions, namely the horizontal spacing between bunds that may be assumed to be the length of the run (L_p), can be determined by the following equation:

$$L = \left(A / RKCPS \right)^m \tag{3.30}$$

where

$$L = \left(L_p / 22.13 \right)^m$$

Equation 3.30 can also be used for assessment of effectiveness of a soil conservation programme on a watershed, by assessing the loss of soil before and after the implementation of the programme.

3.3.1.3 Sediment Delivery Ratio

A long-term average annual sediment yield can be predicted by application of a delivery ratio to the estimate gross erosion. This technique is very convenient and is adequate for problems that require an estimate of average annual sediment yield, namely sediment pools for reservoirs, etc.

The following formula is used for determining the delivery ratio (DR):

$$DR = \frac{\text{Measured sediment yield at the watershed outlet}}{\text{Estimated gross soil erosion } (A)} \tag{3.31}$$

where

A = soil erosion estimated by USLE.

3.4 SEDIMENT YIELD MODELS

3.4.1 Modified Universal Soil Loss Equation

The USLE of Wischmeier and Smith (1978), developed in 1965, can predict average annual sediment yield by application of a delivery ratio in big watersheds, but generally estimating the delivery ratios accurately is problematic at several places, because of non-availability of observed data. USLE is also not considered an appropriate model for modelling water quality problems. Such modelling needs shorter time increments than one year, and also consider both runoff and sediment parameters as carriers of pollutants.

Modification of USLE by Williams (1975) by replacement of its rainfall energy factor with runoff factor is known as Modified 'USLE'. Estimation of sediment yield by the new model is done on a per storm basis, rather than average soil erosion per year as done by USLE. The MUSLE is stated as

$$Y = 11.8(Qq_p)^{0.56}KLSCP \tag{3.32}$$

where

Y = sediment yield from a singular storm, metric tonnes

Q = volume of storm runoff

q_p = peak runoff rate

$KLSCP$ = factors of the USLE.

Applying the delivery ratio is not needed with MUSLE, and Equation 3.32 is valid to singular storms on agrarian watersheds up to area of 70 km^2.

3.4.2 Routing Model of MUSLE

Estimation of sediment yield by the MUSLE equation from very big agrarian watersheds is inaccurate because of changing climatic factors, watershed hydraulics, soil characteristics, crop management, land slope, and erosion control practices inside watershed area. More accurate estimation of sediment yield from large watersheds can be obtained if these watersheds are subdivided into smaller watersheds with area less than 25 km^2, for compensating the non-uniform distribution of sediment sources and routing sediment yield from smaller watersheds to outlet of entire watershed for including impact of sediment particle size and watershed hydraulics. The following routing model was proposed by Williams (1975) to estimate the sediment yield from large watersheds:

$$RY = \sum_{i=1}^{n} y_i e^{-BT_i}\sqrt{D_{50_i}} \tag{3.33}$$

where

RY = sediment yield from whole watershed, in metric tonnes
y_i = sediment yield i from subwatershed i in metric tonnes, as determined by MUSLE
B = coefficient of routing
T_i = travel time from subwatershed i to outlet of watershed (h)
D_{50_i} = median particle diameter of sediment for subwatershed i (mm)
n = number of subwatersheds

In Equation 3.33, value of every factor is known apart from the coefficient of routing B, which is only a function of watershed hydraulics, and its value is found taking into consideration a uniform distribution of USLE factors (*KLSCP*) and D_{50} on whole watershed. Assuming this, Equation 3.33 is reduced to subsequent formula, from which B is computed.

$$RY = (Q \times q_p)^{0.56} = \sum_{i=1}^{n} (Q_i q_{p_i})^{0.56} e^{-BT_i} \sqrt{D_{50_m}} \tag{3.34}$$

where D_{50m} is the sediment median diameter (about 0.001 mm) and it is kept identical for all small watersheds. Value of B is calculated by an interactive solution, which is same for all small watersheds.

Das (1982) and Das and Chauhan (1990) observed that B is equivalent to $1/k$, where k is the storage coefficient when applied to sediment wash load. Since B is determined through an iterative process of which D_{50} is a multiple, and as determination of the accurate value of D_{50} for each subwatershed is very difficult, the product of the factors B and $\sqrt{D_{50}}$ can be considered together, and only one coefficient ($1/k$) can be used without sacrificing accuracy.

3.5 BED LOAD MODELS

Transportation of sediment load in a stream takes place through the processes of contact, saltation, and suspension, depending on the distribution of sediment particle size and the transport capabilities of streams. The sediment transported by saltation and contact processes constitutes the bed load. The total sediment load in a storm is the sum of the bed load and the suspended load (wash load), it is difficult to draw any demarcation line between these two types of sediment materials. The bed load in a stream varies considerably and lies in the range of 5–25% of the suspended load. Mostly, the bed load is computed by empirical relationships, because of the difficulty of making field measurements of bed load in larger streams.

3.5.1 Estimation of Bed Load

The three empirical relationships commonly used to estimate the bead load in streams are described below.

Schoklitsch formula: The Schoklitsch formula for calculating the bed load is as follows:

$$G = 143.98B\sum_{i=1}^{n}\frac{437.9}{(D_i)^{0.5}}(S)^{3/2}(10.76q - q_{oi}) \tag{3.35}$$

where
- G = bed load, t/day
- B = width of the stream, m
- D_i = geometric mean diameter of each individual fraction, mm
- S = hydraulic gradient
- q = discharge per unit width, m³/s-m

$$q_{oi} = \frac{0.00021}{(S)^{4/3}}D_i \tag{3.36}$$

N = number of individual size fractions in the bed material.

3.5.1.1 Meyer-Peter Muller Formula

The Meyer-Peter Muller formula for calculating the bed load is as follows:

$$G = 5.1844B[\left(\frac{10.844Q_B}{Q}\right)\left(\frac{D_{90}^{1/6}}{n_p}\right)^{1.5} dS - 0.627D_m] \tag{3.37}$$

where
- G = bed load, t/day
- B = width of the stream, m
- Q_B = water flowing just over the bed load area, m³/s
- Q = water discharged, m³/s
- D_{90} = particle size of stream bed load such that 90% of the material is finer than this size
- n_p = Manning's n for stream bed material
- D_m = effective bed material size (weighted mean diameter), m
- D = mean channel depth, m
- S = hydraulic gradient

Nomographs (Babu et al., 1979) are used to compute the values of n and Q_B. In the absence of nomographs, the following formulae are used:

$$n_p = n_m[1 + \frac{2d}{B}\left(1 - \frac{n_w^{1.5}}{n_m}\right)]^{2/3} \tag{3.38}$$

$$Q_B = \frac{Q}{1 + \frac{2d}{B}\left({n_w}/{n_m} \right)^{3/2}} \quad (3.39)$$

where

n_m = Manning's n for total channel section

n_w = Manning's n for channel side walls.

3.6 CONTROLLING SOIL EROSION

Soil erosion is mainly caused by two natural agents: wind and water. Soil erosion by wind is controlled by measures which either reduce the erosive power of wind or increase the ability of the soil surface to resist the blowing away of soil particles. Similarly, soil erosion by water is controlled by reducing the erosive power of raindrops and runoff and increasing the resistivity of soil from getting detached and transported by water.

3.7 SOIL CONSERVATION PRACTICES

Mechanical and vegetative practices are applied on mild slopes to conserve soil, by farming across the land slope (Das 2009). The basic principle of this method is reducing the influence of slope on runoff velocity and, thus, reducing soil erosion. On steep slopes, mechanical measures are adopted as well as construction of mechanical structures for reducing effect of slope on runoff velocity. The most common mechanical measure is grading (levelling) of land surface, but on very steep slopes, deeper cuts and fills are encountered which tend to expose the subsoil and subsequently cause reduction in the fertility level of soil. In such cases, the land is divided into strips and laid across the slope in the forms of terraces and bunds. Basically, except for bench terraces, in all these strips, the soil from the top of the strip erodes and settles down at the bottom of the strip, and eventually levels up itself with a grade that is not very erosive. The following categories of mechanical and vegetative practices are being employed at present.

3.8 VEGETATIVE PRACTICES

i. Contouring
ii. Strip cropping
iii. Tilling operation
iv. Mulching

3.8.1 Contouring

Contouring is the practice of cultivation on contour lines, placed across the prevalent slope of land where all agricultural processes, such as cultivation, ploughing, planting, sowing, etc., are approximately carried out on contours. The interculture operations produce furrows, which in conjunction with plant stems function as appropriate barriers to water flowing down the gradient. Also, the ridges detain water for an extensive time period, which results in increasing opportunity time for runoff water for infiltrating into soil, and in that way increase soil moisture. First contour guide lines are laid and simultaneously tillage operations are carried out for laying a contour farming system. In soils with low filtration rates, a non-erosive grade of 0.1–0.2% is provided to the contour guide line leading towards the outlet.

The degree of control of soil erosion by this technique varies with soil texture, crop cover, and land slope. It has been found that soils on lands with medium slopes and with good infiltration capacity are the best.

3.8.2 Strip Cropping

The practice of growing strips of crops with poor ability for controlling erosion, such as cereals, root crops (intertilled crops), etc., alternating with strips of crops having good ability for controlling erosion, such as grasses, fodder crops, etc. that are close-growing crops, is known as strip cropping. It is a more rigorous farming practice compared to farming on contours only. The farming practices comprised in strip farming are contour strip farming, cover cropping, farming with conservation tillage, and suitable crop rotation. Crop rotation with a mixture of intertilled and close-growing crops, farmed on contours, provides soil moisture, food and fodder. Close-growing crop acts as barrier for flowing and reducing runoff velocity generated from the strips of intertilled crops, and eventually reducing soil erosion. Strip cropping systems are laid out by the following three methods:

1. Buffer strip cropping
2. Field strip cropping
3. Contour strip cropping

3.8.2.1 Buffer Strip Cropping

This type of cropping is practised where uniform strips of crops are necessary to be laid out for smooth operations of the farm machineries, while farming on a contour strip cropping arrangement. Buffer strips of grasses, legumes, and other similar crops are laid out between contour strips as improvement strips (Figure 3.1). These strips provide very good security and efficient soil erosion control from strips of intertilled crops.

3.8.2.2 Field Strip Cropping

In a field layout of strip cropping, uniform width strips are placed across prevalent slope, while shielding soil from erosion caused by water. For protecting soil from wind erosion, strips are placed across prevalent wind direction. Generally, such practices are followed in regions with very irregular topography, and contour lines are too curvy for placing farming plots.

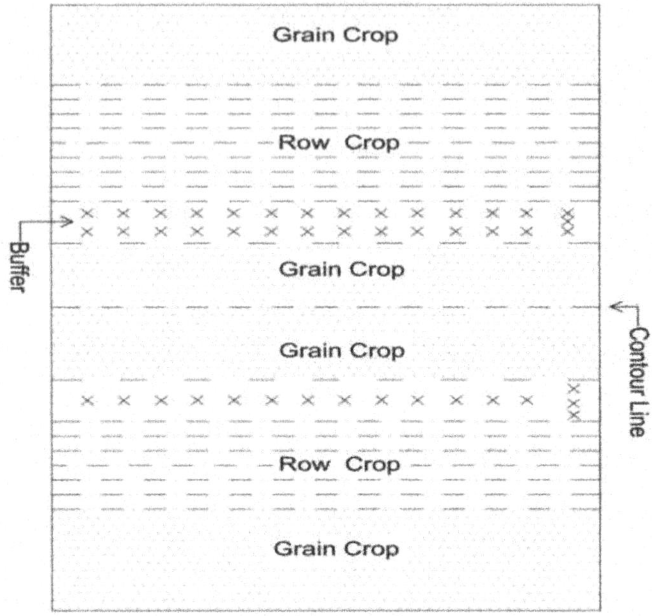

FIGURE 3.1 Buffer strip cropping.

3.8.2.3 Contour Strip Cropping

In contour strip cropping, alternating strips of crops are sown (planted) more or less following the contours (Figure 3.2). Suitable tillage operations and rotation of crops are followed in the course of farming procedures.

Strip cropping systems when laid out on contours follow the same procedure as that for contouring, i.e. contour guide lines are laid out first. Strip cropping system are also done on terraced or bunded lands, where terrace ridges or bunds are taken as the guidelines for the strip layout, and the strip widths conform to horizontal spacing of these mechanical practices.

3.8.3 Tillage Operations

Tillage operations are carried out to make a seedbed conducive to plant growth and good crop yield. Tillage operations that are carried out for conservation of soil are a bit different from the commonly carried out operations. The following are the special benefits of tillage operations:

1. Increase soil infiltration capacity
2. Improve resistivity of soil to erosion
3. Improve coarseness of land surface and reduce evaporation of moisture
4. Improve moisture-retaining capacity of the soil
5. Create a rough land surface for protecting it against water or wind erosion
6. Improve the humus content of the soil

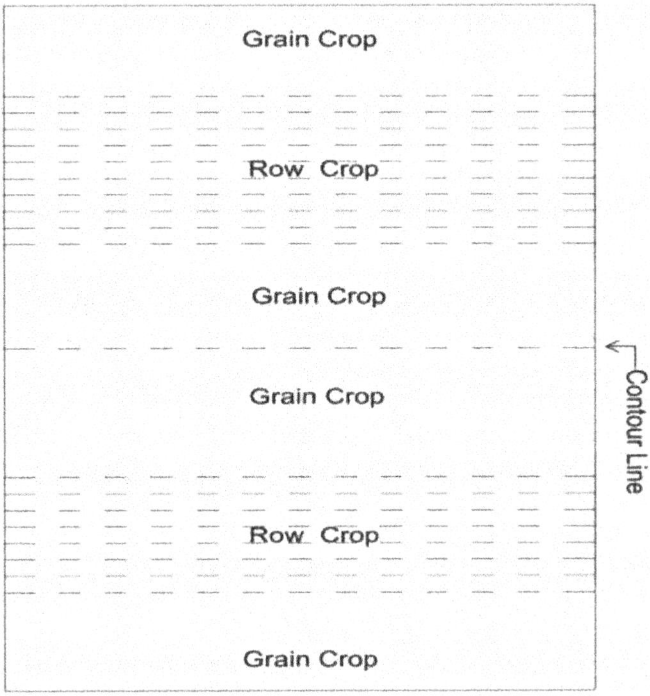

FIGURE 3.2 Contour strip cropping.

7. Use plant residues as mulches to reduce evaporation and increase organic content of soil
8. Regulate the water regime of soil for better plant growth

The following are the commonly performed tillage operations:

1. Deep tillage
2. Removal of stubbles
3. Land smoothening
4. Harrowing

The following types of tillage operations are carried out for conservation of soil (Wischmeier, 1970):

Minimum tillage: It is a combination of tillage and sowing in one operation. It is also called the minimum tillage. Such operations create fine lumps of soil and a coarse soil surface between rows. Porous texture of soil creates a good infiltration capacity. This operation leads to reduction of soil erosion by about 40% and surface runoff by about 35%.

No-tillage: Soil surface is not disturbed much and left more or less intact in no-tillage system of operation. Operations performed are loosening, under cutting, and drying of upper layer of soil, with purpose to stop the growth of weeds and whiskers of preceding crops persist as such in the field. Under cutting, operations are not necessary as there are no weeds, and seeds are sown straight into soil by distinctive kinds of seed drills. Reduction in soil erosion up to 70% has been obtained following this tillage process.

Strip tillage: Strip tillage is a development over no-tillage system. Generally, in this type of cultivation, narrow strips of approximately 0.1 m depth and 0.2 m width are laid following the contour, and land in-between the strips is left uncultivated. Also, these are known as loosening strips. In constructed narrow strips, there are no stubbles, which aids in sowing operation and enables better growth of plant.

The following parameters are sought to be achieved by a conservation tillage operation, in addition to creating a soil surface conducive to good plant growth:

1. To reduce the turning up of the fertile top soil
2. To minimize the breaking up of the surface area
3. To make judicious use of the stubbles of previous crops for increasing the soil organic matter and using it as a mulch

The farm implements generally utilized for conservation tillage operations are arrows-shaped blades, V-shaped sweeps for controlling erosion instigated by water, and chisel-like blades for controlling erosion instigated by rod weeders, wind, etc.

3.8.4 Mulching

Mulches are utilized to reduce evaporation, minimize rain splash, control weeds, allow temperature which is conducive to microbial activity, and reduce temperature of soil in hot climates. Stubbles, different types of vegetation, trash, and polyethylene are some of the most common varieties of mulches used. These materials are spread over the land surface. Mulches assist in breaking energy of rainwater, avoid splash and dissipation of soil structure, and block the flow of runoff for reducing their velocity and stop inter-rill (sheet) erosion, which in return helps to improve the infiltration capacity by sustaining a conducive soil structure at top surface of the land.

3.9 MECHANICAL PRACTICES

Mechanical practices are engineering measures used to control soil erosion from sloping land surfaces. In the design of such particles, the basic approach is (i) to increase the time of stay (opportunity time) of runoff water in order to increase the infiltration time for the water, (ii) to decrease the effect of land slope on runoff velocity by intercepting the slope at several points so that the velocity is less than the critical velocity, and (iii) to protect the soil from erosion caused by the runoff water.

On undulating land surfaces, water gets concentrated along the watershed valley lines. On sloping terrains, where good soil depths are available, gully erosion starts owing to scour of runoff. In such types of land surfaces, land slopes can be reduced by land-grading (levelling) operations. Such operations are recommended where the height differences are not more than 10 cm. In large projects, terrains with 15–20% slopes are also graded with the help of heavy machinery. During land-levelling operations, it is always advisable to scrape the top soil and replace it during final operations to maintain the fertility of soil.

3.9.1 Terraces

The terrace is an earthen embankment, built across slope, for controlling runoff and minimize soil erosion. Terrace acts as an intercept to slope of land distributing surface of land slope into strips. Length of run of the runoff is reduced in limited widths of strips. It has been established that soil loss is proportional to square root of the slope length, i.e. by limiting run length, there is reduction of soil erosion. Soil eroded by raindrop splash and runoff scour streams down the slope and gets blocked up by terraces. Scour of soil surface due to runoff water is started by runoff at a velocity above critical value, achieved during a flow on a long length of sloping run. By restricting the run length, runoff velocity remains less than critical value and hence there is prevention of soil erosion due to scour. Terraces are classified into two major types: broad-base terraces and bench terraces.

3.9.1.1 Terrace Spacing

By using the USLE, the horizontal spacing, that is the length of slope of each strip, can be determined for an admissible soil loss. If in the USLE (3.16) it is assumed that the admissible soil loss is A_{adm}, and the admissible factor for the length of slope is L_{adm}, then

$$L_{adm} = A_{adm} / RKSCP \quad (3.40)$$

From L_{adm} the length of run (L_p) can be determined by using Equation 3.16. The US Soil Conservation Service has recommended spacing for graded terraces, where the land slope is a function of vertical and horizontal intervals, respectively. The space between two adjacent terraces should not be very wide, because the length of run would increase, leading to soil erosion and the flow of eroded soil to the terrace channel. Many empirical relationships for determination of the vertical interval have been developed for different areas, but the following formula is widely used with the adjustment of 25% for local conditions:

$$VI = 0.3(as + b) \quad (3.41)$$

where

VI = vertical interval
a = a constant dependent on geographical location
s = average land slope above the terrace
b = a constant for soil erodibility and cover conditions during critical periods

The USLE is also used to determine the value of vertical interval, based on the admissible soil loss limit. By using the USLE – Equation 3.16 – the value of LS factor can be obtained:

$$LS = \frac{A}{RKCP}$$

When the values of parameters A, R, K, C, and P are known, the value of LS factor can be determined, or when the values of length of run (L_p), land slope (s), and soil loss (A) are known, then by using the above equation, the value of LS can be determined. If the admissible soil loss is given by

$$LS_{adm} = LS \times \frac{A_{adm}}{A}$$

From LS_{adm}, the values for L_{adm} and S_{adm} (%) are determined and the vertical interval is calculated as

$$VI = S_{adm}(\%) \times \frac{L_{adm}}{100}$$

Generally, broad-base terraces are not recommended for field which have slopes beyond 10–12% and should not in any case be constructed on slopes which are more than 20%. Level terraces are primarily meant for conservation of moisture and prevention of soil loss in arid and semi-arid areas, where rainfall is low. In such areas, the spacing between two consecutive terraces is dependent on infiltration rate, runoff, and period of stay of runoff in the channel. The terrace ridge should have sufficient height to protect itself from being over-toppled by the accumulated runoff in the channel.

3.9.1.2 Terrace Grade

Terrace grade is the gradient provided to the base of the terrace channel towards its outlet to remove the runoff water at a non-erosive velocity. The gradient in respect of graded terraces could be either uniformly sloped or variably graded. In level terraces, the grade is 0% because no removal of runoff is required. Graded terraces, constructed to remove ponded water, have longer lengths and pass through varying topographic levels of the field. In such situation, it is recommended not to have uniform grades for terraces, but have variable grades up to the outlet to obtain better control on runoff flow. The channel capacity should progressively increase towards the outlet, which can be ensured by increasing the channel slope gradually up to the outlet. The channel grade is kept minimum at the upper end of the terrace and maximum at the outlet end. This method also permits more infiltration of the runoff water at the upper reaches.

In uniformly graded terraces, though the gradient may have a value from 0.1 to 0.6%, 0.4% is the most common figure. Steeper grades are provided in relatively impervious soil and shorter terraces.

3.9.1.3 Terrace Length

The design length of a terrace depends on the shape of the field, the type of outlet, the runoff rate, and the channel capacity. Very long terrace generally leads to soil erosion and should be avoided. In some cases, however, long terraces are provided with non-erosive gradient on both of their sides from the middle and the runoff is allowed to flow out from both the ends into two separate outlets. The important criterion for the design is that the flow through the channel should be non-erosive.

The length of the terrace generally varies from 300 to 500 m, depending on the requirements of the area. Sometimes, long terraces in permeable soils are provided with check dams, at about 150-m interval to conserve moisture and prevent any breach in the terrace owing to high accumulation of runoff at the downstream end. As regards level terraces, at times, the ends are left opened or partially opened to prevent any over-toppling of ridges when there is a high runoff.

3.9.1.4 Terrace Cross Section

The cross section of terrace should have the required capacity for runoff and wide side slopes for farming operation. It should get easily constructed at reasonable cost. The design of terrace which has parabolic shapes of channel and ridges under natural conditions needs either a triangular or a trapezoidal approach.

During construction, the soil on the terrace ridge is loose and it settles down only after some times. A provision for settlement of 10–20% of soil is made. The terrace channel depth is kept in between 30 and 40 cm and the minimum cross-sectional area, about 1 km^2, of terrace as a function of land slope. It is recommended that the ridge back slope and the terrace ridge should be kept under sod or grasses.

3.9.1.5 Terrace Outlet

Terrace has to be provided with safe and proper outlet to dispose of the excess runoff. The outlet could be natural streams or drains, constructed drains, sod flumes, pastures, wastelands, stabilized gullies, vegetated airways, etc. The outlet for pastures, wetlands, etc., which are not channel flows, should be carefully planned and staggered, to avoid erosion in these areas.

3.9.1.6 Terrace Location

The location for the layout of a terrace system depends on the requirements of the watershed characteristics and the type of farming to be practised in the area. The primary consideration is the ease of farming operations and reduction of maintenance costs. The ridges are recommended to be placed just above the erosion-affected areas in the field. Generally, the topmost terrace is laid out first, by starting its construction from the outlet point, because in the event of any breach in the system, the resulting sudden outflow of runoff through it will not be able to damage any constructed terrace, as there will be none below it. Breaches occur as sometimes the fills of soft soil are not able to hold on and give away during a heavy rainfall.

In places where the area above the top terrace has a pointed top, it is recommended that the vertical be kept 1.5 times the design interval. In places, where the area above the terrace cannot be included under the terracing system, the runoff outflow from this area should be drained to the outlet through an interceptor drain, so that this load of water does not damage the other terraces.

3.9.1.7 Terrace Construction

Machinery used for the construction of terraces can be divided into two groups: bullock-driven equipment and power-driven machinery. Bullock-driven equipment includes mould-board ploughs, buck scrapers, and V-ditchers. Power-driven machinery includes bulldozers, rotary scrapers, motor patrollers, elevating graders, mould-board, disc ploughs, etc. Before undertaking construction, the area should be cleared off from all types of vegetation, stubbles, etc. The ploughing is done either from both sides or from the upper sides only. Though mould-board ploughs are commonly used, bulldozers are employed for large-scale terracing work.

3.9.1.8 Terrace Maintenance

Utmost care is required for the upkeep of terraces as the fill soil is initially in a loose condition. Some important tips for good maintenance of terrace are given below:

1. As far as possible, all farming operation should be carried out parallel to the terraces.
2. Crops should be preferably planted parallel to the terrace, and soil-conservation-oriented rotation of crops should be adopted.
3. Any break or breach in the terrace system should be repaired promptly, as it may widen with another rain and create serious flood and erosion problem.
4. Sometimes strip cropping is combined with terracing as it is a very effective method of soil conservation.

3.9.2 Broad-base Terraces

Broad-base terraces were developed by the farmers in the western countries (particularly in the United States), out of their experience only. These types of terraces consist of a ridge which has a flatter slope and a fairly broad base, so that farming equipment can simply pass over the ridge area. On broad-base terraces, ridge area is also cultivated and these are no loss of land for agricultural processes owing to terracing. These are of two types:

1. Graded terraces
2. Level terraces

3.9.2.1 Graded Terraces

Drainage or graded terraces are built in high (>600 mm) precipitation regions, where excess runoff needs to be safely and quickly removed for protecting the crops from waterlogging. These terraces are provided with either a constant or a variable grade along the terrace length to convey the excess runoff safely into a vegetated outlet channel. In drainage-type terraces, a channel is normally provided, along with the ridge on the upstream side, to convey the excess runoff to the outlet. In some cases, the entire strip of terrace system is utilized as a channel for transmission of runoff to the outlet. This system is practised where the excess runoff produced is less, and water can be allowed more time for getting absorbed into the soil and thus increase its soil moisture. Thus, graded terraces can be of two types: (a) terraces with a proper channel, (b) terraces without a proper channel.

3.9.2.1.1 Graded Terraces with a Channel

In case of graded terraces with a channel system, the ridge of the terrace is provided with a drainage channel on its uphill side. The terrace design involves the design of the channel and the ridge as a combined unit. These channels are shallow and wide and have a slope of low gradient. They intercept and infiltrate the water running down the land surface. They do not hinder cultivation, and can be crossed over by the farm machinery and sown with crops. Through graded terraces, the excess runoff is removed at a velocity which does not cause erosion of soil. It is called the non-erosive velocity. The design of graded terraces basically involves the design of three parameters: spacing, capacity, and cross section.

3.9.2.2 Level Terraces

Ridge or level terraces are similarly of two types: narrow-based and wide-based. These are based on the width of the channel and the ridge. Narrow-based terraces have widths ranging from 1.2 to 2.5 m, and are inappropriate for operation of farm machinery. Narrow-based terraces are common in India and are known as contour bunds.

3.9.3 Bench Terraces

Bench terraces are constructed for reducing the gradient of slopes of hillsides to help impede runoff paths, and consequently prevent soil erosion. In hilly regions, bench terraces are broadly utilized as agricultural field plots; however, these are costly in construction. Bench terraces cause extensive interventions in the landscape and are objected to by environmentalists, in the present-day context of environment consciousness. These are constructed on hillsides which have slope gradients in the range of 15–30% or more and land surface has a medium to high degree of hazard of soil erosion. The dimensions of these terraces depend on the gradient of the slope, depth profile of the soil, volume of the earthwork involved, and the crops to be grown. Bench terracing transforms a steep land surface into a series of nearly levelled steps across the slope of the land. These terraces convert the erodable sloping lands into farmlands that are safe for cultivation. Though these terraces are costly to construct, they are being used as no other better land is available for cultivation in such areas.

3.9.4 Bunds

Though bunds are similar to narrow-base terraces, no farming is done on bunds except at some places, where some types of stabilization grasses are planted to protect the bund. Generally, two types of bund systems are practised: graded and contour. The choice of the type of bund is dependent on land slope, rainfall, soil type, and the purpose of making the bund in the area. Graded bunds are used for safe disposal of excess runoff in areas with high rainfall and relatively impervious soil, while contour bunds are recommended for areas with low rainfall (<600 mm), agricultural fields with permeable soils, and having a land slope of less than 6%.

3.9.4.1 Graded Bunds

They are constructed in regions where land is vulnerable to water erosion, permeability of soil is less, and the area has waterlogging complications. Design of graded bund system is done for disposing of excess runoff securely from agricultural fields. It is laid out with a longitudinal slope gradient leading to the outlet. Gradient can be either variable or uniform. Uniformly graded bunds are appropriate for areas where runoff is low and bunds need smaller lengths. Variable-graded bunds are necessary where bunds require longer lengths, because of which the cumulative runoff keeps getting higher towards the outlet. In this type of bund, variations in grade are given at various divisions of the bund to keep the velocity of runoff within desirable limits, so as to prevent the occurrence of any soil erosion or landslides.

3.9.4.2 Contour Bunds

Contour bunds are laid out in those areas which have less rainfall and less permeable soils. The major requirements in such areas are prevention of soil erosion and conservation of rainwater in the soil for crop use. To maximize the conservation of rainwater in the soil, no longitudinal slope is provided to the field strip. In such a system of bunding, the bunds are designed to be laid out on contours with minor adjustments, wherever necessary. The layout of bunds on contours makes no provision for any longitudinal slope to the field strip. Therefore, the rainwater in each strip collects at the down end of the horizontal slope, i.e. behind the up-slope side of the lower bund. This ponded water slowly infiltrates into the soil and increases the soil moisture. Because of the impact of rainfall and the runoff scour, soil erosion takes place and the eroded soil from the up-slope areas flows down along with the runoff water and settles behind the lower bund. This fills the down-slope area with soil particles, and eventually decreases the horizontal slope gradient of the strip. In course of time, the lower end gets filled up with soil and the strip gets more or less levelled. The process of deposition of soil near the lower bund also decreases the effective height of the bund. The height of the lower bund, therefore, either needs to be increased periodically, or a provision needs to be made at the time of the design itself. Soil loss on this account can be estimated by using the USLE, or any other soil loss equation, and the lower bund designed accordingly.

3.10 CONCLUSION

The main source of the sediment was identified as the soil erosion from the settlement areas within a watershed. From survey it is found that the volume of annual soil erosion is estimated at around 90% of the annual sediment inflow into the watershed. This may be due to the poor land management and agricultural development of local farmers. This chapter provides a clear description on different types of soil erosion and its estimation through different scientific formulae. Also, controlling of soil erosion through different approaches has been described. Similarly, different vegetative practices like contouring, strip cropping, tilling operation, and mulching are described to minimize soil erosion.

REFERENCES

Bagnold, R. A. 1941. The Physics of Blown Sand and Desert Dunes, Methuen, London.

Chepil, W. S. 1953. "Field structure of cultivated soils with special reference to erodibility by wind." *Soil Science Society of American Journal*, 17, 185–190.

Chepil, W. S. 1960. "How to determine required width of field strips to control wind erosion." Journal of Soil & Water, 15 (2).

Chepil, W. S. and N. P. Woodruff. 1963. "The physics of wind erosion and its control." *Advances in Agronomy*, 15, 211–302.

Das, G. 1982. Runoff mind sediment yield from upper Ramganga catchment, Ph.D. thesis, G.B. Pant University of Agriculture and Technology, Pantnagar, India.

Das, G. 2009. Hydrology and Soil Conservation Engineering, PHI Learning Private Limited, New Delhi.

Das, G. and H. S. Chauhan. 1990. "Sediment routing for mountainous Himalayan region." *Transactions of the ASAE*. 33 (1), 95–99.

Dhruvanarayana, V. V. 1993. Soil and Water Conservation Research in India, ICAR.

Dvorak, J. and L. Novak (Eds.). 1994. Soil Conservation and Silviculture, Elsevier, Amsterdam.

Elwell. 1978. "Modelling soil losses in southern Africa." *Journal of Agricultural Engineering Research*, 23, 117–127.

Finkel, J. H. and Noveh. 1986. Semi-Arid Soil and Water Conservation, CRC Press Inc., Boca Raton, Florida.

Foster, G. R., D. R. 'McCool, K. G. Renard, and W. C. Mobdenhaur. 1981. "Conversion of universal soil loss equation to SI metric units." Journal of Soil Water Conservation.

Franken Berger, E. 1951. Quoted by WMO.

Gillette, D. A., I. H. Blifford, Jr., and C. R. Fenster. 1972. "Measurement of aerosol size distributions and vertical fluxes of aerosols on land subject to wind erosion." *Journal of Applied Meteorology and Climatology*, 11, 977–987.

Gurmel Singh, Ram Babu, and Subhash Chandra. 1981. "Soil loss prediction research in India. Bulletin No. T-12/D-9, Central Soil and Water Conservation Research Training Institute, Dehra Dun.

Hsu. S. A. 1973. "Computing eolian sand transport from shear velocity measurement." *Journal of Geology*, 81, 739–743.

McCool. D. K., G. R. Foster, C. K. Mutcheler, and L. D. Meyer. 1989. "Revised slope length factor for the Universal Soil Loss Equation." *Transactions of the ASAE*, 32 (5): 1571–1576.

Musgrave. G.W. 1947 "The quantitative evaluation of factors in water erosion, a first approximation." *Journal of Soil Water Conservation*. 2, 133–138.

Pasak. V. 1974. "Determination of potential wind erosion of soil." Transactions of the 10th Congress hit. Soc. Soil. Sci.

Ram Babu, K. G. Tejwani, M. C. Agarwal, and I. S. Bhusan. 1978. "Determination of erosion index and isoerodent map of India." *Indian Journal of Soil Water Conservation*, 6 (1).

Ram Babu, K. G. Tejwani, M. C. Agarwal, and I. S. Bhusan. 1979. "Rainfall: intensity duration return period equations and nomographs of India." Proceedings of the 1st International Conference on Statistical Climatology, Hachioji (Tokyo), Elsevier Scientific Publishing Company, Amsterdam, 1979.

Rattan Lai. 1990. Soil Erosion in the Tropics, McGraw-Hill, New York.

Skidmore, E. L. and W. P. Woodruff. 1968. "Wind erosion forces in the United States and their use in predicting soil loss." Agriculture Handbook Number 346, Agricultural Research Service, USDA.

Smith, D. D. and W. H. Wischmeier. 1962. "Rainfall erosion." *Advances in Agronomy*. 14, 109–148.

Williams, J. R. 1975. "Sediment yield prediction with universal equation using runoff energy factor." In: Sediment Yield Workshop, Oxford, November 1972. Present and Prospective Technology for Predicting Sediment Yield and Sources, USDA Sedimentation Lab, Oxford.

Williams, J. R. and H. D. Berndt. "Determining the universal soil loss equation's length-slope factor for watersheds." In: Proceedings of the National Soil-Erosion Conference, May 25–26, 1976, Akeny, Iowa.

Wischmeier, W. H., "Relationship between soil erosion, plant growing and soil management" (in Czech)." Proceedings of the International Symposium on Water Erosion, ICID, Prague, 1970.

Wischmeier, W. H. and D. D. Smith. 1958. "Rainfall energy and its relationship to soil loss." *Eos: Transactions American Geophysical Union*, 39 (2), 285–291.

Wischmeier, W. H. and D. D. Smith. 1965. "Predicting rainfall erosion losses from cropland east of the rocky mountains." USDA Handbook No. 282, USDA, Washington DC.

Wischmeier, W. H. and D. D. Smith. 1978. "Predicting rainfall erosion losses: a guide to conservation planning." USDA Handbook No. 537, USDA, Washington DC.

Zingg, A. W. 1940. "Degree and length of land slope as it effects soil loss in runoff." *Agricultural Engineering*, 21, 59–64.

4 Water Harvesting

4.1 INTRODUCTION

At the rate in which the Indian population is increasing, it is said that India will surely replace China from its number one position of the most densely populated country in the world after 2020–2030. These will lead to high rate of consumption of most valuable natural resource 'water' resulting in augmentation of pressures on the permitted freshwater resources. Ancient methods of damming rivers and transporting water to urban areas has its own social and political issues. In order to conserve and meet our daily demand of water requirements, we need to develop alternative cost-effective and relatively easier technological methods of conserving water. Rainwater harvesting is one of the best methods fulfilling those requirements. Rainwater harvesting is the accumulating and storing of rainwater for reuse before it reaches the aquifer. It has been used to provide drinking water as well as water for livestock, irrigation, and other typical uses. Rainwater collected from the roofs of houses, tents, and local institutions can make an important contribution to the availability of drinking water. In arid and semi-arid regions, there is a scarcity of both surface water and groundwater. Hence, efforts are made in these areas for collecting and preserving rainwater to maximum possible level. This collection of rainwater is called water harvesting, which is defined as collection of rain and runoff water primarily for irrigation and human and livestock consumption. Mostly, water harvesting refers to direct collection of rainwater. The water collected can be stored to use in future and again recharged into groundwater. Rainfall is the principal source of water, whereas lakes and rivers are secondary sources. As water is necessary for irrigation, there are two ways by which it is preserved. The first method involves collection of rainwater or storing in the soil itself for immediate use. In the second method, rainwater is stored in reservoirs and ponds. For human and livestock consumption, water also needs to be treated to make it safe for consumption.

Design of a storage structure depends on area of the catchment, amount of runoff which is to be stored, and amount of water required for irrigation of a specific size of cultivated area. The design should take account of losses during storage. The ratio of catchment area to cultivated area is used, as a rule of thumb, for the design of reservoir capacity. The relationship is expressed as follows:

Catchment area × Expected rainfall × Runoff coefficient and an efficiency factor = Cultivated area (Crop water requirement − Depth of rainfall)

$$\text{Therefore, } \frac{\text{Catchment area}}{\text{Cultivated area}} = \frac{(\text{Crop water requirement} - \text{Depth of rainfall})}{(\text{Expected rainfall} \times \text{Runoff coefficient and efficiency factor})} \quad (4.1)$$

Wherever data are available, consumptive use model, curve number method, and recurrence interval model of storm can be utilized for determining some of the parameters with fair accuracy. If the area ratio is 1:8, it would mean that 1 ha of cultivated land would require runoff from 8 ha of catchment area for its irrigation requirement.

Human beings have been reliant on adequate supply of water for their well-being, food, and security. Water is considered as a principal regulating aspect for human life and is a universal need. Devastation of natural catchments has led to serious water scarcities affecting massive population and areas. Different means helping to ensure sufficient water supply for domestic, agrarian, and other usages are available to communities and farms. This process is known as water harvesting. Practice of collection and storage of water from different sources for beneficial usage is a part of water harvesting process. Water collected from watershed and directed to ponds for storing purpose can significantly improve availability of water for garden livestock watering, irrigation, aquaculture, and other domestic necessities.

4.2 TYPES OF WATER HARVESTING STRUCTURES

4.2.1 Contour Furrows

An imaginary line on ground along which all points are at equal elevation is called a contour line. Contour ridges and furrows are built along contour lines, and this method is known as contour farming. Contour farming helps in reducing soil erosion, supporting runoff water infiltration into soil, and can also be relatively applied by small-scale farmers. Strip cropping or cultivation is a process involving growth of a cultivated crop (e.g. corn) in striping pattern with sod-forming crop (e.g. hay) in alternate strips in an organized manner for following an estimated land contour and minimizing erosion of soil.

4.2.2 Contour Bunding

Contour bunds are built in moderately low precipitation areas with annual precipitation less than 600 mm, mainly in regions containing lightly textured soil. These are basically meant to store rainwater received in the course of a 24-hour period at 10 years return period. Maximum water depth to be allowed, design flow depth over waste weir, and preferred freeboard are major considerations in constructing contour bunding.

4.2.3 Micro-catchments

Finkel and Naveh (1986) have described different types of bunded micro-catchment, which are popular in some African and Middle East countries. Micro-catchment is a specifically planned area consisting of slope for increasing runoff from precipitation and collects the resulting runoff in a planting basin. Here, infiltration of runoff occurs and water is made available to plants for its subsequent usage. It has many advantages:

i. Moisture is provided to plants with a little effort by micro-catchments and is easily developed utilizing manpower.

ii. They can be used to grow many plants and crops.
iii. Improvement in fertility of soil by trapping organic matter impending from catchment.

They can be of various shapes and sizes conditional to effective rainfall, slope, plant type, soil texture/structure etc. Shapes can be rectangular, square, triangular, semi-circle, and V-shape (Figure 4.1).

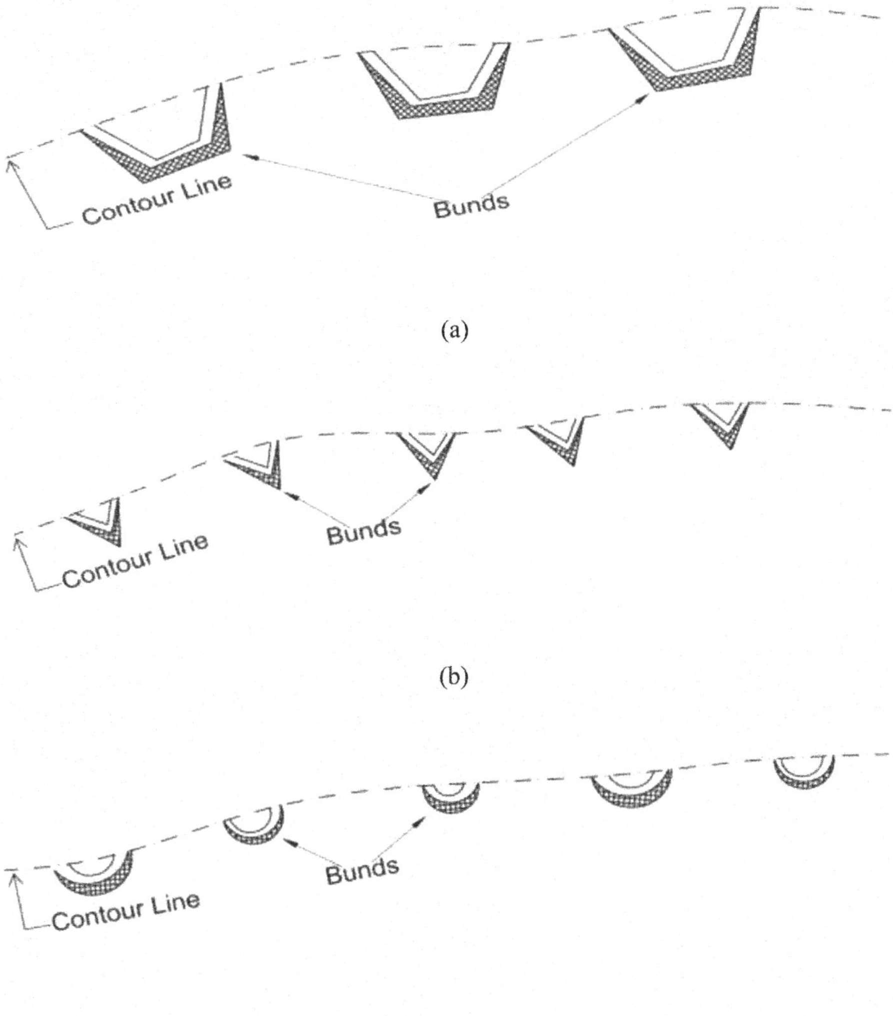

FIGURE 4.1 (a) Trapezoidal, (b) triangular, (c) semicircular bunded micro-catchment for water storage.

4.3 WATER YIELD FROM CATCHMENTS

4.3.1 Vegetation Management

The amount of water yield from a catchment depends on physiographic and climatic factors. Climatic factors involve precipitation, rainfall intensity, rainfall duration, direction of prevailing wind, rainfall distribution, etc., whereas physiographic factors include shape, size, slope and orientation of watershed, soil type, soil moisture, land use, drainage density, and topographic characteristics. Strips of grass planted along a contour can reduce soil erosion and water yield. They can also be used in conjunction with other water-harvesting techniques. Forestry is the best land cover for regulating seasonal flows, maximizing water yield, and ensuring high quality of water. Based on this assumption, conservation of forest land in upstream catchments was considered as the most efficient step for enhancing availability of water for domestic, industrial, and agriculture practices, along with preventing flood occurrences in downstream regions.

4.3.2 Land Alteration

Any action altering the topography of land or removing vegetation from land by clearing, grubbing, grading, tree removal, excavating, or filling, except for activities carried out for maintaining existing ground conditions, is known as land alteration. The land alteration of a catchment can increase the surface runoff yield. This method is recommended for those places, where land surface is undulating and a flatter surface is needed to enable runoff to flow smoothly into storage tanks.

4.3.3 Treatment of Soil

Treatment of soil with chemicals has been found to be a promising technique for reducing infiltration and making soil water repellent or hydrophobic, and thus increasing runoff rate. These chemicals (sodium salts) act as soil dispersants, which reduces permeability of soil, and also as soil sealants (fuel oil, paraffin wax, methyl silanolate, etc.) making the soil impermeable. Physical treatments of soil, which can retard infiltration and increase runoff, comprise soil covers of impermeable materials, such as fibreglass, concrete, plastic films, asphalt, polythene matting saturated with asphalt, artificial rubber membrane, sheet metal, etc.

4.3.4 Small Water Storage Structures

Rainwater harvesting fundamentally aims at collection and storage of water in abundance days and can be used in lean periods. Rainwater can be stored in two ways: (i) storing in the soil media as groundwater and (ii) in an artificial storage. Roof water harvesting is one kind of artificial storage and is somewhat a transitory measure which focuses on human necessities providing instant relief from scarcity of water, whereas groundwater storage has potential for providing sustainable relief from scarcity of water, addressing necessities of every living class in nature. Rainwater or runoff in the form of a stream or spring can be harvested in ferrocement, RCC, fibre tanks, plastic, or different kinds of low-cost lined ponds to utilize in lean days. Experiments showed that collected rainwater could be utilized for irrigating winter season vegetables using micro-irrigation arrangement for enhancing profitability and productivity.

4.4 RUNOFF DIVERSION

Runoff water from seasonal stream is diverted to field either directly or using a network of field/canals channel. Appropriate actions are taken to prevent silt deposition in conveyance channels and field. Runoff water is sometimes diverted to graded bunds with channels for flood irrigation. Such practices are common in northwestern India, Pakistan, Afghanistan, etc.

4.4.1 Check Dams

Check dams are built in natural waterways to block runoff up to a definite height in the waterway. Surplus water is allowed for flowing over the dam. The blocked water flows gradually and infiltrates into soil, which finally recharge the groundwater aquifers. The water is harvested by digging well very near to infiltration zone. Underground vertical barriers are also occasionally used in some particular area, where valley wall below the surface are comparatively impervious and bedrock under the stream is not too deep (Finkel and Naveh, 1986).

4.4.2 Underground Impervious Layer

Sandy soils, which are very porous, impervious layer of polythene sheet, or emulsified asphalt, after a certain depth, have been found to check infiltration of water to underground layers. Specifically designed machineries are utilized for the layout.

4.4.3 Rooftop Rainwater Harvesting

In semi-arid and arid regions, like Rajasthan, Gujarat in India, rooftops are used for harvesting rainwater. The slope of the roofs to a drainage channel (pipes) collecting rainwater into a reservoir is positioned down below. This water is mostly used for domestic purposes and animal consumption.

4.5 PONDS

Ponds are designed with surplus storing capacity for attenuating surface runoff during rainfall. They comprise a permanent pond area with landscaped banks. They are created by constructing embankments, utilizing an existing natural depression, or excavating a new depression. Prevailing natural water bodies should not be utilized because of risk due to pollution and poor quality of water that might damage/disturb natural ecology of the system. These are constructed to store runoff or spring water for irrigation, domestic, and livestock uses.

Additional storage capacity in ponds can provide both water quality treatment and storm water attenuation by retaining runoff and releasing it at a measured rate.

Design of ponds are done in such a way that runoff from all storms can be controlled by surface drainage storing and discharging it gradually after flooding risk has passed. Runoff is detained from each rainfall occurrence and then treated in pond. They are frequently redesigned with variety of shrubs, grasses, or/and wetland plants for providing aesthetic benefits and bank stability.

4.6 RESERVOIRS

A reservoir is a storage space for water. Usually a reservoir is an enlarged artificial or natural lake, storage pond, or impoundment formed using a lock or dam for storing water. They can be made by regulating a stream which drains a prevailing water body and can also be constructed in river valleys utilizing a dam. In an alternate way, reservoirs can be constructed by flat ground excavation or building retaining walls and levees.

Construction of a dam across a river might create water storage reservoirs, along with appropriate appurtenant constructions. Reservoir size is based on demand for which it is being built. In addition, reservoirs are meant for absorbing a portion of floodwater and discharging surplus water by a spillway. A reservoir basically stores runoff and its size is administered by volume of water which must be put in storage. This is affected by inflow variability accessible for reservoir. The following are several purposes for building reservoirs. The major function of a reservoir is to provide water for fulfilling these multiple purposes, which are called 'Multi-Purpose' reservoirs.

- *Consumption:* Both by humans and for industrial usage.
- *Agricultural purpose:* Generally for supplementing inadequate rainfall.
- *Hydropower systems for pumped storage:* Here water pours from upper end to the lower end of reservoir, producing energy and power at times of heavy demand using turbines that might be alterable, and water is pumped back to the upper end of reservoir at times of extra energy. For meeting peak demands, this cycle is repeated usually once or twice in a day. Inflow to such reservoirs is not vital, provided it is mandatory for replacing losses of water by evaporation and leakage or for generating surplus electricity. In such amenities, conduits, power stations and either or both of the reservoirs can be built underground if it was found necessary for doing so.
- *Hydropower:* For generating energy and power during availability of water or for providing consistent supplies of energy and power at all times when necessary for meeting demands.
- *Controlling flood:* Maintenance of capacity for storage is essential for absorbing probable inflow of flood to reservoirs, as they would cause excess of acceptable discharge spillway opening. Storing of retained floodwater lets it for future usage.
- *Convenience use:* Includes provision for sight-seeing, boating, fishing, and water sports.

For determining catchment area of pond, the method of ratio of catchment area to irrigated area or equivalent can be followed. An allowance for losses due to seepage etc. is involved in the design. Catchment area should be vegetated for preventing eroded soil (sediment) from being transported along with runoff to pond/reservoir.

4.6.1 Components of a Reservoir

Reservoir consists of three basic components:

i. Water storage area
ii. Spillway
iii. Earth embankments

4.6.2 Reservoir Location

The area selected for the construction of a reservoir should be such that it requires the minimum possible earth work, so that it is not expensive. To save the stored water from contamination, sewage disposal drains, septic tanks, soakage pit, etc. should be kept away or interceptor drains should be provided to check the seepage of contaminated water into the reservoir.

4.6.3 Reservoir Storage Capacity

It is calculated on the basis of the amount of water required by different types of users such as human and animal consumption, agricultural and industrial purposes, etc. Losses because of seepage and evaporation account for 40–70% of the storage capacity, depending on the type of soil, season, temperature, etc. This is a very large quantity and should be considered for a design.

Schwab et al. (1993) and Murty (1985 and 2002) have given the following formulae to calculate the volume of storage in a reservoir from the contour map of the site.

$$\text{Average end area formula: } V = d\left(A_1 + A_2/2\right) \tag{4.2}$$

where

v = volume of storage (L^3) between two contours
D = vertical distance between the end area or the two contours (L)
A_1 and A_2 = end area (L^2)

$$\text{Prismoidal formula: } V = \frac{d}{6}(A_1 + 4A_m + A_2) \tag{4.3}$$

where

A_m = middle area halfway between the end area (L^2)

4.6.4 Reservoir Behind a Dam or an Embankment

At the slope end of a watershed, a reservoir if required can be constructed by erecting a dam or an embankment on the stream. The reservoir storage volume is then calculated by crossing the contour line with the centre line of the dam and enclosing the contour lines. The rest of procedure is similar to that of dug-out pond. The calculations for capacity and catchment area are similar for the dug-out and embankment type of reservoir.

4.6.5 Reservoir Spillway Design

Spillways are structures provided in reservoirs to discharge outflow from them, at controlled velocities, for domestic use or for the outflow of excess water during an emergency. Such situation arises in those types of reservoirs which are constructed by putting up a dam or an embankment. In dug-out ponds, invariably water is directly pumped out for use. Such ponds are located in those places where there is no provision for any drainage. The dug-out ponds are constructed below the ground surface on all sides and there is no danger of breach or over-toppling of retaining wall (dam) and its failure. Therefore, no emergency spillways are required to be provided. However, the inflow to the pond is regulated.

The level of emergency spillway crest is kept at the maximum allowable flood level in the reservoir. This level is generally based on the selected storm frequency of flood expected in the area. The size of the outlet is calculated based on the required outflow from it, by using a weir formula after assuming the required depth of flow over the weir. The outflow channel is provided with a 2:1 slope and is recommended to be sodded or vegetated. The permissible flow depth in the channel is 30 cm. Based on the situation, either a drop spillway or a drop-inlet spillway, or a chute spillway is constructed.

4.6.6 Reservoir Seepage Loss

There occurs considerable loss of stored water in ponds owing to seepage. The following measures are adopted to reduce the seepage losses.

4.6.6.1 Soil Compaction

Soil with 70% sand, 20% clay, and 10% silt can be effectively compacted to reduce infiltration.

4.6.6.2 Lining

a. A mixture of a swelling type of clay, namely bentonite and soil, is used for lining the surface of the pond. The soil is applied as a layer. It is necessary for the soil material to have 10–15% of sand for strength.
b. A layer of polythene sheet is laid out on the pond surface with the help of a mildly hot ironing box. Above the sheets, a 15-cm thick layer of soil is placed to protect the polythene from damage.
c. A layer of masonry lining of bricks and cement is the best, but it is expansive and hence generally not adopted.
d. A layer of soil cement is sometimes also used.

4.6.7 Protection of Reservoir Banks

Vegetation planted on the banks of large reservoir improves the stability of banks, provides protection against water waves, converts the water area into a landscape, protects against soil erosion, enhances the quality of water for aquatic animal, and provides woods and timber. The distribution of types of plants depends upon water level in the reservoir, position of side slope, and types of plants.

4.6.8 Damage to Reservoir Sides by Landslides and Abrasion

Reservoir sides are highly susceptible to abrasion because of high slopes and also during runoff and wave action. The side need to be protected fully under such situation.

4.7 EARTHEN EMBANKMENTS FOR WATER CONSERVATION

Embankments built of rock and earthen materials are usually referred to as earthen embankments or fill-type embankments. Earthen embankments are categorized into two major classifications, mainly based on types of soil utilized as construction ingredients, for example earth-fill embankments and rock-fill embankments. The central structure of rock-fill embankments having structural resistance contrary to failure consists of transition zones and rock-fill shell. For minimizing leakage through embankment, core and facing zones play a vital role. Provision of filter zone in any kind of rock-fill dam is essential for preventing loss of soil elements by erosion because of seepage flow via embankments. On the other hand, in an earth-fill dam, dam body should have both seepage and structural resistance contrary to failure with provision of drainage facilities.

In a project, type of embankment is determined by keeping in view several aspects associated with geology and topography of embankment site. It also depends on quantity and quality of available construction materials. Inclined core is implemented instead of centre core. For example, in case of an embankment where its foundation has a sharp inclination along the stream, a blanket zone is provided. Here a pervious foundation is to be linked with an impervious core zone, where various construction procedures are accessible for settlement of rock fill and core materials.

4.7.1 Construction of Earthen Embankments

Based on size and scale of an embankment, specific engineering management might be necessary at critical construction stages. Generally, the extent of specific engineering management necessary for construction of embankment is proportional to threat classification of dam. Anyhow, a properly qualified and skilled engineer must be referred to for helping the future proprietor with necessities for dam construction. Engineering proficiency can be utilized in the course of planning and building of dam, as well as throughout the dam life.

Area to be enclosed by embankment must be pegged out before beginning any kind of work. Embankment and area to be dug should be grubbed and cleared. Top soil should be piled in areas exterior to area enclosed by embankment and all roots, trees, and scrubs eradicated. Placement of top soil should be in stratums not more than 200 mm and grass should be planted if it is to be left for a substantial period (in excess of 6 months). This will preserve reliability of top soil. All waterlogged constituents inside the area of embankment must be taken out far away from the location and should not be utilized in embankment.

All loose material, topsoil, vegetable matter, and silt, should be stripped from the base of embankment and then scratched over its entire area. Loss of water in dams is mainly due to seepage and evaporation. Although for evaporation losses a little can be done, seepage losses can be reduced with good construction techniques. One critical feature is cut-off trench construction. A keyway or cut-off trench should be 1½ times the dam height at bottom of trench. This trench helps in reducing seepage underneath the embankment and stability of dam thus increases.

The keyway should be dug down to a minimum of 600 mm into the impervious soil and rock, backfilled with suitable quality of clay which is compacted thoroughly. Extension of the keyway should be along the embankment length as well as hillside flanks, but extension is not necessary under the spillway where spillway is cut into rocks. Basic functions of filters in embankment dam and their foundation are as follows:

1. Prevention of erosion of soil constituents from soil they are guarding.
2. Allowing seepage water drainage.

Instalment of an outlet pipe is needed at the dam base. Outlet pipe is necessary for enabling the permit owner to pass inflows during winter and summer as per the conditions of water licence. Also, it is necessary to permit water in upstream of structural work construction to be side-stepped at the time of construction. Minimum outlet pipe size is stated in the authorization; nevertheless, if authorization owner desires of installing bigger diameter pipe, it is permissible.

Complete excavation of embankment materials should be kept below full supply level (FSL) of storage area as much as possible. Excavation of appropriate constituents within wetted perimeter of storage area will lead to maximization of total storage capability, or else appropriate materials for constructing the dam may be obtained from outside of wetted perimeter.

Inadequate compaction levels often lead to leakages in earth-filled dams, which results in failures of dam. Hence, it is very significant to achieve effective compaction level. Application of requisite compaction work to heavy clay materials help in attaining proper compaction. This should be done by utilizing a tamper foot roller, usually denoted as sheep foot roller. For all dams, approximately greater than 3 m in height and capacity of 3 megalitres, subsequent steps should be accepted as a minimum thumb rule to achieve the requisite compaction level:

1. Placement of all filling materials for embankment should be layer wise (or lifts), not higher than 150 mm.
2. The biggest size of particle should not be more than one-third the lift height, i.e. 50 mm.
3. Every layer must be compacted thoroughly before placing the following layer. Minimum six passes for achieving requisite compaction strength is often essential for an appropriate machine.

For ensuring that downstream and upstream batters are compacted well and clipped to a slope of 3:1 (*H*:*V*), special care shall be taken. Stability of embankment may be greatly affected if there is a failure in complying with this requirement. Soil bank settlement is common and there must be an allowance to settle the embankment of the dam. Embankment may be allowed to settle to a level where it can avoid overtopping of water that will avoid the resultant failure. Overtime settlement may result in embankment height becoming lesser than spillway. Settlement of clay soil can be in excess of 10% of dam height; however, in case of a compacted and well-constructed clay dam, embankments are improbable of settling more than 5%. Provision of 5% allowance of the embankment height (along its length) is necessary for settlement.

Spreading of topsoil over exposed embankment surface to a minimum depth of 150 mm and sowing of pasture grass as soon as possible for establishing a better cover are essential. Any vegetation bigger than pasture grass should not be allowed to grow or nourish on or near the embankment. Several tree roots, particularly the eucalyptus tree roots, lead to cracking of the embankment core which results in dam failure. As a thumb rule, shrubs and trees should be at least at a distance of 1½ times the height of trees away from embankment of dam.

Spillway is designed with the purpose of passing floodwater without overtopping wall of the dam. Specific consideration must be given to provide suitable depth (freeboard) and width of spillway as per provisions specified in dam authorization. Vertical distance from embankment top to spillway level is known as the freeboard. It is very significant to provide appropriate depth. In case of insufficient depth, flood flow will be overtopping the dam and embankment materials will be taken away at increasingly larger rate and there will be a widespread destruction. Several cases of complete failure of dams have occurred from inadequate provision of freeboard. It must not be less than specifications stated on permit of dam construction. Absolute minimum is 0.50 m, usually with an additional 0.25 m to be taken into account for possible action of waves.

Dimensions of stored water surface are such that prevailing wind causes wave action on embankment. Upstream face of dam should be protected by providing rip-rap or stone pitching. As a thumb rule, wave height can be computed by formula developed by Hawksley (Wood and Richardson, 1975):

$$H = 0.0138(F)^{0.5} \tag{4.4}$$

Here H is the wave height (in metres) and F is the fetch distance over longest exposed water surface (in metres). Minimum depth for wave action is 0.25 m.

Over a period of time, sedimentation will decrease the capacity of the dam. Hence, during the planning procedure, the licence owner must take into consideration the capacity of dam for filling with silt and the measures needed to be undertaken for reducing the possibility of its occurrence. To stop huge volumes of sediment from entering a dam, certain erosion control and drainage measures must be undertaken for reducing the potential of sedimentation process. Once construction of the dam gets over, regular inspections and maintenance are required for ensuring good operational condition of the dam.

4.7.2 Embankment Design

Most failures of earthen embankment occur by overtopping of reservoir water because of loss of freeboard or flooding. Except for this condition, an embankment should not be designed for withstanding erosive exploitation of water flow above the crest. Many case studies in history have revealed that insufficient spillway capacity, i.e. inadequate estimation of amount of floodwater, has repeatedly led to embankment failure. However, tragedies of this kind cannot be a conclusive shortcoming of embankment dams, since collection of accessible accurate hydrology data and improved design methods can resolve the problem readily. Other major aspects causing failures in embankment are earthquake forces, high pore-water pressure, hydraulic erosion, and so on. More than 50% of embankment failures are principally because of hydraulic erosion, and rest is caused by other relevant factors.

Water flows through soil in foundation of an embankment due to its viscosity seepage forces that act on soil particles. If seepage forces acting on soil are large enough in comparison to resistive forces on the basis of effective earth pressure, erosion by quick sand occurs removing soil elements away from the surface, and there is a successive development of piping as erosion progresses gradually.

4.7.2.1 Recommended Safety Measures

The following are the safety measures that should be taken during the design and construction of an embankment:

1. Provision of an emergency spillway and sufficient freeboard height to avoid overtopping of the embankment.
2. Passing of seepage line through downstream face of the structure should be avoided. Seepage line should be well within the embankment.
3. Provision of sufficient upstream slope to evade any destruction and sloughing during drawdown period.
4. Design a steady structure with appropriate side slopes.
5. Foundation shear stress should be less than shear strength of foundation material.
6. No free passage should be allowed for water to flow without any protection, from upstream to downstream face.
7. Velocity of seepage water should be as low as possible.
8. Protection of upstream of the dam from reservoir waves and downstream face from the effect of raindrops.

4.7.3 Seepage Analysis

The line of seepage in an earthen embankment is the line above which there is atmospheric pressure and below it there is hydraulic pressure. It is the uppermost flow line and is called phreatic line. In a dam made of coarse material, the seepage line is the line of saturation where capillarity has no influence (Jumikis, 1967).

In earthen dams if seepage lines intersect outside, downstream fall much above its toe, and then invariably a serious sloughing occurs, which ultimately leads to dam failure. In earth dam of homogeneous material located in an impervious foundation, seepage of water from reservoir is not able to percolate vertically downward owing to impervious foundation layer, and thus it cuts across downstream fall of dams above its base unless a properly designed drainage system is laid. The location of the seepage line within points and dam, at which it cuts the downstream fall, is influenced by type of cross section of dam and not by permeability of the fill of the dam, as long as it is homogeneous. Under these conditions, Casagrande (1937) has observed that seepage line can be approximately replaced by a parabola. However, the parabola deviates from the true seepage line at the upstream and downstream faces of the embankment.

The slope of the seepage line is evaluated by

$$i = -{}^{dy}\!/_{dx} \tag{4.5}$$

The equation of the seepage line is given by

$$y = \sqrt{H^2 - \frac{2qx}{k}} \tag{4.6}$$

4.7.3.1 Seepage Analysis by Flow Net

Studies on seepage are also conducted by the use of flow nets. In earth dams, it is commonly used method. A flow net is a graphical device consisting of two sets of mutually perpendicular intersecting curves: one is the flow line and the other is the equipotential line. The flow net represents the direction of the seepage flow. The equipotential line represents the hydraulic head at any point in a cross section of a soil mass through which the seepage is occurring (Forchheimer, 1914). There are three popular ways of constructing flow nets: (i) mathematically, (ii) experimentally, and (iii) sketching with free hand.

In flow nets, it is a two-dimensional study of the flow in a plane. A flow line represents the direction and pattern of flow and the space between the two adjacent flow lines is called the flow channel, whereas an equipotential line passes through all the points of equal piezometric head.

$$\text{Total head} = \text{pressure head} + \text{geodetic head} + \text{velocity head}$$

The seepage analysis of flow net is based on the following assumptions:

1. Soil is isotropic and homogeneous and its voids are filled with water.
2. There exists a laminar flow which is continuous and steady.
3. Both soil and water are incompressible.
4. Water is ideal and of constant density.
5. Flow is governed by Darcy's law and is not rational.

4.8 CONCLUSION

The efficiency of rainwater harvesting scheme lies in its capability to satisfy the end-use preferences and site requirements. Although these systems are simple, they are very site-specific and detailed planning is needed before implementing the system. It prevents travelling far distances for obtaining water and can assist in overall health and community growth. With a decrease in water availability at many locations, rainwater harvesting presents the best selection for future times. In general, this chapter provides a clear idea about the structure affecting water harvesting, like reservoir, earth embankment, seepage analysis, etc., which will be helpful for construction of new structure towards water harvesting.

REFERENCES

Casagrande, Arthur. 1937. "Seepage through dams: New England Water Works Association", LI (2), 131–172.

Finkel, J. H. and Naveh. 1986. Semi-arid Soil and Water Conservation, CRC Press, Boca Raton, FL.

Forchheimer, P. 1914. Hydraulik, B.G. Teubner, Leipzig.

Jumikis, A. R. 1967. Introduction to Soil Mechanics, Van Nostrand Reinhold Inc., 436 pp.

Murty, V. V. N. 1985. Land and Water Management Engineering, Kalyani Publications, Ludhiana, India.

Murty, V. V. N. 2002. Land and Water Management Engineering, Kalyani Publications, Ludhiana, India.

Schwab, G. O., D. D. Fangmeier, W. J. Elliot, and R. K. Frevert. 1993. Soil and Water Conservation Engineering, John Wiley and Sons, Inc., New York.

Wood, A.D. and Richardson, E.V., 1975. *Design of small water storage and erosion control dams* (Doctoral dissertation, Colorado State University. Libraries).

5 Water Quality Management in Watershed

5.1 INTRODUCTION

The quality of water is influenced by both human and natural influences. We can consider various factors for the determination of quality of water, such a human activities may be considered one of the major criteria. How we use the different water bodies such as rivers determine the water quality. The causes of water pollution may be classified into point and non-point sources. Bad water quality is a direct threat to the society as it cannot be used for any purpose and decrease the usability of water in that water.

5.2 WATER POLLUTANTS AND THEIR SOURCES

5.2.1 Point Sources

Both domestic and industrial sewages can be considered as point sources. As both the domestic and industrial wastes go through pipes and are collected as a single point by the receiving water, they are point sources. Domestic waste includes wastes from schools, colleges, household, etc.

5.2.2 Non-point Sources

Agricultural and urban runoff can be considered as non-point sources. They are considered as non-point sources because urban and agricultural wastes have multiple discharge points. These wastes generally flow through combined sewers during urban storm. Runoff from agriculture contains various toxic materials such as insecticides and pesticides, which are carried during soil erosion.

5.2.3 Oxygen-Demanding Materials

The materials that oxidize receiving water with the intake of dissolved molecular oxygen are known as oxygen-demanding materials. These generally include biodegradable organic materials but also consist of inorganic materials. The aquatic life is threatened by dissolved oxygen; however, critical DO level fluctuates between species. In domestic sewage, oxygen-demanding materials is mainly due to human waste and food residue.

5.2.4 Nutrients

Nitrogen and phosphorous are nutrients which are very important for growth. But too much of nitrogen and phosphorous can be harmful. They must be existing in water bodies to maintain the food chain. Major difficulties arise when nutrient level increases and food chain is highly disturbed causing some organisms to flourish at expense of others. Few major causes of nutrients are phosphorous-based detergents, food processing waste, and fertilizers.

5.2.5 Pathogenic Organisms

Microorganisms in wastewater consist of protozoa, bacteria, and virus excreted by unhealthy animals or persons. When these microorganisms get discharged into surface water, they make the water frail for drinking purposes; if present in excess amount, the water may likewise be unfit for fishing and swimming.

5.2.6 Suspended Solids

Organic and inorganic materials that are carried by wastewater into receiving water are called suspended solids. When water flow into a lake or pool, the speed of water gets reduced and the suspended solids settle at the bottom as sediments.

5.2.7 Salts

All water comprises some salts. These salts are always measured by vaporization of a filtered sample of water. Salts and other solids that do not evaporate at all are termed as total dissolved solids. A point occurs when excessive amount of salts threatens the natural population of plants and animals and is no longer useful for public supply.

5.2.8 Toxic Metals and Toxic Organic Compounds

Agricultural runoff consists of pesticides and herbicides that have been used on crop. A major source of zinc in water bodies is urban runoff. The zinc emanates from tire wear. Much industrial wastewater contains either toxic organic compounds or toxic metals (Peavy et al. 1986). If these are discharged in huge quantities, many can make a water body almost useless for a long period of time. Many toxic compounds are concentrated in the food chain, making fish and shellfish unfit for human consumption.

5.2.9 Endocrine Disrupting Chemicals

The classes of chemicals identified as endocrine disrupters (EDCs) alter the standard physiological function of endocrine system affecting the hormone synthesis. EDCs can as well target the tissues where hormones exercise their effect.

5.2.10 Heat

Even though heat is not usually accepted as a pollutant, people in electric power industry are very well alert of complications caused by waste heat disposal. Similarly, water released by several industrial procedures is very warmer compared to receiving water. In certain environment, a rise in temperature of water can be advantageous (Barthwal 2012, Canter 1996). For instance, production of oysters and clams can be augmented in certain areas by heating the water. In contrast, increase in temperature of water have adverse impacts. Many significant game and commercial fish, such as trout and salmon, survive in cool water only.

5.3 WATER QUALITY INDICATOR AND STANDARDS

Water is completely colourless, transparent, odourless, and tasteless liquid at standard temperature; it is chemically neutral and a universal solvent for numerous compounds in reactions (Goel and Sharma 1996). If all these properties are met in practice, then we can say that it is pure in quality, otherwise not. Just by evaluating the chemical, biological, and physical characteristics of water, the quality of water can be determined. The physical properties of water like colour, odour, temperature, presence of solid, turbidity, oil, and grease contents are natural indicators of water quality. The presence of organic and inorganic quality of water makes its quality very much different than its pure form. There are some gases like hydrogen sulfite, methane, and oxygen which also change its quality. Biological constituents of animals, plants, protists, etc. also affect water quality adversely.

Water quality standard has been prescribed by most of the developed and developing countries like India as well as the World Health Organization (Table 5.1). In India, three important agencies prescribing water quality standard are the Bureau of Indian Standard (BIS), the Indian Council of Medical Research (ICMR), and the Ministry of Works and Housing. Drinking water quality have also been specified by the European countries.

5.4 WATER QUALITY MANAGEMENT IN RIVERS

The main aim of water quality management in rivers is to keep the discharge of contaminants within limit so that water quality does not go below the accepted level. However, regulating wastes discharge is a quantitative effort (Assar 1971). We must be capable of measuring the contaminants, predicting the effect of the contaminant on the quality of water, determining the related water quality which would be present in the absence of human interference, and deciding the acceptable levels for projected utilization of water. For most people, the dipping mountain stream, icy cold and crystal clear, nourished by snow melt and safe for drinking, is personification of high quality of water.

The impact of river pollution depends on both the contaminants and features of specific rivers (Morris et al. 1996). The significant criteria are discharge and velocity of water, the flowing depth, the channel slope, and the nearby crops and vegetation. Water quality management of a specific river must take into consideration all these aspects which include the land use patterns, climatic factors of the region, availability of mineral in the watershed, and nature of aquatic life in the aquifer. Therefore, certain rivers are vastly vulnerable to pollutants such as heat, sediments, and salt.

TABLE 5.1
US Water Quality Standard of Maximum Contaminant Level in Community Water System

Contaminant Category	Maximum Contaminant Level	Contaminant Category	Maximum Contaminant Level
Primary Standard			
Inorganic Chemicals		**Organic Chemicals**	
Arsenic	0.05 mg/L	Endrin	0.0002 mg/L
Lead	0.05 mg/L	Methoxychlor	0.1 mg/L
Silver	0.05 mg/L	Chlorophenoxys	0.005 mg/L
Mercury	0.002 mg/L	Lindane	0.004 mg/L
Selenium	0.01 mg/L	Toxaphene	0.005 mg/L
Barium	1.0 mg/L		
Nitrate	10.0 mg/L		
Cadmium	0.010 mg/L		
Chromium	0.05 mg/L		
Fluoride	4.0 mg/L		
Secondary Standard			
Iron	0.3 mg/L	Chloride	250 mg/L
Manganese	0.05 mg/L	Corrosity	Non-corrosive
Zinc	5 mg/L	Odour	3 Ton
Aluminium	0.05–0.2 mg/L	Copper	1.0 mg/L
Sulfate	250 mg/L	Foaming agent	0.5 mg/L
pH	6.5–8.5	Silver	0.1 mg/L
Colour	15 CU	Total dissolved solid	500 mg/L
Fluoride	2.0 mg/L		

5.4.1 Total Maximum Daily Load (TMDL)

A TMDL can be described as quantity of pollutant a water body can accept and still meet the water quality standard. The TMDL allocates the pollutant loading that might be contributed between point and non-point sources. TMDL is calculated as

$$\text{TMDL} = \sum \text{WLA} + \sum \text{LA} + \text{MOS} \tag{5.1}$$

where

WLA = waste load allocations allocated to prevailing and upcoming point sources
LA = load allocations allotted to prevailing and upcoming non-point sources
MOS = margin of safety

Some contaminants, predominantly oxygen-demanding nutrients and wastes, are very common and have such a deep effect on nearly every type of river that they deserve superior importance. However, this cannot be said that they are most substantial pollutant in any one river.

5.4.2 Effect of Oxygen-Demanding Waste on Rivers

Introduction of materials demanding oxygen, either inorganic or organic, into a river leads to reduction of dissolved oxygen in river. If oxygen concentration drops less than a critical point, it would be life-threatening to fisheries and other advanced forms of marine life. For predicting the magnitude of oxygen reduction, it is essential to identify what quantity of waste is being discharged and for degrading the waste how much oxygen will be necessary. Nevertheless, complete replenishment of oxygen from atmosphere and from photosynthesis takes place by aquatic plants, algae, in addition to consumption by organisms. Oxygen concentration in river is obtained by relative proportions of the competing process. Generally, oxygen is measured by finding quantity of oxygen consumed in the course of degradation of natural water. Factors which affect oxygen consumption in the course of organic matter degradation is considered in this section and then inorganic nitrogen oxidation has been discussed.

5.4.3 Chemical Oxygen Demand (COD)

Organic matter present in wastewater can be measured in many ways. Most often organic matter is assessed in terms of oxygen necessary for completely oxidizing organic matter to CO_2, H_2O, and other oxidized species.

Oxygen necessary for oxidizing organic matter present in a specified wastewater can be computed theoretically, if organics existing in wastewater are known. Therefore, if concentrations and chemical formulas of chemical compounds present in water are known, theoretical oxygen demand (ThOD) of each compound can easily be calculated by writing balanced reaction for compounds with oxygen for producing CO_2, H_2O, and other oxidized inorganic components.

Therefore, if organic compounds and their concentrations are given, ThOD of water can be computed accurately, but it is virtually not possible to know this about organic compounds available in any natural raw water or wastewater. By performing laboratory test on a given water sample with strong oxidant like dichromate solution, COD of raw water or wastewater is determined.

5.4.4 Dissolved Oxygen (DO)

Determination of DO present in sewage is very significant. This is because while treated sewage is discharged into some river, it is essential to guarantee at least 4 ppm of DO in it, else fish will probably die, causing trouble near disposal vicinity. To ensure this, DO tests are performed during sewage disposal treatment processes. The DO concentration of sewage is generally determined by Winkers technique, which is an oxidation-reduction procedure carried out chemically for liberating iodine in quantity equivalent to quantity of DO initially present.

5.4.5 Biochemical Oxygen Demand

The organic matter is of two types: (i) biologically oxidized, i.e. oxidized by bacteria, and is called biologically active or biologically degradable and (ii) that cannot be biologically oxidized and is known as biologically inactive. While testing a wastewater, the main interest lies in finding out quantity of biologically active organic matter available in it, while COD test gives total biologically inactive as well as biologically active organic matter. Therefore, we need to determine biochemical oxygen demand (BOD) which directly gives the amount of biologically active organic matter present.

If availability of oxygen is adequate in wastewater, valuable aerobic bacteria will keep flourishing causing aerobic biological decomposition of wastewater that will continue until completion of oxidation. BOD is the amount of oxygen consumed in this procedure. BOD is defined as the oxygen requirement of bacteria to decompose the organic matter under aerobic conditions. For many months, contaminated water will endure to absorb oxygen and practically it is not feasible to determine this ultimate oxygen demand in laboratory. Hence, the BOD of water during 5 days at 20°C is commonly considered as standard demand and is nearly 68% of total demand. A 10-day BOD is around 90% of total.

This standard 5-day BOD, written as BOD_5, or simply as BOD, is found in laboratory by diluting or mixing a given volume of a wastewater sample with a given volume of aerated pure water and computing DO of this diluted sample. The diluted sample is then incubated for 5 days at 20°C. DO of diluted sample after this incubation period is again computed. Difference in initial value of DO and final value of DO will specify consumption of oxygen by the sample in 5 days (used in causing aerobic decomposition of wastes).

Then BOD is calculated in ppm by utilizing the following equation:

$$\text{BOD or } BOD_5 = \text{DO consumed in the test by the diluted sample} \times \frac{\text{Vol. of the diluted sample}}{\text{Vol. of the undiluted sample}} \quad (5.2)$$

The factor on the right-hand side is the dilution factor.

5.4.6 Dissolved Oxygen Sag Curve

The concentration of dissolved oxygen in a river is indication of the health of river. All the rivers have certain ability of self-purification. As long as discharge of oxygen-demanding waste is well within the capacity of purification, DO level will remain high and a diverse population of animals and plants involving game fish can be found. As quantity of wastes exceeds the self-purification capacity, it can cause detrimental change in plants and animals. The streams loses its ability for cleansing itself and DO increases.

5.4.7 Dilution in Rivers and Self-purification of Natural Streams

Discharge of sewage into a natural water body leads to pollution of receiving water because of waste product present in effluents of sewage. However, these conditions do not continue forever, since natural purification forces such as sedimentation, dilution, oxidation-reduction in sunlight, etc. act upon elements causing pollution, thus bringing back water into its original form. In due course, this automatic purification of polluted water is known as self-purification phenomenon. But, if self-purification is not successfully achieved either because of discharging too much of pollution into it or because of other reasons, river water itself will become contaminated, which in return may contaminate the sea also where river falls.

Various natural forces of purification helping in process of self-purification are summarized below:

1. Physical forces include dilution and dispersion, sedimentation, and sunlight.
2. Chemical forces assisted by biological forces (known as biochemical forces) include oxidation (bio) and reduction.

These forces are described below:

i. *Dilution and dispersion:* When putrescible organic matter is discharged into a river or stream, it gets rapidly dispersed and diluted. This diminishes organic matter concentration, and thus reduces the potential nuisance of sewage.
ii. *Sedimentation:* If settleable solids are present in sewage effluent, they will settle down into the bed of the river, near the sewage outfall, thus helping in the process of self-purification.
iii. *Sunlight:* The sunlight has a bleaching and stabilizing effect on bacteria. It also helps certain microorganism to derive energy from it and convert them into food for other forms of life, thus absorbing carbon dioxide and releasing oxygen by a process known as photosynthesis.
iv. *Oxidation:* Oxidation of organic matter present in sewage effluent will begin once the sewage outfalls into river water comprising DO Atmospheric oxygen will fill up the created oxygen deficiency. Oxidation process will endure until complete oxidization of organic matter.
v. *Reduction:* Owing to hydrolysis of organic matter settled at bottom either biologically or chemically, reduction occurs. The complex organic constituents of sewage is split into liquid and gases with the help of anaerobic bacteria, thus paving the way for their final stabilization by oxidation.

5.5 WATER QUALITY MANAGEMENT IN LAKES

With time, the impact of global climate change will be felt more; it is of paramount importance to preserve water and make wise use of water resources for sustaining aquatic ecosystems. The health of a lake is greatly dependent on the quality of inbound water. Human activity, soil erosion, soil fertility, watershed topography, and vegetation all affect quality of water, and, consequently, the lake. The objective of watershed management is to avoid toxic waste from entering the lake by moving off the land. For preservation and restoration of all lakes, effective watershed management is essential.

Lake pollution sources are distributed into two extensive categories – point and non-point. Point source is a concentrated discharge which enters a water course at a solitary point, most often a pipe. Discharge from a wastewater treatment plant is an example of a point source. Point sources have an impact on few lakes only in Connecticut (a US state in southern New England) and impoundments on rivers receiving discharge of treated wastewater. Firm criteria of state and central water pollution laws standardize these sources for preventing pollution. Non-point sources are rambling, arising mainly as storm water discharge from terrestrial areas in the watershed. Non-point pollution sources affect every lake and pond in Connecticut. Even if a solitary non-point pollution source may not remarkably affect a lake, collective influence of several non-point sources can have a greater impact in the watershed. Most of these sources in Connecticut are related to human activities and growth. They may consist of agriculture, septic systems, gardens and lawns, storm water drainage system, and development, which lead to land erosion. Other sources may comprise household pets, recreational beaches, waterfowl, and poorly managed timber harvesting operations. Distinctive non-point pollutants involve the following features:

- *Solids:* Both floatable wastes and sediments are included in solids. Solid sources consist of stream banks, eroding construction sites, agricultural lands, and roadway runoff. In a watershed, sediments from eroding land increase organic matter and nutrient in a lake. Deposition of solids makes the lake shallower and destroys natural bottom habitation. Water is clouded by suspended solids, which interfere with digestion and respiration of organisms.
- *Nutrients:* The most significant are nitrogen and phosphorus as they stimulate algae growth and aquatic weed. Nutrient sources consist of agricultural runoff, septic systems, and pets, suburban/urban landscape runoff, and waterfowl.
- *Pathogens:* Viruses and bacteria from animal trashes, deteriorating septic system, and other sources pose health hazards.
- *Hydrocarbons:* Substances affiliated to petroleum, comprising grease and oil, are poisonous to subtle aquatic species. Runoff from roadways, marinas, and parking lots is a distinctive pollutant source of hydrocarbons in a watershed.
- *Heavy metals:* Copper, lead, zinc, cadmium, chromium, and mercury can come from automobiles, marinas, atmospheric pollution, and industrial waste. Deposits in a lake can be deadly to aquatic organisms.
- *Humic substances:* Leaves, grass clippings, and other plant materials dumped from landscaping practices in a lake results in problems after they are decomposed. DO is lowered affecting respiration of aquatic species and allows nutrients to be free from sediments. These organic constituents break down which forms muck on bottom of the lake.
- *Pesticides:* These include garden, lawn, commercial, and household pest-controlling chemicals. These contaminants can be poisonous to several creatures in a lake.

Often sources of non-point contamination which cause eutrophication are produced from land that drains into lake. This area of land is called drainage basin and is similarly known as watershed. In general, most plans for abating eutrophication of lake need watershed management for reducing contaminants in storm water runoff.

5.5.1 Eutrophication

It is a type of water pollution that is caused by the extreme presence of sediments, plant nutrients, and organic matter. These contaminants are often known as non-point source contamination as they originate from verbose sources. Symptoms of eutrophication include nuisance weed beds, depletion of oxygen in bottom waters, and dense algae blooms.

Different control methods that are being used for controlling pollution in lakes include:

- herbicide treatments,
- harvesting,
- sediment dredging,
- bottom barriers.
- sterile grass, and
- winter drawdown.

Experimental controlling techniques being examined consist of aquatic weevils and mixtures of aquatic herbicides.

5.5.2 Constituents of Managing Lake Water Quality

5.5.2.1 Monitoring from Baseline

Monitoring of synoptic water quality is conducted for assessing degree of eutrophication. Baseline monitoring is repeated over time for evaluating trends in conditions of water quality.

5.5.2.2 Diagnostic investigation

Conduct intensive water quality monitoring, typically over a year, for characterizing conditions of water quality and for identifying particular water quality complications which require consideration.

5.5.2.3 Watershed Assessment

An assessment of significant watershed characteristics is done in detail, such as soil types and land uses; it is conducted for identifying potential or active sources of pollution which must be addressed for protecting and improving quality of lake water.

5.5.2.4 Management Plan

Results of diagnostic investigation and assessment of watershed are utilized for evaluating alternative techniques for remediating unwanted lake conditions and for managing contamination sources in a lake watershed (Morris and Therivel 2009). Planning recognizes maximum cost-effective approaches for achieving water quality.

5.5.2.5 Implementation

One or more variety of technologies, including weed harvesting, sediment dredging, aquatic herbicide, and artificial aeration treatments, may be involved in remediation of undesirable lake conditions. Implementation of best management practices for non-point pollution sources are invariably involved in watershed management. Improved lawn fertilization practices, installation of storm water treatment technology, and routine catch basin cleanouts are few examples. For impoundments of river, watershed management may also include upgradation of treatment for point source discharges of wastewater.

5.6 WATER QUALITY MANAGEMENT IN ESTUARIES

Freshwater and seawater meet and mix in estuaries. The estuaries are habitat to a variety of wildlife and fisheries, and also provide areas for several recreational activities. Changes and development in land affect estuaries as they are fragile and sensitive to ecosystems. In estuarine areas, declination of water quality could endanger aquatic life. Recreational activities like bird watching and fishing are also dependent on quality of water; therefore, poor quality of water can have an effect on economy of surrounding areas because of decrease in tourism. Due to subtle balance between mixture of freshwater and saltwater, estuaries are vulnerable to destruction from humans. They can be destroyed or damaged from land draining, development, or even toxins and chemicals washing into the habitats. For ecological reasons, it is important to safeguard this delicate habitat.

5.6.1 Assessment of Estuary Water Quality

Dynamic water bodies are influenced by both freshwater from catchment and marine environment estuaries. Conditions inside estuaries differ both diurnally (a result of temperature, tide, and sunlight) and seasonally.

Throughout the state, water quality monitoring programs in estuaries is coordinated by Water Department. Parameters that are characteristically measured comprise temperature, chlorophyll, salinity, light, nutrients, and dissolved oxygen. These parameters assist in monitoring natural phenomena such as tidal flushing and stratification and also impacts produced by human actions and from watershed Rau and Wooten, 1985. The location and frequency at which these measurements are done fluctuate based on estuary (pressures and location) and requirements for current reporting. A specific problem, for example a phytoplankton bloom, may need monitoring of water quality in estuary more often over bloom duration.

The following five steps are used for monitoring estuarine water quality:

Step 1: Definition of estuaries, description of their location, and presenting how collection of real-time water quality measurements is done. Analysing water temperature data and identifying daily and seasonal temperature patterns by researchers or scientist from more National Estuarine Research Assets.

Step 2: Analysis of dissolved oxygen data from estuaries is done by researcher across the United States for identifying daily and seasonal patterns.

Researcher will utilize data for examining relationship between dissolved oxygen and water temperature.

Step 3: Salinity varies over space and time within the work station. Investigations are done through collected data set and prepare a brief report supporting or disproving a hypothesis.

Step 4: Water quality data will be obtained and evaluated in estuaries for identifying best timing of spring time depositing migrations of Atlantic sturgeon.

Step 5: Water quality data is collected and analysed through SWMP data tool. Designing their own investigations, collecting and analysing data, and constructing an argument will be done, which sensibly demonstrates how data supports their conclusions.

5.7 GROUNDWATER QUALITY MANAGEMENT

Groundwater by its location has a large measure of protection from containments that are found in surface waters. If groundwater once becomes polluted or contaminated, it becomes very difficult to replace it with freshwater and return to its previous state. Two major sources of groundwater pollution are: excessive interference of biological and chemical contaminants and saltwater produced over pumping of wells. The excessive release may occur under the following conditions:

I. When located septic systems are improperly operated, the discharge from them can cause contamination.
II. Leakage of various storage tanks that are located underground.
III. When chemical wastes and other toxic wastes are disposed improperly.
IV. When any transportation accident occurs, there is spills from pipelines.
V. When contaminated surface water recharges the groundwater.
VI. When landfills are leaked.
VII. Leaking retention ponds or lagoons.

Groundwater quality is important than its quantity. Quality necessary for supply of groundwater is dependent on the purpose of its usage. It can be widely utilized as irrigation water, industrial water, domestic water, etc. Salts in form of solution which are derived from location and previous movement of water are present in all groundwater. For establishing quality standard measures of physical, chemical, and biological characteristics of groundwater, radiological components must be quantified and also standard method to report and compare outcomes of water analysis be specified. If presence of dissolved gases goes unrecognized in groundwater, it can create hazard. The regularity of groundwater temperature is beneficial for industrial purposes and water supply and primary saline groundwater are significant since it offers potential advantages.

5.7.1 Measures of Water Quality

To specify quality features of groundwater, physical, chemical, and biological analyses are generally necessary. A comprehensive chemical study of a groundwater

TABLE 5.2
India: BIS Standards for Drinking Water IS 10500-1983

Substance or Characteristic	Desirable Limit	Substance or Characteristic	Desirable Limit
Turbidity	10 JTU	Taste	Agreeable
Magnesium	30 mg/L	Copper	0.05 mg/L
Chlorides	250 mg/L	Sulfates	150 mg/L
Mercury	0.001 mg/L	Fluorides	0.6–1.2 mg/L
Arsenic	0.05 mg/L	Phenolics	0.001 mg/L
Zinc	5.0 mg/L	Cadmium	0.01 mg/L
Pesticides	Absent	Selenium	0.01 mg/L
Colour	10 Hazen unit	Lead	0.10 mg/L
pH	6.5-8.6	Anionic detergents	0.2 mg/L
Calcium	75 mg/L	Residual-free chlorine	0.2 mg/L
Iron	0.3 mg/L	Alpha emitters	10^{-8} mCi/ml
Nitrates	45 mg/L	Beta-emitters	10^{-7} mCi/ml
Cyanide	0.05 mg/L	Total hardness	300 mg/L
Chromium	0.05 mg/L	Manganese	0.1 mg/L
Mineral oil	0.1 mg/L	Odour	Unobjectionable

sample involves determining concentration of inorganic components present; radiological and organic parameters are usually of concern only at places induced by pollution created by humans affecting quality. Normal salinity dissolved salts present in groundwater occur as separated ions; whereas other trivial components exist in elementary form. The study also involves specific electrical conductance and measurement. Based on purpose of a water quality analysis, partial analysis of specific components will suffice only occasionally (Hammer and Mackichan 1981). Illustrative chemical analyses of groundwater from a variety of geologic formations are shown in Table 5.2. Properties of groundwater assessed in a physical study involve colour, temperature, turbidity, taste, and odour. Biological study comprises tests to detect existence of coliform bacteria indicating hygienic quality of water for consumption by humans. Since certain coliform bacteria are usually present in human and animal intestines, existence of such organisms in groundwater is equivalent to its interaction with sewage sources. Standard methods of water analysis are specified by the American Public Health Association (APHA, 2017) and other organisations most laboratories conducting water analyses follow these procedures (United States Environmental Protection Agency, 1994).

5.7.2 Chemical Analysis

After analysis of a groundwater sample has been done in a laboratory, techniques to report water analysis must be taken into consideration. From an understanding of expressions and units to describe quality of water, standard measures can be established for interpreting the analyses in terms of definitive purpose of water supply.

In a chemical analysis of groundwater, concentration of various ions are given by weights or by chemical equivalence. Moreover, measurement of total dissolved solids can be done in terms of electrical conductance.

5.7.3 Physical Analysis

In a physical analysis of groundwater, temperature is reported in degrees Celsius and it must be necessarily measured. Colour in groundwater may be because of minerals present or organic matter in solution which is reported in mg / L by assessment with standard solutions. Turbidity is a measure of colloidal and suspended matter in water, such as silt, clay, microscopic organisms, and organic matter. Often measurements are dependent on length of a light path through water causing a flame's image of a customary candle to disappear. Unconsolidated aquifers producing natural filtration largely eradicate turbidity, but other forms of aquifers can produce turbid groundwater. Derivation of odours and tastes may be done from dissolved gases, bacteria, phenols, or mineral matter. These features are subjective sensations which can be described only in terms of knowledge of a human being. Quantitative determination of taste and odour has been made on the basis of maximum dilution degree which can be distinguished from odour-free and taste-free water.

5.7.4 Biological Analysis

As mentioned before, bacteriological study is significant to detect biological pollution of groundwater. Almost all pathogenic bacteria present in water are indigenous to intestinal tract of humans and animals; however, separating them from natural water is problematic in the lab (Needham and Needham 1962). Because coliform group of bacteria are comparatively simple for isolating and identifying, standard tests for determining their absence or presence in a sample of water are taken as a straight signal of the water safety for the purpose of drinking. In a given volume of water. coliform test results are stated as most probable number (MPN) of coliform group organisms. By analysing a number of discrete parts of a water sample, MPN for this purpose is calculated from probability tables.

5.7.5 Groundwater Samples

To prepare samples of groundwater for analysing its quality, polyethylene bottles or pyrex glass are mostly suitable. Volume of 1–2 L are generally adequate for a standard routine chemical analysis. After the bottle is rinsed with water, the sample is then collected and sealed securely. Sample water should be kept in a cool place and promptly moved for analysis to a laboratory. Samplings should be collected from a well once it has been pumped for a while, or else non-representative samples of polluted, stagnant water may be acquired. Considering each sample, a record should be kept of the depth of sample, well location, date, colour, water turbidity, temperature, size of casing, odour, and working conditions of the well instantaneously prior to sampling. For analysing organic and radiological components, special storage techniques and sampling are necessary.

The shorter the time elapse between a sample collection and its study, usually the more consistent the analytical results. For some components and physical values, instantaneous analysis is required in field for obtaining reliable outcomes; thus, determination of temperature, pH, dissolved gases, and alkalinity should always be done in field itself since variations are inevitable by the time samples reach a test centre. Storing of samples before analysis can also affect results. Cations such as Mn, Cu, Al, Fe, Zn, and Cr are subject to adsorption loss or exchange of ions on walls of glass containers.

Finally, it should be kept in mind that samples collected from wells penetrating stratified aquifers can yield solute concentrations that differ significantly from those occurring in the individual layers. Under these conditions, it is possible for obtaining water meeting quantified quality criteria, whereas in case of individual strata concentrations, it could be totally unacceptable.

5.8 GROUNDWATER QUALITY SCENARIO IN INDIA

Indian subcontinent is gifted with varied geological formations starting from old Achaeans to recently formed alluviums. These are characterized by changing climate in various regions of the country. Subsurface geological formations through which groundwater continue to interact and the depth of soils influence the natural chemical content of groundwater. Generally, in most part of India, groundwater is of good quality and appropriate for human consumption as well as industrial or agricultural purposes. Generally in shallow aquifers, groundwater is suitable to use for various purposes and is primarily of calcium bicarbonate and mixed form. Yet, additional forms of water are available as well involving sodium chloride water. In deeper aquifers, quality also differs from one place to another and is usually observed to be appropriate for collective usages. However, there are salinity problems in coastal territories.

In major parts of country, groundwater is drinkable. Yet, some issues related to water quality are informed from secluded pockets from different regions of the country. Higher level of components like fluoride, arsenic, salinity, and iron in groundwater is because of the normal geological phenomena. These are geo-genic pollutants found in groundwater. Human-made actions like disposal of industrial wastes, untreated domestic wastes, and mining activity are the cause for pollution like nitrate and heavy metals.

5.9 CONCLUSION

This chapter covered some basic concepts of water quality indicator, hydrologic science, and a brief description of water quality impact on river, lake, groundwater etc. As there is a growth in water consumption, wastewater increases considerably and due to lack of proper measures for treatment and management, pollution of existing freshwater reserves increases. Consequently, pollution of water has emerged as one of the nation's severest environmental threats. The study delivers numerous interesting conclusions which should be considered by academicians and policymakers. A significant finding specifies that policymakers should note that total pollution coefficients should be considered as alternative environmental management policies and not just direct pollution coefficient. This area of study is highly specialized and technical in nature. Also, it sources of pollutant affected towards water are described. Similarly, physical, chemical, and biological analysis is done for groundwater quality management.

REFERENCES

Assar, M. 1971. Guide to Sanitation and Natural Disasters, WHO Publication, Geneva.

Barthwal, R. R. 2012. Environmental Impact Assessment, 2nd edition, New Age International Publishers, Chapter 5.

Canter, L. W. 1996. Environmental Impact Assessment, 2nd ed. McGraw-Hill, New York.

Goel, P. K. and K. P. Sharma. 1996. "Environmental Guidelines and Standards in India, Techno Science Publications, Jaipur, India." Chapters 2 and 3.

Hammer, M. J. and K. A. Mackichen. 1981. Hydrology and Quality of Water Resources, John Willy & Sons, Inc., New York.

Indian Standard Specification For Drinking Water: 10500–1983.

McGauhey, P. H. 1968. Engineering Management and Water Quality, McGraw Hill Book Co., New York.

Needham, J. C. and P. R. Needham, 1962. A Guide to Study of Fresh Water Biology, Holden Day, San Francisco.

Peavy H. S., Rowe D. R., and Tchabanaglous, G. 1986. Environmental Engineering, McGraw Hill International Edition.

Peter Morris, and Jeremy Biggs. 1996. Water, Chapter 10. Taylor & Francis Group, London and New York.

Peter Morris and Riki Therivel (Eds.). 2009. Methods of Environmental Impact Assessment, 3rd ed., UCL Press.

Rau, J. G. and D. C. Wooten. 1985. Environmental Impact Analysis Hand Book, Chapter 6. McGraw-Hill, New York.

Standard Method of Water and Sewage Analysis, 2017. American Public Health Association Publication, 23rd edition, Washington, DC, America.

Water Quality Standards: Regulations and Resources, 1994. United States Environmental Protection Agency, 2nd edition, Washington, DC, America.

6 Groundwater

6.1 INTRODUCTION

Maximum improvement of groundwater resources for advantageous usage includes planning in terms of a complete groundwater basin. Recognizing that a basin is a big natural reservoir located underground, it follows a rule that use of groundwater by one land owner affects supply of water to all other land owners. Selection of management objectives must be done for developing and operating the basin in a proper manner. Objectives consist not only of hydrologic and geologic considerations but legal, economic, financial, and political aspects as well. A combined approach coordinating the usability of both surface water and groundwater resources is very much essential for optimum economic development of water resources in an area. After assessment of complete water resources and preparing alternative management plans, decisions on action to be taken can be made by fitting public agencies or bodies.

6.2 GROUNDWATER DEVELOPMENT

Development of groundwater can be carried out at a small capital cost and time consumed for this development is also very less. Generally, chemical quality of groundwater is found to be good and can be utilized for drinking, industrial, and agriculture purposes. A programme organized by UNICEF known as hard rock drilling has dropped the unit cost (per capita) significantly to make supply of safe water available; usually water is struck at 30–60 m depth in hard rock formations and no treatment is required. Tapping of location of groundwater, spacing, and yielding in a well field must be so phased that recharge and discharge of aquifer on annual basis are very nearly balanced without causing over usability in the area. As there is an increase in groundwater developments, problems related to management of well field will become critically dangerous in several places and investigations on optimum spacing of wells will be necessary for minimizing mutual intrusion among pumped wells (Water Resources Development Centre, 1960).

6.3 PROPERTIES AFFECTING GROUNDWATER

6.3.1 Aquifers

Aquifer is a formation containing adequate saturated permeable material for yielding substantial amount of water to wells (Das et al., 2019). It has the capability of storing and transmitting water, unconsolidated gravels, and sands. Furthermore, it is commonly understood that an aquifer involves unsaturated part of permeable unit.

6.3.2 Porosity

It is defined as the ratio of the volume of interstices to total volume:

$$\alpha = {}^{v_i}\!/_{V} \tag{6.1}$$

Where
α = porosity
v_i = volume of interstices
V = total volume

Further, it can be expressed as

$$\alpha = {}^{\rho_m - \rho_d}\!/_{\rho_m} \tag{6.2}$$

where
ρ_m = density of mineral particle
ρ_d = bulk density

Effective porosity denotes amount of interconnected pore spaces accessible for fluid flow.

6.3.3 Soil Classification

According to size and distribution, unconsolidated geologic materials are classified into various types. It is classified based on particle or grain size such as fine gravel, fine sand, silt, clay, coarse gravel, medium sand, etc. Assessment of size distribution is done by mechanical study involving sieving elements coarser than 0.05 mm and measurement of settlement rate for smaller suspended elements. The distribution of particles is characterized by uniformity coefficient U_c as follows:

$$U_c = {}^{d_{60}}\!/_{d_{10}} \tag{6.3}$$

where
d_{60} = 60% finer than value.

6.3.4 Specific Surface

Surface area has direct impact on water retentive property of a soil. This depends on particle size, shape, and types of clay minerals present. This term denotes area per unit weight of the material (m^2/g). Alternately specific surface measurement is based on retaining a polar organic molecule like ethylene glycol. Derived from statistical calculations of surface area, these have been related to absolute values.

6.3.5 Vertical Distribution of Groundwater

Subsurface of groundwater is divided into two types:

a. Zones of aeration
b. Zones of saturation

Zone of aeration comprises interstices partially occupied by air and water. In saturation zone under hydrostatic pressure, all interstices are filled with water. On most land masses of earth, a single aeration zone overlies a single saturation zone and spreads upwards towards ground surface.

6.3.5.1 Zone of Aeration

6.3.5.1.1 Soil-Water Zone

When excess amount of water reaches ground surface due to rain or irrigation, water exists in soil-water zone. The zone spreads from ground surface down through the main root zone. Its thickness differs with vegetation and soil type. Owing to the agricultural significance of soil water in providing moisture to roots, soil scientists and agriculturists have investigated distribution and movement of soil moisture comprehensively. Amount of water available in soil-water zone is dependent mainly on fresh exposure of soil to moisture. Under hot and arid conditions, water-vapor balance tends to become established between the ambient air and the surfaces of fine-grained soil particles.

6.3.5.1.2 Intermediate Vadose Zone

It spreads from lower edge of soil-water zone to upper limit of capillary zone. The thickness may differ from zero, where leaping zones combine with a higher water table impending ground surface, to more than 100 m beneath deep water table circumstances. This zone primarily assists as a relating zone close to ground surface that is close to water table through which water moving perpendicularly downwards need to pass.

6.3.5.1.3 Capillary Zone

This zone spreads from water table till capillary rise limit of water. Idealizing the pore space to characterize a capillary tube, rise in capillary can be determined from equilibrium amid weight of water raised and surface tension of water. Thus,

$$h_c = \frac{2\tau}{ry}\cos\lambda \tag{6.4}$$

where

τ = surface tension
y = specific weight of water
r = radius of tube
h = angle of contact between meniscus

6.3.5.2 Zone of Saturation

Here groundwater fills all interstices; therefore, porosity gives a directive measure of water contained per unit volume. A part of the water can be detached from subsurface strata through drainage or through pumping of a well. Yet, surface tension and molecular forces hold the remainder of water in its place.

6.3.6 Specific Retention

Ratio of volume of water against gravitational force to its own volume is known as specific retention.

$$S_r = {}^{w_r}\!/_{v} \tag{6.5}$$

where
w_r = volume occupied by retained water
v = bulk volume of the soil or rock.

6.3.7 Specific Yield

The specific yield is the ratio of volume of water after saturation to its own volume. It depends on grain shape, size, and compaction of the stratum, distribution of pores, and drainage time.

$$S_y = {}^{w_y}\!/_{V} \tag{6.6}$$

where
w_y = volume of water drained

6.4 BASIN MANAGEMENT

Management of a groundwater basin suggests a platform for improvement and use of subsurface water for certain specified objective, generally of an economic or social nature (ASCE, 1972; Burt, 1967). Generally, desired aim is for obtaining maximum water quantity for meeting predetermined quality necessities which is at least budget. Since visualization of a groundwater basin can be made as a big natural underground reservoir, it holds that withdrawal of water by wells at one site affects the quantity of water accessible at other sites inside the basin. Extraction of water from ground is done just like other minerals such as gas, gold, or oil. Typically water carries a distinctive parameter which is considered as a renewable natural resource. Hence, when drilling of a water well is completed, the public assume that production of water will indefinitely continue with time. This can simply come into effect if there is an existence of equilibrium between recharge of water from surface sources to basin and pumping of water by wells from inside the basin. Improvement of water supplies from groundwater typically begins with a few pumping wells which are distributed over a basin. With time passing by, there is drilling of more wells which results in increase of water extraction rate. Since the number of wells goes on increasing, there is an extension in basin development, which finally exceeds its natural recharge ability (Ambroggi, 1977). Continuous development without a management strategy could ultimately diminish the groundwater resource.

6.5 DARCY'S LAW

A French hydraulic engineer Henry Darcy in 1856 studied the flow of water through horizontal beds of sand to be utilized for filtration of water. He attempted to determine the law of flow of water through filters by precise experiments. These experiments positively demonstrate that the volume of water passing through a sand bed of a given nature is directly proportional to pressure whereas inversely proportional to thickness of traversed bed; thus, in calling the surface area of a filter, k – coefficient based on nature of sand, e – thickness of sand bed, $P - H_o$ pressure below the filtering bed, $P + H$ – pressure of atmosphere summed with water depth on filter. Universally, Darcy's law can be defined as the flow rate through porous media which is proportional to head loss and is inversely proportional to length of flow path. More than any other contribution, it assists as the base for present-day information regarding flow of groundwater. Study and explanation of difficulties related to movement of groundwater and well hydraulics started only after Darcy's experimental study.

Verification of Darcy's law on experimental basis can be conducted with water flowing at a rate Q through a cylinder of cross-sectional area A crammed with sand and has piezometers separated by distance L, as presented in Figure 6.1. Fluid potentials or total energy heads above a datum plane may be articulated using Bernoulli's equation:

$$\frac{p_1}{\gamma} + \frac{v_1^2}{2g} + z_1 = \frac{p_2}{\gamma} + \frac{v_2^2}{2g} + z_2 + h_L \tag{6.7}$$

where p is the pressure, y is the specific weight of water, v is the flow velocity, g is the gravitational acceleration, z is the elevation, and h_L is the head loss. Subscripts denote points of identified dimension in Figure 6.1. Since velocities are usually low in porous media, velocity heads may be neglected devoid of considerable error. Therefore, head loss can be rewritten as

$$h_L = (\frac{p_1}{\gamma} + z_1) - (\frac{p_2}{\gamma} + z_2) \tag{6.8}$$

Thus, resultant head loss is described as probable loss inside sand cylinder and by frictional resistance debauched as heat energy, this energy is lost.

At this moment measurements done by Darcy revealed that proportionalities $Q \sim h_L$, and $Q \sim 1/L$ exist. A proportionality constant K is introduced leading to following equation:

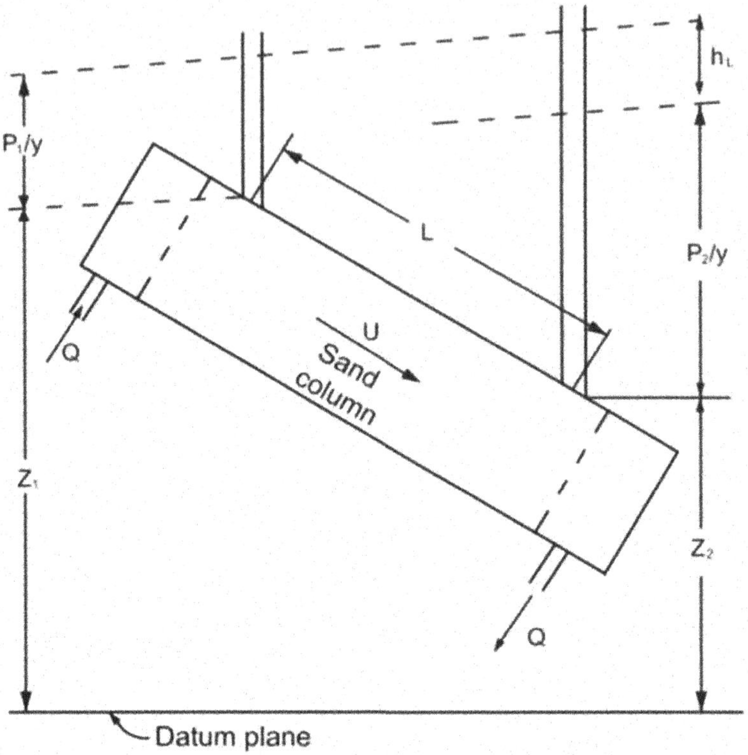

FIGURE 6.1 Distribution of pressure and head loss while flowing through sand.

$$Q = -KA\frac{h_L}{L}$$

Expressed in general terms:

$$Q = -KA\frac{dh}{dl}$$

or simply:

$$v = \frac{Q}{A} = -K\frac{dh}{dl} \tag{6.9}$$

where

v = specific discharge or Darcy's velocity
K = hydraulic conductivity (permeability of porous medium)
dh/dl = hydraulic gradient

Here negative sign specifies that flow of water is in decreasing head direction. The simplest form of Darcy's law is stated in Equation 6.9, which denotes that flow velocity v equals the product of constant K, known as hydraulic conductivity, and hydraulic gradient.

Actually, the flow is limited to the age interstitial velocity pore space only so that the OVPF

$$v_a = \frac{Q}{aA} \tag{6.10}$$

where

a is the porosity. This specifies that for a sand having porosity 33%, $v_a = 3v$.

6.6 HYDROLOGIC EQUILIBRIUM

For groundwater management, knowledge regarding quantity of water that can be developed is a prerequisite. Available water determination within a catchment requires evaluation of constraints constituting the hydrologic cycle. In a hydrologic cycle for a particular basin, an equilibrium must exist between the amount of water supplied to the basin and leaving the basin (Bear and Levin, 1967). The following expression is the equation for hydrologic equilibrium:

$$S_i + Ss_i + P + I_w + SS_d + G_d = S_o + Ss_o + C + E_w + SS_i + G_i \tag{6.11}$$

S_i = Surface inflow
Ss_i = Subsurface inflow
I_w = Imported water
C = Consumptive use
E_w = Exported water
SS_d, G_d = Surface and groundwater storage (decrease)
SS_i, G_i = Surface and groundwater storage (increase)
P = Precipitaion

Above equation can be applied for area of any size, it may be groundwater basin, river valley, aquifer etc.

For theoretical analysis, hydrologic equilibrium equation must be balanced after evaluation of all items in actual condition. It occurs due to inaccuracies of measurement, incorrect approximation, or lack of adequate dataset. The amount of unbalanced should not be more than accuracy of basic dataset. Adjustment should be done in terms of error to achieve balance equation. Further investigation is required if amount of unbalanced exceeds the limit for accuracy of basic dataset. In this context, equation requires careful analysis of hydrology and geology of particular area, adequate dataset, good judgement, etc. The quantity of water can be evaluated for existing conditions as well as any future condition. Also, we can calculate the unknown constraints if all other datasets are known. So inaccuracies of one or more quantities may exceed the magnitude of unknown quantity.

6.7 INVESTIGATION OF GROUNDWATER BASIN

An investigation of underground water resources is required before groundwater development within a basin (Revelle and Lakshminarayana 1975). Investigation is seldom concerned with simply locating groundwater supplies. Basically, it focuses on quality and quantity of groundwater sources. Usually these studies are taken by local government agencies. It consists of the following four steps:

1. *Preliminary examination:* To find or define the specified area a meeting is organized in presence of well-experienced expertise.
2. *Reconnaissance:* For water management, a possible alternative is formulated to meet the defined need for an area (Samantaray et al. 2019, Ghose and Samantaray 2019). This includes the estimation of benefit and cost of the project. Generally, investigation drawn from available datasets requires minimum numbers of new data collection.
3. *Feasibility:* Here a detailed analysis of hydrologic, engineering and economic is done in context to cost and benefit estimation for ensuring project have an optimum benefit. Basically, this investigation includes preparation and approval recommendations of reports and funding requisition of projects.
4. *Define project:* For defining specific features of the project, a detailed planning of the project is necessary. The complete report includes final design, preparation of plan, and specifications.

For groundwater management, the following are the data types and physical portion of reconnaissance:

Topographic data: Basic requirements are aerial photographs, contour map, benchmark levelling network, etc., which are applicable for identifying and locating well, plotting areal data, measuring groundwater level, etc.

Geologic data: For movement and occurrence of groundwater, surface and subsurface geologic mapping is essential. Subsurface information includes geophysical survey, analysis, and classification of well logs, pumping test of wells, etc. Pumping test of well is done to calculate the transmissivity and storage coefficient of aquifer (Chun et al. 1964). Location of dikes, faults, and other structures and interpolation of geologic data are also a part of this section.

Hydrologic data: Basic purpose of this dataset collection is to determine equation of hydrologic equilibrium. The required dataset and analysis of methods are summarized below.

These constraints are determined from hydrographic and hydraulic procedures where all the datasets are not easily available.

Precipitation: If gauges are provided in a basin, then annual precipitation is evaluated by using isohyetal or Thiessen method. Supplemental station is established if gauges are not located.

Consumptive use: By the process of evaporation and transpiration, water released into atmosphere from surface and subsurface is consumptive use or evapotranspiration. Aerial photography is needed to evaluate this discharge from the given catchment.

Changes in surface storage: This is evaluated from changes in water level of reservoir or lakes.

Changes in soil moisture: Neutron probe is used to measure the moisture content of the soil. Accurate measurement is difficult with respect to time and place. The amount of water at the beginning is nearly same as at the end in unsaturated storage, through selecting periods of storage change, so that we can minimize these problem.

Changes in groundwater storage: Groundwater storage change is determined from pumping record, artificial recharge, antecedent moisture content, groundwater level, geologic datasets, etc. Similarly, specific yield and storage coefficient are evaluated from laboratory and pumping test of well, respectively.

Surface inflow and outflow: Determination of both the parameters are difficult as they cannot be measured directly. Sometimes one of them is known, only unknown can be evaluated from the equation. Often after study both the items are cancelled as evaluated subsurface inflow is equal to the subsurface outflow.

6.8 BASIN MANAGEMENT BY CONJUNCTIVE USE

Maximum benefit can be achieved by conjunctive usage that includes planned and conditioned operation for surface and groundwater resources, while considering development in water resources (Bittinger, 1964). The simple difference between traditional surface water development with its related groundwater development and an integrated operation of surface water and groundwater resources is that separate firm yields of former can be substituted by more economic and larger joint yields of latter. The basic concepts of conjunctive use of groundwater and surface water are predicted based on surface reservoir impounding stream flow that is transferred to groundwater storage with an optimal rate.

Surface storage is the primary supplier to the reservoir for fulfilling annual water requirements, while groundwater storage retained for cyclic storage for annual subnormal precipitation. Thus, groundwater is lowered during dry season and is raised in monsoon season. Figure 6.2 shows variation of groundwater levels (Odisha, India) under such a system of conjunctive usage (Samantaray et al., 2020). During the monsoon period, surface water is artificially recharged to the groundwater storage and thus groundwater level rises. Conversely, during drought season, surface water are supplemented by pumping groundwater, hence groundwater level decreases. Possibility of conjunctive usage approach is based on operating a groundwater basin over water level. Conjunctive use management is essential for artificial recharge, water distribution and pumping purposes, etc.

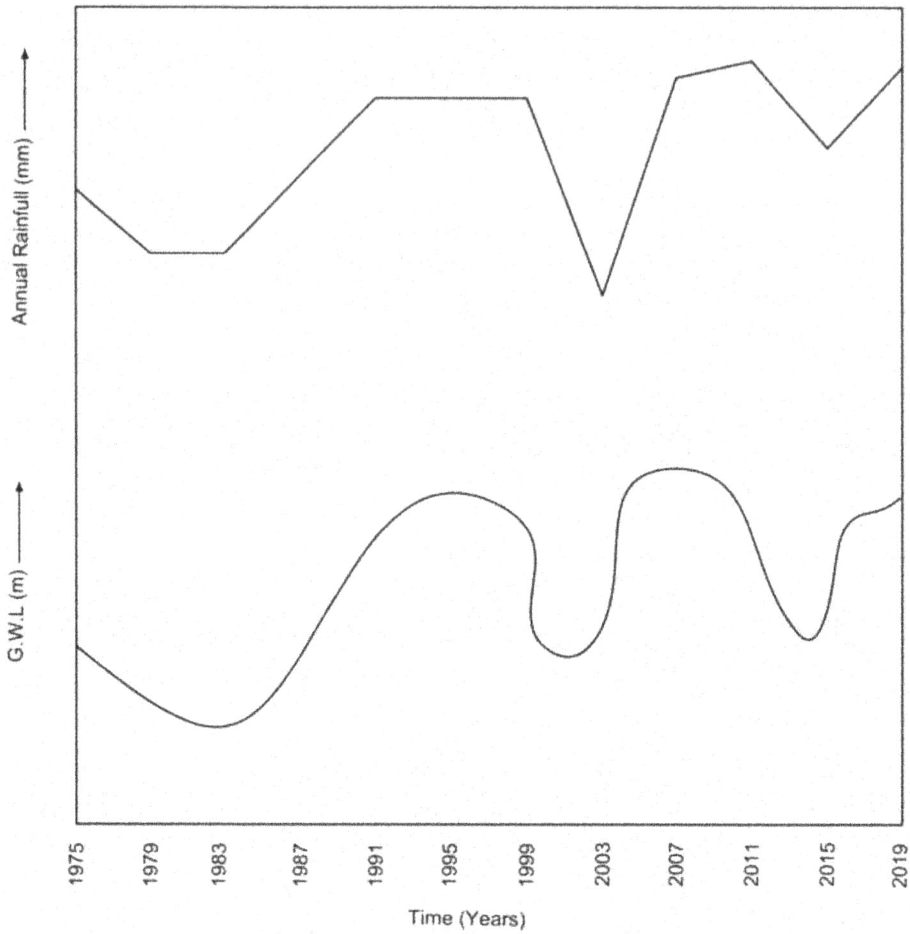

FIGURE 6.2 Groundwater fluctuation with respect to rainfall.

The procedure required careful management of optimal use of surface and groundwater resources (Samantaray et al., 2017; Rayner, 1972; Brown et al. 1978). Whole operation is complex in nature and highly technical; so it requires personnel with detailed knowledge regarding hydrogeology of basin, records of recharge and pumping rate, and continual updating of datasets of groundwater quality and level. Geological condition of distribution system, water use, wastewater disposal data are also required for analysis of consumptive use.

The following steps show the details analysis of conjunctive use:

Step 1: Identification of recent problem.
Step 2: Find out various constraints affecting problem.
Step 3: Specify objectives.
Step 4: Modelling through statistical, hybrid ML approaches.

Step 5: Establish model by training/testing/validation phases.
Step 6: Check whether it satisfies or not.
Step 7: If not, then repeat it by changing hidden layer/membership function/ learning rate until it satisfies.
Step 8: Choose the optimum model.

6.9 WELL HYDRAULIC: STEADY STATE

The primary objective is to determine the quantity of groundwater that can be safely drawn perennially from the basin. It was first developed by Dupuit (1863), and later modified by Theim (1906). The drawdowns stabilize after pumping from well, when the cone of depression spreads to steady state and shape conditions have been developed.

Unconfined aquifer (Figure 6.3): The yield from the well is

$$Q = KiA = K\frac{dy}{dx}(2\pi xy) \tag{6.12}$$

$$Q\int_{r_1}^{r_2}\frac{dx}{x} = 2\pi K\int_{h_1}^{h_2} y\,dy \tag{6.13}$$

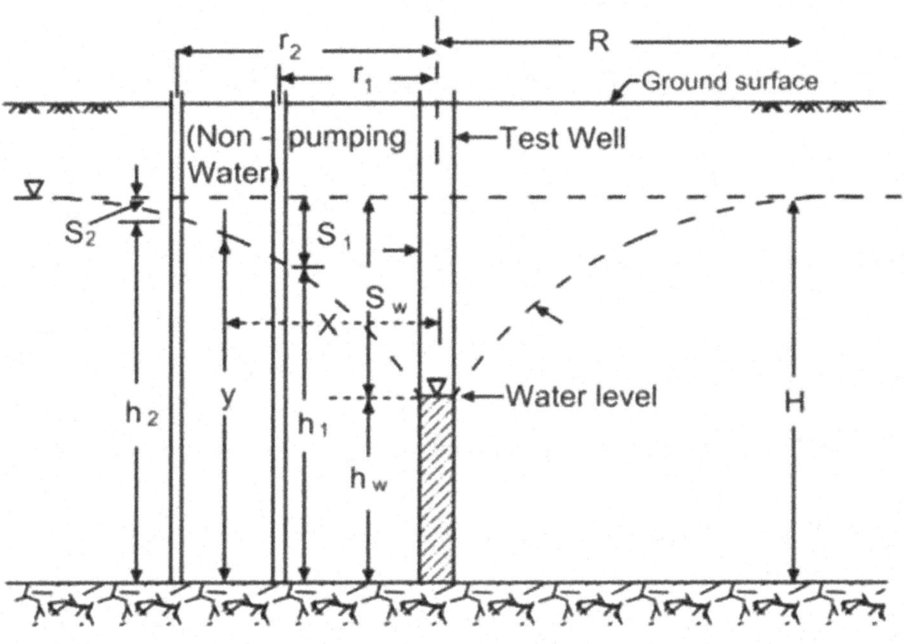

FIGURE 6.3 Flow in unconfined aquifer.

$$Q = \frac{\pi K(h_2^2 - h_1^2)}{2.303 \log_{10} \frac{r_2}{r_1}} \tag{6.14}$$

$$Q = \frac{\pi K(H^2 - h_w^2)}{2.303 \log_{10} \frac{R}{r_w}} = \frac{\pi K(H + h_w)(H - h_w)}{2.303 \log_{10} \frac{R}{r_w}} = \frac{\pi K 2H(H - h_w)}{2.303 \log_{10} \frac{R}{r_w}}$$

$$Q = \frac{2.72T(H - h_w)}{\log_{10} \frac{R}{r_w}} \text{ (assuming } T = KH\text{)} \tag{6.15}$$

where

T = transmissibility
Q = yield from well
H = saturated thickness
R = radius of influence
r_w = radius of well
h_w = depth of water

Confined aquifer (Figure 6.4): The yield from well is

$$Q = KiA = K\frac{dy}{dx}(2\pi xb) \tag{6.16}$$

$$Q\int_{r_1}^{r_2} \frac{dx}{x} = 2\pi Kb \int_{h_1}^{h_2} dy \tag{6.17}$$

$$Q = \frac{2.72Kb(h_2 - h_1)}{\log_{10} \frac{r_2}{r_1}} \tag{6.18}$$

S_1 and S_2 are drawdown in the observation well, $h_2 - h_1 = S_1 - S_2$, and assuming $T = Kb$

$$Q = \frac{2.72T(H - h_w)}{\log_{10} \frac{R}{r_w}} \tag{6.19}$$

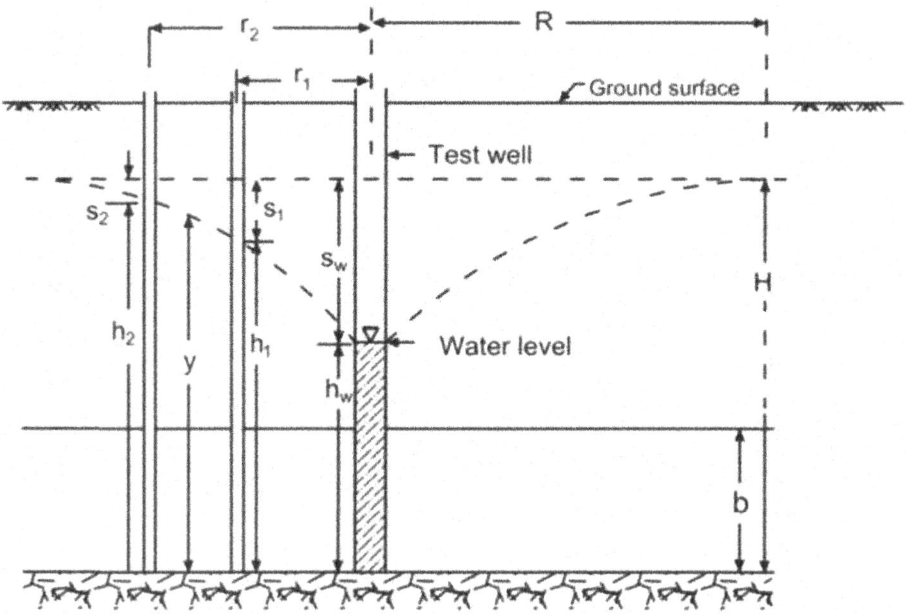

FIGURE 6.4 Flow in confined aquifer.

The assumptions of the equation are as follows:

a. Complete penetration of well
b. Radial flow into the well
c. Constant thickness of aquifer
d. Stabilized drawdown

6.10 WELL HYDRAULIC: UNSTEADY STATE

For confined aquifer with a constant rate of well penetrating, influence of discharge extends outward with time. Storage coefficient over an influencing area is same as discharge, as water comes from reduction of storage in an aquifer. The aquifer is infinite, so transient and unsteady flow exists. The differential equation for above conditions with respect to polar coordinate is

$$\frac{\partial^2 h}{\partial r^2} + \frac{1}{r}\frac{\partial h}{\partial r} = \frac{S}{T}\frac{\partial h}{\partial T} \tag{6.20}$$

where

S = storage coefficient
r = radial distance
h = head
T = transmissivity
T = pumping time

Assuming boundary condition, $h = h_0$, at $t=0$ and $s=$drawdown, $Q=$well discharge, then

$$s = \frac{Q}{4\pi T}\int_u^{\infty}\frac{e^{-u}du}{u} \text{ (where } u = \frac{r^2 S}{4Tt}) \tag{6.21}$$

The above equation is known as Theis equation. This equation permits determination of the formation constant S and T through pumping test of well. The equation is widely applied and preferred over the equilibrium equation.

Assumptions of the equations are as follows:

a. Instantaneously water removed from storage is discharged.
b. Piezometric surface is horizontal before pumping.
c. Well diameter is infinitesimal, hence storage can be neglected.
d. Aquifer is isotropic, has uniform thickness, and is homogeneous.
e. Pumping well penetrates entire aquifer and the flow is horizontal.
f. Always well is pumped with a constant discharge rate.

6.11 GROUNDWATER POLLUTION

It is defined as the man-made or artificially produced degradation of groundwater quality. Pollution can damage the usability of water and induce hazards to public health by toxicity. Basically, pollution emerges from disposal of wastewater. Usually boron, calcium, chloride copper, iron, lead, cadmium, detergent, and phenols are some pollutants found in groundwater. It is classified according to physical, biological, inorganic, organic, and chemical types. Human-used water is the cause and sources of groundwater pollutants. The major cause and sources of groundwater pollution are given below.

6.11.1 Municipal Sources

In general, leakages of sewage from the old sewer is one of the reasons of groundwater pollution. It occurs due to breakage of tree root, settlement of manhole, loss in foundation, infiltration, soil slippage, effective sewer pipe, weak workmanship, etc. Sewage leakage can produce COD, BOD, organic chemical, bacteria, nitrate, etc., into groundwater. Heavy metals like manganese lead, cadmium, cobalt, mercury, chromium, and iron may come into groundwater if sewer is present in industrial area. Similarly, in municipal area, liquid waste may come from industrial and domestic use, storm runoff, etc. Municipal liquid waste may produce viruses, bacteria, and organic chemicals and enter into the groundwater. Solid waste is another important pollutant of groundwater. A landfill can pollute groundwater if water passes through the fill material. Primary pollutants produced from landfill are nitrate iron, chloride, BOD, COD, hardness, manganese, etc. Gases like ammonia, hydrogen sulfide, carbon dioxide, and methane are the groundwater pollutants generated from landfill.

6.11.2 Industrial Source

Water is essential in industrial plants for sanitation, cooling, and manufacturing process. It depends upon the type and use of industry. Industrial water discharged into lagoons, pits, and ponds affect groundwater directly. For industrial installation, underground storage and transmission of gas, fuel, and chemicals are common practices. If structural failure occurs, leakage of fuel and chemical from tank pollutes groundwater. Gasoline, home fuel oil, and petroleum and its products are also primary pollutants of groundwater which directly go downward though permeable soil until they reach underground water level (Thomas, 1951). Subsequently, they spread into all zones of soil, which have an indirect pollution of groundwater. Also, they produce wastewater in forms of bines, which consist of ammonia, chloride, sulfate, sodium, boron, and total dissolve solid, which may have an indirect contact with groundwater. Metallic ores like phosphate, coal, iron, and uranium mines produce zinc, copper, iron, lead aluminium, and sulfate which are present at underground surface of soil. These often directly contact with groundwater table, which may cause pollution of groundwater (Sridharam et al., 2020; Sarkar et al., 2020). Sometimes, leaching of old mine and settling pond causes groundwater pollution.

6.11.3 Agricultural Source

Large amount of agricultural waste is produced from animal and its products after decomposing into the ground. Basically, animal waste needs 120–150 days to decompose into the ground. Bacteria, organic-inorganic contaminant, salts, nitrate-nitrogen pollutant produced from animal waste may reach groundwater (Casey, 1972). About 30–45% of water is consumed by evapotranspiration from applied irrigation water called irrigation return flow. This causes increase in salinity from the applied water. The addition of salts is caused by dissolution of salts added as fertilizer or soil ingredients. Basically, it includes nitrate, bicarbonate, chloride, and sulfate, which constitute the primary pollutants of groundwater. Phosphate and potassium fertilizers are also readily absorbed soil particles and so cause pollution problem. Sometimes nitrogen solution is used by plants and absorbed by soil, which may cause groundwater pollution.

6.11.4 Miscellaneous Sources

6.11.4.1 Roadway De-icing

Application of de-icing salts to street and highway may cause of pollution of groundwater. After spreading on roadway and also from stockpiles, soil reaches the groundwater through solutions. Solution consists of calcium and sodium chloride which have direct effect on pollution.

6.11.4.2 Interchange Through Well

Aquifers are interconnected vertically, which serve as avenues for groundwater pollution. Pollution occurs when incomplete hydraulic separation within a well and a vertical difference exist between two aquifers.

6.11.4.3 Cesspool and Septic Tank

Huge amount of sewage is discharged from residence directly into the ground across the world. Some commercial places like industrial plants, hospitals, and resorts have septic tank where community sewer system is not available. Liquid discharged from domestic sewage dissolves and passes into soil through subsurface percolation system like seepage bed, sad filer, etc.

6.11.4.4 Stockpiles

Raw solid material are stockpiled near the construction site, industrial plant, and agricultural farm for future use purposes. In monsoon season, stockpiles have direct contact with rainwater, which causes leaching and directly move into the soil. It consists of organic and inorganic material, silts, and heavy metal which cause groundwater pollution.

6.12 CONCLUSION

The importance of groundwater to well-being of humans is well documented. Groundwater is a major source of freshwater for public consumption, industrial uses, and irrigation of crops. Groundwater protection and management practices must be based on an understanding of groundwater sources; the manner in which groundwater is distributed below the earth's surface; geologic, topographic, and soil characteristics in the region; and interconnection between groundwater and surface water sources. The discussion in the preceding sections indicates the vast scope of groundwater assessment, encompassing a range of quality issues, purposes, types, scales, and levels of assessment. Investigations are made to determine the availability and suitability of groundwater for beneficial use and to provide groundwater information needed to plan, design, and construct works of improvement. Because problems associated with groundwater are complex, a flexible pattern of investigational procedures is needed. Identifiable groundwater characteristics help to determine the best procedures to use.

REFERENCES

Ambroggi, R. P. 1977. "Underground reservoirs to control the water cycle." *Scientific American*, 236 (5), 21–27.

ASCE. 1972. "Groundwater management." Manual Engineering Practice, 40, 216 pp.

Bear, J., and O. Levin. 1967. "The optimal yield of an aquifer." *International Association of Scientific Hydrology Bulletin*, 72, 401–412.

Bittinger, M. W. 1964. "The problem of integrating ground-water recharge and surface water use." *Ground Water*, 2 (3), 33–38.

Brown, R. F., et al. 1978. "Artificial ground-water recharge as a water-management technique on the southern high plains of Texas and New Mexico." Texas Department of Water Resources Report, 220, p. 32.

Burt, O. 1967. "Temporal allocation of groundwater." Water Resources Research, 3, 45–56.

Casey, H. E. 1972. Salinity Problems in Arid Lands Irrigation, Office of Arid Lands Studies, University of Arizona, Tucson, 300 pp.

Chun, R. Y. D, et al. 1964. "Ground-water management for the nation's future: optimum conjunctive operation of ground-water basins." *Journal of the Hydraulics Division*, 90 (HY4), 79–95.

Das, U. K., S. Samantaray, D. K. Ghose, and P. Roy. 2019. "Estimation of aquifer potential using BPNN, RBFN, RNN, and ANFIS." In Smart Intelligent Computing and Applications. 569–576, Springer, Singapore.

Dupuit J. (1863) *Etude Th´ eoriques et Pratiques Sur le Mouvement Des Eaux Dans Les Canaux D´ ecouverts et à Travers Les Terrains Permeables*, Dunot: Paris.

Ghose, D. K. and Samantaray, S. 2019. "Integrated sensor networking for estimating ground water potential in scanty rainfall region: challenges and evaluation." In Computational Intelligence in Sensor Networks. 335–352, Springer, Berlin.

Rayner, F. A. 1972. "Ground-water basin management on the high plains of Texas." *Ground Water*, 10 (5) 12–17.

Revelle, R. and V. Lakshminarayana. 1975. "The Ganges water machine." *Science*, 188, 611–615.

Samantaray, S., Rath, A., and Swain, P. C. 2017. "Conjunctive use of groundwater and surface water in a part of Hirakud Command Area." *International Journal of Engineering and Technology*, 9 (4), 3002–3010.

Samantaray, S., A. Sahoo, and D. K. Ghose. 2019. "Assessment of groundwater potential using neural network: a case study." In International Conference on Intelligent Computing and Communication, 655–664, Springer, Singapore.

Samantaray, S., A. Sahoo, and D. K. Ghose. 2020. "Infiltration loss affects toward groundwater fluctuation through CANFIS in arid watershed: a case study." In Smart Intelligent Computing and Applications, 781–789, Springer, Singapore.

Sarkar, B. N., S. Samantaray, U. Kumar, and D. K. Ghose. 2020. "Runoff is a key constraint toward water table fluctuation using neural networks: a case study." In Communication Software and Networks. 737–745, Springer, Singapore.

Sridharam, S., Sahoo, A., Samantaray, S., and Ghose, D. K. 2020. "Estimation of water table depth using wavelet-ANFIS: a case study." In Communication Software and Networks. 747–754, Springer, Singapore.

Theim, G. 1906. Hydrologische Methoden. Gebhardt: Leipzig, 56.

Thomas, H. E. 1951. The Conservation of Groundwater, McGraw-Hill, New York, 327 pp.

Water Resources Development Centre. 1960. Large-scale Groundwater Development, United Nations, New York, 84, pp.

7 Flood and Drought

7.1 INTRODUCTION

Flood is an unusual high-water stage in a river, generally the level at which water overflows its natural and artificial banks and submerges land which is usually dry. Occurrence of floods results in significant damages in terms of property loss, loss of lives, and financial loss because of disturbance in economic activities, which are all very well known. Every passing year crores of rupees are spent on flood forecasting and flood control. For hydrologic design purposes, hydrograph of extreme flood events and stages in correspondence to peak flood provides significant data. Furthermore, amongst different flood hydrograph characteristics, probably the most vital and broadly applied parameter is flood peak. Flood peak fluctuates year by year and its magnitude comprises a hydrologic series that enables one to allocate a frequency to a specified value of flood peak. Practically for designing all hydraulic structures, expected peak flow with a given frequency (1 in 100 years) is of major significance for effectively proportioning the structures for accommodating its influence. Designing of spillways for dams, culvert waterways, and bridges and estimating scour depth at any hydraulic structure are few instances in which flood peak values are necessary. For estimating a flood peak magnitude, alternative approaches are given below:

1. Rational formula
2. Empirical formula
3. Flood frequency analysis

Utilization of a specific technique depends on (i) desired objective; (ii) availability of data; and (iii) significance of project. In addition, rational formula is simply appropriate for small-size (<50 km^2) watersheds and unit-hydrograph approach is usually limited for moderate-sized (< 5000 km^2) watersheds.

7.2 RATIONAL FORMULA

A rainfall of very long duration and uniform intensity happening over a basin is considered. Gradually there is an increase in runoff rate from '0' to a constant value as specified in Figure 7.1.

As more and more flow from faraway areas of watershed reach watershed outlet, runoff increases. Time required for a drop of water flowing from the farthest point of watershed to reach its outlet is designated as t_c = time of concentration. It is apparent that if rainfall endures beyond t_c, runoff will attain peak value and remain constant. Peak runoff value is given by a basic equation as specified below:

$$Q_p = CAi \text{ for } t \geq t_c \tag{7.1}$$

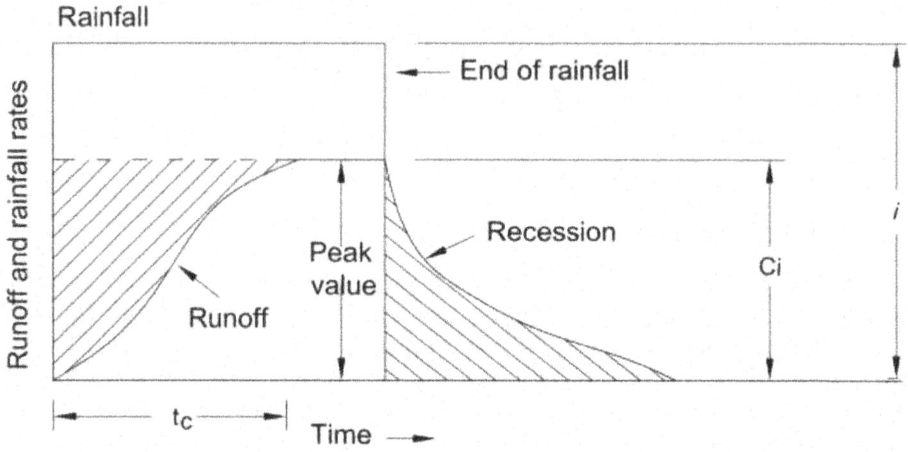

FIGURE 7.1 Runoff hydrograph due to uniform rainfall.

C = runoff coefficient
A = catchment area
i = rainfall intensity

Applying the regularly utilized unit, Equation 7.1 for field application is written as

$$Q_p = \frac{1}{3.6} C\left(i_{tc,p}\right) A \tag{7.2}$$

where
Q_p = peak discharge $\left(m^3/s\right)$
C = runoff coefficient
$i_{tc,p}$ = mean precipitation intensity for duration t_c and exceedance probability P
A = drainage area km^2
t_c = time of concentration

There are several available empirical equations for estimating t_c, of which two are defined below:

US practice: It is assumed that t_c is equal to lag time of peak flow for small drainage basins. Hence,

$$t_c = t_p = C_d\left(\frac{LL_{ca}}{\sqrt{S}}\right)^n \tag{7.3}$$

t_c = time of concentration
t_p = basin lag in hours
C_d and n = basin constants

L = basin length measured along water course from basin divide to gauge site in km

L_{ca} = distance along main water course from gauge site to a point opposite to the centroid of watershed in km

S = basin slope

Kirpich equation (1940): Universally utilized formula linking t_c of travel length and catchment slope is given by

$$t_c = 0.01947\ L^{0.77} S^{-0.385} \tag{7.4}$$

L = maximum travel length of water (m)

S = catchment slope = $\Delta H/L$

ΔH = difference in elevation amid most distant point on catchment and its outlet

Somewhere Equation 7.4 can be written as

$$t_c = 0.01947 K_1^{0.77} \tag{7.4a}$$

where $K_1 = \sqrt{L^3/\Delta H}$

7.2.1 Rainfall Intensity

Rainfall intensity in correspondence to a duration and desired exceedance probability P (i.e. return period $T = 1/P$) is obtained from relationship between frequency, rainfall, and duration for known area of catchment. Usually this will be a relationship represented in the form of

$$i_{tc,p} = \frac{KT^X}{(t_c + a)^n} \tag{7.5}$$

where K, x, n, and a are coefficients with respect to specific area.

7.2.2 Runoff Coefficient (C)

Coefficient C denotes the combined effects of catchment losses and therefore rely on rainfall intensity, surface slope, and nature of surface. Impact of rainfall intensity is not taken into consideration in existing table of C values. A homogeneous catchment surface is assumed in Equation 7.2. However, if the catchment is not homogeneous, it can be subdivided into separate areas each consisting of a different runoff coefficient and then runoff from each small area is discretely computed and combined in appropriate sequence of time. Occasionally, a non-homogeneous catchment may comprise constituent subareas dispersed in such a complicated way that distinctive subzones cannot be parted. In such circumstances, a weighted equivalent runoff coefficient (C_e) is used as shown below:

$$C_e = \frac{\sum_1^N C_i A_i}{A}$$

where

A_i = areal extent of subarea i
C_i = runoff coefficient
N = number of subareas in catchment

This formulae is observed to be appropriate for prediction of peak flow in small catchments (area < 50 km^2). It has substantial applications in design of urban drainage and in designing small bridges and culverts.

7.3 EMPIRICAL FORMULAE

Empirical formulae utilized for estimating flood peak are basically provincial formulae on basis of statistical correlation of important catchment properties and observed peak. For simplifying the given arrangement of equation, a small number of parameters among many other affecting peak flood are only utilized. Such as, catchment area is utilized as a parameter affecting flood peak by almost all formulae and most of these formulae do not consider flood frequency as a constraint. Keeping these in sight, empirical formulae are appropriate simply in area from which they are established. These formulae when employed to some other regions, can provide only rough values.

7.3.1 Relationship Between Flood Peak and Area

Considerably the most simple among the empirical relationships are those that associate drainage area with flood peak. Maximum flood discharge Q_p from a watershed area A is specified by formulae given in Equation 7.6:

$$Q_p = f(A) \tag{7.6}$$

Although there exist many formulas of this type projected for different areas around the world, a limited prevalent formulas utilized in different regions of India are provided herein.

7.3.2 Ryves Formula (1884)

$$Q_p = C_R A^{2/3} \tag{7.7}$$

where

C_R = Ryves constant=6.8–8.5 for east-coast
= 10.2 for hilly area

Originally this was developed for Tamil Nadu state and is now used in Tamil Nadu and some parts of Andhra Pradesh and Karnataka.

7.3.3 Dickens Formula (1865)

$$Q_p = C_D A^{3/4} \tag{7.8}$$

Q_p = maximum flood discharge
C_D = Dickens constant (6–30); (coastal Odisha and Andhra Pradesh = 22–28, central India=14–28, hilly region of Northern India=11–14, North India plains=6)
A = catchment area

For using in actual, local experience will be of help in properly selecting C_D. This is utilized in northern and central parts of India.

7.3.4 Inglis Formula (1930)

This formula works on the basis of flood data obtained from Western Ghat catchments located in Maharashtra:

$$Q_p = \frac{124A}{\sqrt{A + 10.4}} \tag{7.9}$$

7.3.5 Other Formulae

There are several empirical formulae established in different portions around the world. Certain empirical formula exists relating peak discharge to area of basin and includes flood frequency as well. Fuller's formula (1914) derived for catchments in the United States is typically one of the kind and is represented by

$$Q_{Tp} = C_f A^{0.8} \left(1 + 0.8 \log T\right) \tag{7.10}$$

where

Q_{Tp} = maximum 24-h flood for T year frequency
C_f = A constant (0.18–1.88)

7.3.6 Envelope Curves

The envelope curve method can be utilized for developing a relationship amid maximum flood flow and drainage area in regions having similar climatic characteristics and availability of flood data are insufficient. In this technique, accessible peak flood data from a large number of catchments that don't vary from one another significantly in terms of topographical and meteorological characteristics are composed. Then plotting of the collected data are done between flood peak and catchment area on a log-log paper resulting in a scattered plot representation. If an envelope curve is drawn encompassing all data points that are plotted, it can be utilized for obtaining maximum peak discharge for whichever specified region. Thus, the obtained enveloping curves are very beneficial in making rapid rough peak value estimations. If equations are fixed to the envelope curves, they give empirical flood formula as $Q = f(A)$. Figure 7.2 illustrates an enveloping curve for southern, northern, and central Indian rivers developed by Subramanya (2008).

Baird and Millwright (1951) related maximum flood discharge (Q_{mp}) having catchment area A on basis of maximum flood recorded all over the world as follows:

$$Q_{mp} = \frac{3025A}{(278 + A)^{0.78}} \quad (7.11)$$

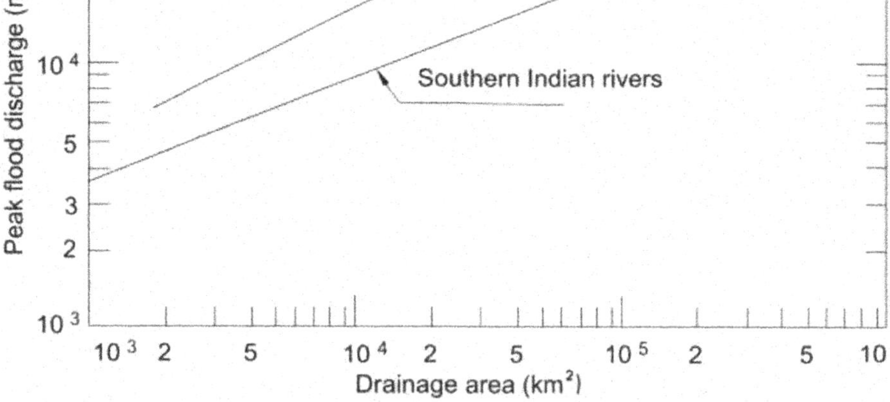

FIGURE 7.2 Enveloping curves for Indian rivers.

7.4 FLOOD FREQUENCY STUDIES

Floods are a type of natural hydrological process which are extremely complex in nature. They are very difficult to model analytically due to a number of constituent parameters resulting in occurrence of flood. For example, floods in a catchment is dependent on rainfall antecedent conditions and characteristics of catchment; in turn, each one of these factors is based on a host of component constraints. As a result, flood peak estimation becomes a very difficult problem leading to several different methodologies. Unit hydrograph and empirical formulae techniques shown in prior sections are few among them. An additional approach for flood flow prediction, and also valid for other hydrological processes, for example precipitation etc., is the statistical technique of frequency study (UNO/WMO, 1967).

Annual peak flood values from a specified area of catchment for many successive years comprise a hydrologic series of data known as the annual series (Central water commission India, 1973). Then data are organized in declining order of magnitude and probability P of every event being equated to or exceeded (plotting position) is computed by plotting location formula:

$$P = \frac{m}{N+1} \tag{7.12}$$

where

m = event order number

N = total number of data in event

Recurrence interval T can be computed as

$$T = \frac{1}{P}$$

Occurrence probability of the event in n successive years for r number of times is given by

$$P_{rn} = {}^{n}_{0}C_{r} P^{r} q^{n-r} = \frac{n!}{(n-r)!r!} P^{r} q^{n-r} \tag{7.13}$$

where $q = 1 - p$

Chow (1964) revealed that maximum frequency distribution functions appropriate for hydrological analysis can be articulated using subsequent equation called the generalized equation of hydrological frequency study:

$$X_T = \bar{x} + K\sigma \tag{7.14}$$

where

X_T = value of variate x of an arbitrary hydrologic series of T return period
$\overline{x}$ = mean of variate
σ = standard deviation (SD) of variate
K = frequency factor

Some frequently utilized frequency distribution function to predict extreme flood values are as follows:

1. Gumbel's distribution
2. Log-Person type III distribution
3. Normal distribution

The first two distribution functions are discussed here.

7.4.1 Gumbel's Distribution

Gumbel (1941) established this extreme value distribution and is generally called Gumbel's distribution. For extreme hydrological and meteorological studies such as prediction of flood, wind speed, etc., it is one of the most widely used probability distribution functions.

Based on his extreme events theory, probability of occurrence of an event equivalent to or greater than x_0 is

$$P(X \geq x_0) = 1 - e^{e^{-y}} \tag{7.15}$$

y is a dimensionless variable as given below:

$$y = \alpha(x - a) \quad a = \overline{x} - 0045005\ \sigma_x \quad \alpha = 1.2825 / \sigma_x$$

Hence, $y = \dfrac{1.285(x - \overline{x})}{\sigma_x} + 0.577$

where

$\overline{x}$ = mean
σ_x = SD of variate X

Practically, it is value of X for a specified P which is necessary and Equation 7.15 can be expressed as

$$y_p = -\ln\left[-\ln(1 - P)\right] \tag{7.16}$$

7.4.2 Log-Pearson Type III

This distribution is excessively utilized in the United States for projects supported by their government. Here, first the variate is converted into logarithmic form (base 10) and then converted data is analysed. X is the variate of an arbitrary hydrologic series of Z variates. Here

$$z = \log x \tag{7.17}$$

are obtained first. For this series of Z and for any T recurrence interval, Equation 7.14 provides

$$z_T = \overline{z} + K_z \sigma_z \tag{7.18}$$

where

K_z = a frequency factor that is a function of recurrence interval T and coefficient of skewness C_s

σ_z = SD of sample of Z variate = $\sqrt{\Sigma(z-\overline{z})^2 / (N-1)}$

C_s = coefficient of skewness of Z variate = $\dfrac{N\Sigma(z-\overline{z})^3}{(N-1)(N-2)(\sigma_z)^3}$

$\overline{z}$ = average of z values

N = sample size

Once z_T is found by Equation 7.18, equivalent value of x_T is found by Equation 7.17 as

$$x_T = antilog\ (z_T)$$

7.5 DESIGN FLOOD

At the time of designing the hydraulic structures, considering economic factors is not practical for providing safety to the system as well as structural safety at maximum potential flood in a catchment. Design of small structures such as storm drainage and culverts for lesser severe floods as consequence of a higher magnitude flood in comparison to design flood might not be too serious. Temporary inconvenience like very rare severe damage to property, loss of life, and disruption of traffic can be caused. Then again, storage structure, such as dam, demands more consideration to flood magnitude utilized in design. Catastrophe of these constructions leads to huge loss of life and excessive damage to property on downstream of these structures. From this it can be concluded that design criteria for choosing flood magnitude dictates the type and importance of structural and economic growth of neighbouring areas. Processes adopted to select flood magnitude for designing of several hydraulic structures are highlighted in this section.

7.5.1 Design Flood

IT is assumed flood for designing a structure.

7.5.2 Spillway Design Flood

To design a spillway of stone structure, design flood is used for this specified purpose. This word is often utilized for denoting maximum discharge that can be allowed to pass in a hydraulic structure devoid of any harm or critical danger to firmness of the structure.

7.5.3 Standard Project Flood (SPF)

It is a flood resulting from an austere amalgamation of hydrological and meteorological factors which are sensibly appropriate to the area. Exceedingly exceptional amalgamations of aspects are omitted.

7.5.4 Probable Maximum Flood (PMF)

It is extreme flood which is possible physically in an area resulting from severe most amalgamations, inclusive of exceptional amalgamations of hydrological and meteorological features. PMF is utilized in conditions where structural failure would result in catastrophic damage and loss of life and intrinsically whole safety is sought from possible flooding. SPF, on other hand, is frequently utilized where structural failure would cause fewer severe harms. Characteristically, SPF is around 40–60% of PMF for identical drainage basin.

'IS: 11223-1985: Guidelines to Fix Spillway Capacity' is presently utilized for selecting design floods for dams in India. In these guiding principles, dams are categorized in accordance to size by utilizing gross storage and hydraulic head behind the dam. Difference between maximum level of water on upstream and standard annual average flood level on downstream is known as the hydraulic head.

7.6 DESIGN STORM

Design storm is necessary for estimating design flood for a project by using a unit hydrograph. This can be the storm producing probable maximum precipitation (PMP) to derive a standard project storm (SPS) for computations of SPF or PMF. Calculations are done by skilled hydro-meteorologists by utilizing hydro-meteorological data. Different techniques ranging from very complex hydro-meteorological techniques to simplified study of past precipitation data are used subject to accessibility of consistent applicable data and proficiency.

A process followed in India is described in the following steps:

- Selection of duration of critical rainfall. If flood peak is of concern, this will be basin lag. Longest storm duration experienced in basin is selected if flood volume is of major concern.
- Selection of previous major storms occurred in the basin under study of that region. Perform DAD analysis and enveloping curve which represents relation between maximum depth and duration for basin under study is obtained.

- Scaling of rainfall depths for suitable intervals of time from envelope curve. Arrangement of these increments for getting a critical order producing maximum flood peak when employed to pertinent unit hydrograph of basin.

Critical order of rainfall growths can be found by trial and error method. In an alternative way, precipitation increments are first organized in a table of appropriate ordinates of unit hydrograph:

i. Increment of maximum rainfall is contrary to maximum ordinate of unit hydrograph.
ii. Second top increment of rainfall is in contrary to second largest ordinate of unit hydrograph, and so on.
iii. Now reverse the order of increments of rainfall organized above, with first item last and last item first. This novel order provides design storm.
iv. Then combine design storm with hydrological abstractions most favourable to high runoff, i.e. primary loss is low and rate of infiltration is lowest for getting hyetograph of rainfall excess for operating on unit hydrograph.

7.7 RISK, RELIABILITY, AND SAFETY FACTOR

7.7.1 Risk and Reliability

Designer of a hydraulic structure constantly faces a niggling uncertainty regarding risk of failures in the structure. This is due to estimation of hydrologic design values (such as river stage and design flood discharge in the course of design flood) involving an in-built or natural uncertainty and also a hydrological risk of failure. Hence, risk is given by

$$\bar{R} = 1 - (1 - p)^n \tag{7.17}$$

$$\bar{R} = 1 - \left(1 - \tfrac{1}{T}\right)^n \tag{7.18}$$

where

P = probability $P(x \geq x_T) = \tfrac{1}{T}$
T = return period

The reliability R_e is defined as

$$R_e = 1 - \bar{R} = \left(1 - \tfrac{1}{T}\right)^n \tag{7.19}$$

It can be observed that design of a structure based on return period depends upon allowable risk level. Practically, acceptable risk is administered by policy and economic considerations.

7.7.2 Safety Factor

As mentioned above, in addition to hydrological uncertainty, a water-resource development project is inclusive of many other uncertainties. These may arise from construction, structural, environmental, and operational causes in addition to non-technological aspects such as political, sociological, and economic causes. As such any water resource development project will have a safety factor for a given hydrological constraint M as defined below:

$$SF_M = \frac{\text{Actual value adopted in the design of the project}}{\text{Value of the parameter M obtained from hydrological considration}} = \frac{C_{am}}{C_{hm}}$$

Parameter M comprises such items as flood discharge maximum river stage, magnitude, freeboard, and reservoir capacity. Difference $(C_{am} - C_{hm})$ is known as margin of safety.

7.8 DROUGHT

Lack of precipitation, such as snow, rain, or sleet, for a prolonged time period causes shortage of water which leads to occurrence of drought. Though droughts are natural phenomena, human activities, such as improper usage and management of water, can aggravate dry conditions. Consideration of drought conditions vary from region to region and is largely based on precise weather patterns of a specific area. On the tropical island of Bali, threshold for drought may be attained just after six rainless days, whereas in the Libyan desert, annual rainfall would need to fall below 7 inches for warranting a similar declaration.

Particularly, developing nations are susceptible to climate change impacts, involving drought. From 2005 to 2015, developing nations suffered more than 80% of drought-induced economic loss and damage which was related to fisheries, crops, and livestock. Approximately some $29 billion economic toll suggests only part of the entire scenario. In these nations, drought becomes notorious for causing water and food scarcity and aggravating already existing difficulties such as famine and civil unrest. It also can lead to mass migration which results in dislocation of entire population.

Most people think drought as a phase of remarkably dry weather which continues long enough for causing problems like water supply shortages and crop damage. However, due to development of dry conditions for various reasons, drought can be defined in more than one way.

'Occurrence of drought conditions is not only due to high temperatures and lack of precipitation but also due to overpopulation and overuse of water', said David Miskus, a meteorologist and drought expert at National Oceanic and Atmospheric Administration's (NOAA) Climate Prediction Centre.

In the 1980s, two researchers discovered more than 150 published definitions of drought, which they published in *Water International* journal. In an effort for bringing some order to measure drought, these researchers assembled definitions into four fundamental classifications: hydrological, meteorological, socio-economic, and agricultural. First three explanations classify drought as a physical phenomenon. Last classification pacts with drought as problem related to supply and demand, through effects of water deficits.

These descriptions commonly stipulate the beginning, end, and degree of drought severity by associating rainfall over a definite period of time with a historic average. Researchers take account of both snow and rain in rainfall measurements, as some of the US regions, such as mountainous West, depend upon winter snow for plenty of their annual water.

7.8.1 Meteorological Drought

When an area's precipitation drops far short of expectations, a large band of parched, cracked earth is visible and you're probably visualizing the impact of meteorological drought which occurs in that area.

7.8.2 Agricultural Drought

When availability of supplied water are not able to meet need of livestock or crops at a specific period, agricultural drought may follow. It may be resulting from reduced access to water supplies, meteorological drought, or just bad timing; for instance, for hydrating crops, runoff is most needed before melting of snow occurs.

7.8.3 Hydrological Drought

Once deficiency of rainfall endures for long enough to diminish surface water such as streams, reservoirs, or rivers, the supply of groundwater is also reduced, which results in a hydrological drought.

7.8.4 Socio-economic Drought

This type occurs when water demand surpasses the supply. Too much irrigation or reduction of energy production by hydroelectric power plant operators due to low flow in rivers are some examples of socio-economic drought.

7.8.5 Causes of Drought

7.8.5.1 Natural Causes

Droughts have afflicted mankind considerably throughout our history, and until in recent times they were frequently natural phenomena activated by recurring patterns of weather, like amount of heat and moisture in sea, land, and air.

7.8.5.2 Fluctuating Ocean and Land Temperatures

Global weather patterns are largely dictated by ocean temperatures, including wet and dry conditions on land, and also minute variations in temperature can have enormous ripple impact on climatic systems. Study reveals that prolonged and dramatic temperature variations in North Atlantic Oceans and North Pacific directly correspond to severe weather patterns on land, involving tenacious droughts in North America and east Mediterranean – second of which has been defined as region's vilest drought in almost 900 years. Fluctuation of ocean temperature is also the reason for La Niña and El Niño weather phenomena. In drying out southern United States, La Niña played a very notorious role. In the meantime, warmer surface temperature on land results in more evaporation of moisture from ground, increasing the effect of drought conditions.

7.8.5.3 Varying Patterns of Weather

Rainfall distribution all over the world is affected by the way air circulates through atmosphere. When irregularity in surface temperature occurs predominantly over sea-air interface, patterns of air circulation are changed, altering where and how rain drops all around the world. Water supply and demand can be thrown out of sync by new weather pattern. It is also the situation when earlier snowmelt than standard diminishes quantity of water availability for harvests in summer.

7.8.5.4 Reduced Soil Moisture

Cloud formation can be affected by soil moisture, and therefore results in precipitation. When evaporation of water takes place from wet soil, it donates to rain cloud formation, which returns water back to earth. When land is dry than normal, moisture still vaporizes into atmosphere, although not at an adequate volume for forming rain cloud. Effectively, the land bakes, eliminating extra moisture and in addition aggravating drier conditions.

7.8.5.5 Man-made Causes

Though drought is a natural phenomenon, human activities such as use of water and emission of greenhouse gas, have an increasing effect on their possibility and intensity.

7.8.5.6 Climate Change

Global warming and climate change precisely impact drought in two simple methods: rise in temperature usually makes dry regions drier and wet regions wetter. In wet areas, warm air absorbs more water, which leads to greater rain events. However, in arid regions, warmer temperature leads to evaporation of water more rapidly. Additionally, climate change results in variation of larger scale patterns of atmospheric circulation, shifting storm tracks off their characteristic routes. In turn, this can increase extreme weather conditions, which is the purpose of climate models that predict previously parched United States. Southwest and Mediterranean will endure to become drier.

7.8.5.7 Excess Demand of Water

Most often drought reveals a disparity in supply and demand of water. Booming of regional population and intensive use of water for agricultural purposes can put a stress on water resources management, even tipping the scale sufficiently for making danger of drought a certainty. A study makes assessment that from 1960 to 2010, water consumption by humans increased occurrence of drought by 25% in North

America. Moreover, when rainfall declines and drought conditions take grip, tenacious demand of water in form of augmented pumping from reservoirs, rivers, and groundwater can diminish valuable water resources which might need years for replenishing and enduringly affect availability of water for future use. In the meantime, demand for water supplied by upstream rivers and lakes, predominantly in the form of hydroelectric dams and irrigation, can result in drying out or diminishing of downstream sources of water contributing to drought in other regions (Khushalani & Khushalani 1971).

7.8.5.8 Deforestation and Soil Degradation

When moisture is released into atmosphere by plants and trees, formation of clouds occur and return moisture as rain back to earth. When vegetation and forests disappear, availability of water becomes less for feeding water cycle, making the whole region extra susceptible to drought. In the meantime, poor land use practices, like intensive farming, and deforestation can reduce quality of soil which in turn decreases the ability of land of absorbing and retaining water. Consequently, soil quickly becomes dry (inducing agricultural drought), and replenishment of groundwater is less (contributing to hydrological drought).

7.9 HOW IS DROUGHT MONITORED AND ASSESSED?

US Drought Monitor (USDM) is a fundamental monitoring constituent in National Integrated Drought Information System (NIDIS), developed in 2006 by Congressional Act for implementing a combined drought monitoring and forecasting scheme at local, federal, and state levels. It involves monitoring, forecasting, research, response, and educational constituents related to drought as part of its early warning system. All these constituents are highlighted in US Drought Portal.

USDM is a weekly merchandise providing a common summary of present drought conditions. Several drought indicators, comprising different outlooks, indices, news accounts, and field reports, are studied and synthesized. Additionally, consultation with many specialists from different offices and agencies across the country is done. Result is consensus valuation represented on USDM map: http://droughtmonitor.unl.edu; http://drought.unl.edu

7.9.1 Regression Analysis

One of the broadly adopted forecasting approaches and early candidates applied for predicting time series is the regression analysis. It is a statistical technique for examining relation among variables (Sykes, 1993). Performance of this technique vastly depends on the shape of regression line, type of dependent variables, and number of independent variables. Essentially, broad range of regression analysis utilized to forecast time series involves log-linear regression and logistic regression.

7.9.1.1 Logistic Regression (LR)

For statistically downscaling results from simulations of a universal climatic model with coarse resolution, logistic regression technique is used. Certain descriptions are necessary to clarify about questions on how a LR can be utilized as a downscaling method. In simple LR, dependent variable is generally binary or dichotomous, i.e. it can assume value of 0 (zero) with a probability of P for failure and value of 1 (one) with a probability of P for success. On the other hand, in cases where dependent variable can take more than two values known as polytomous or multinomial, LR can also be protracted (Prasad et al., 2010; Preisler and Westerling, 2007; Kleinbaum and Klein, 2002; Tabachnick and Fidell, 1996; Aldrich and Menard, 1995; Menard, 1955). LR is more authoritative compared to other regression models, and no assumption is made by it on probability distribution of predictors. There is no need of probability distribution to be linearly related, have an equal variance within each group or normally distributed; these limitations are required in customary linear regression models. Derivation of the relationship is done from supposed logit-transformation by successive probability of:

$$P(x) = \frac{\exp(b_0 + b_1x_1 + b_2x_2 + \ldots + b_nx_n)}{1 + \exp(b_0 + b_1x_1 + b_2x_2 + \ldots + b_nx_n)} \tag{7.20}$$

where $x_i\,(i = 1, 2, \ldots, n)$ is predictor and $b_i\,(i = 0, 1, 2, \ldots, n)$ is coefficient of LR. As LR estimates probability of success above probability of failure, outcomes of the study could be positioned as odd ratios in subsequent expression.

Structurally LR is analogous to familiar multivariate linear regression, where logit performs as predictor and predictors are natural logarithm of success probability (that is the odds).

7.9.1.2 Log-linear Models

Processes for predicting drought class applying log-linear regression are given below:

1. First calculate observed frequencies in accordance to hydrological or meteorological drought class time series.
2. Develop log-linear regression model and using maximum likelihood method estimate the model parameters.
3. Calculate odds and confidence intervals. An odd is a ratio of expected frequencies representing number of times that it is less, more, or equally possible to occurrence of a particular event instead of another (Moreira et al., 2008).
4. Predict drought class transition

7.9.2 Stochastic Modelling: ARIMA and SARIMA

For many scientific applications, stochastic models have been broadly utilized which include investigating and modelling of hydrologic time series. Benefits of stochastic models involve improved selection of sequential linear correlation feature of time series and proficiency of systematically searching for identifying, estimating, and diagnostic checking for development of model. Two significant and prevalent categories of stochastic models are autoregressive integrated moving average (ARIMA) and seasonal ARIMA (SARIMA). Both variants of stochastic models comprise three vital constraints: moving average order of q, autoregressive order of p, and dth difference of time series z_t, where iterative fine-tuning should be carried out for generating a robust model. Using defined parameters, models are generally defined as ARIMA (p, d, q) for ARIMA and ARIMA $(p, d, q)\ (P, D, Q)s$ for SARIMA, where (p, d, q) is non-seasonal part of model and (P, D, Q)s is seasonal part of model.

7.9.2.1 ARIMA

For time series forecasting, time series (ARIMA) or stochastic models are an approach. These models give a systematic empirical technique to forecast and analyse hydrological time series data (Rath et al., 2017). Hence, Box-Jenkins approach for development of ARIMA permits a suitable therapy for non-stationary points in historic time series data. Integration of parts of AR and MA is the cause of this ARIMA model property. They let every variable to be specified by stochastic error terms and its own lagged values. Multiplicative ARIMA $(p, d, q)\ (P, D, Q)s$ model is created by the AR and MA, where $(P, D, Q)s$ is the seasonal part and (p, d, q) is the non-seasonal part of the model. Non-seasonal portion of ARIMA model (AR) can be articulated as follows:

$$\varnothing(B)\ \nabla^d\ Z_t = \theta\ (B)\ a_t \tag{7.21}$$

where $\varnothing\ (B)$ and $\theta\ (B)$ are polynomials for p and q in respective order.

7.9.2.2 Seasonal ARIMA Models

An ARIMA model was generalized for dealing with seasonality, and describe a common multiplicative seasonal ARIMA (SARIMA) model (Box et al., 1994). A characteristic benefit of SARIMA model is that only little model parameters are essential to describe time series exhibiting non-stationary both across and within seasons. In brief, SARIMA model can be clarified as ARIMA $(p, d, q)(P, D, Q)s$, which is stated below:

$$\varnothing_p(B)\phi_P\left(B^S\right)\nabla^d\ \nabla_S^D Z_t = \theta_q(B)\Theta_Q\left(B^S\right)a_T \tag{7.22}$$

where p is the order of non-seasonal autoregression, d is the number of regular differencing, q is the order of non-seasonal MA, P is the order of seasonal autoregression, D is the number of seasonal differencing, Q is the order of seasonal *MA*, s is the season length, U is the seasonal AR constraint of order P, and H is the seasonal *MA* constraint of order Q.

7.9.3 Probabilistic Modelling: Markov Chain (MC)

It is a memory-less random procedure wherein if a current state has been given or known, past and future are independent of one another Yang et al. (2012). It is a mathematical method for obtaining probabilities of system utilizing a set of transition probabilities from one state to other. In general, when transitional probability is reliant on conditions in preceding m time period, it is called an m th-order MC.

They are generally utilized for assessing probability of drought occurrence and for evaluating and predicting occurrence time of a drought condition. A standard class of stochastic models for representing a time series of separate variables is identified as MC. It is dependent on a collection of system states, with first-order MC as most general form based on present system state only, and not on preceding states. Contrarily, a first-order MC is a stochastic procedure (random variable), such that X_{t+1} is provisionally not dependent on $X_0, X_1, X_2, \ldots, X_{t-1}$, given X_t, for any time t. Probability that X_{t+1} takes a specific value j is then found as follows (Çinlar, 1975):

$$\Pr\{X_{t+1} = j \mid X_0 + X_1, \ldots, X_t = \Pr\{X_{t+1} = j \mid X_t = i\} \; \forall i,\, j \in S,\, t \in T \qquad (7.23)$$

Thus, an MC is categorized by a set of states S and by transition probability p_{ij} between states. Transition probability p_{ij} is the probability that MC is at subsequent point of time in state j, provided that it is at present point of time in state i (Paulo and Pereira, 2007).

7.9.4 Artificial Intelligence Based Models

7.9.4.1 Artificial Neural Network (ANN)

ANNs are non-linear, flexible models resembling arrangement of a nervous system. They can become familiarized to inserted data and study and determine patterns from it. In theory, by providing a suitable quantity of non-linear processing units, ANNs are capable of gaining experience and learn for estimating any multifaceted function relationship precisely (Mishra and Singh, 2010). The neural networks learn on basis of a black-box model. Major aspects influencing model performance are network architecture, input capability, and validation of model. Construction of ANN network include three major constituents: input layer, hidden layer, and output layer (Sahoo et al., 2019). For generating an ANN model, tuning of parameters by researchers is essential: number of hidden layer neurons, learning rate (training parameter controlling weight size and bias variations to learn training algorithm), and momentum (update weightage of model input to the following input). Therefore, a strong advantage of ANNs is that they do not necessitate for defining the processes or procedures between the input and output. Moreover, flexibility in architecture of ANN also permits cases to be easily protracted from univariate to multivariate ones. Because of variations in architecture of neural network, there are several alternatives in ANN models and multilayer perceptron feed forward model is most prevalent architecture of ANN. Yet, discussions regarding application of ANNs is not limited to any specific ANN variant.

7.9.4.2 Fuzzy Logic (FL)

Zadeh (1965) conceptualized fuzzy logic and is described as a convenient way for mapping an input space to an output space (Sandya et al., 2013; Prasad and Sudha, 2011). Amid many advantages of utilizing FL, the most applicable for the present study is the information that it is based on a natural language and can model nonlinear functions and inaccurate data of arbitrary complex processes. In traditional (Crisp or Boolean) set concept, membership function (MF) of an element x in a set A is described by a distinctive function assigning a value of either 1 (true) or 0 (false) to every single function in universal set X. It can be said that 'every proposition is either true or false'. However, according to Klir and Yuan (1995), FL infringes both 'contradiction' and 'excluded middle' laws. In traditional sets, the membership is not a 'true-false' answer, but FL is based on fuzzy sets in that where the answer is 'not quite true or false'. Therefore, common theory behind FL is that a set of predefined rules (if-else statements) are employed in parallel for interpreting certain values in input vector and then allocate these values to output vector. For achieving the output vector, a fuzzy MF curve is utilized for defining the way for mapping points into input space to a membership value (or grade/membership degree) between 0 and 1.

7.9.4.3 Support Vector Machine (SVM)

Vapnik (2000) introduced SVM for describing properties of learning machines so as to make them capable for simplifying unseen data (Kisi and Cimen, 2011). Learning procedure is not responsive to comparative number of training samples in positive and negative categories. SVM aims to minimize a bond on generalized error of a model in high-dimensional space, termed as structural risk minimization, whereas other empirical risk minimization based learning algorithm (e.g. ANN) classifies only positive class appropriately for minimizing error over the dataset. In brief, SVMs pursue for minimizing generalization error; however, ANNs and other empirical risk minimization based learning algorithms pursue for minimizing training error. SVMs are characterized into two categories: support vector regression (SVR) and support vector classification (SVC). Among them, SVR is mostly preferable for forecasting problems. Vital parameters to tune SVMs comprise kernel-type parameter (categories of algorithms for analysing patterns), regularization parameter (trade-off amid accomplishing a low testing error and a low training error), Gamma parameter (model intricacy), and boundary of error acceptance (Samantaray et al., 2019). Using an iteration procedure, scholars are capable of developing a robust SVM model with accessibility to various kinds of kernel and with aid of tuning above-mentioned parameters.

7.9.4.4 Hybrid Models

Hybrid model is an innovative class of hydrological modelling which was developed in last decade. Based on knowledge of authors, Mishra et al. (2007) first introduced hybrid model for drought forecasting in the hydrological field. In accordance to papers reviewed, authors witnessed that hybrid models can be congregated into two different groups: firstly, combination between the machine learning models and, secondly, hybridization between machine learning models and data preprocessing techniques (Mohanta et al., 2020).

7.9.5 Dynamic Modelling

In dynamic modelling approach, real-time data is utilized for describing a phenomenon over time. Owing to fast growth of remote sensing (RS) in monitoring and assessment of impact of drought, accessibility to real-time variables related to drought has increased as well. Hence, over the years, there is an increase of studies on dynamic drought forecasts. Dynamic drought modelling greatly depends on real-time RS data, whereas statistical drought forecasting models utilize long-term conventional observations from gauging stations. Remote sensing is a method to obtain consistent information regarding areas, objects, or phenomenon from faraway or/ and deprived of making any physical contact, i.e. typically from a satellite (NOAA, 2017) or an aircraft. For research on droughts, RS observations can be utilized for monitoring climatological variables related to drought and quantify impact of drought from perspective of an ecosystem (AghaKouchak et al., 2015). Satellite RS has been utilized to monitor Earth's climate or weather since the accomplishment of Television and Infrared Observation Satellite (TIROS-1) mission in 1960. For instance, quantifying temporal terrestrial water storage irregularities by computing distance amid two spacecraft; volumetric water content of soil of 2–5 cm depth from ground surface can be transformed from active microwave backscattering and passive microwave brightness temperature using empirical relationship (Njoku et al., 2003). The rate of rainfall can be transformed from visible images and satellite infrared of cloud top temperature utilizing empirical statistical relationship (Joyce & Arkin, 1997; Arkin et al., 1994).

7.10 CONCLUSION

Disasters due to flood and drought are predictable in terms of place and time of their occurrences. In addition, with assistance of developed techniques for predicting the occurrence of flood and drought conditions, their magnitude, and intensity, it has become plausible for managing the hazards to certain degree (Sahoo et al., 2020 a; Sahoo et al., 2020 b; Samantaray and Sahoo, 2020). Based on the above discussion in the chapter, it can be concluded that disasters can be results of human activities or can be natural, and all menaces need not turn into catastrophes as it is challenging to eradicate disasters, predominantly the natural disasters. However, due to high vulnerability of large population residing in the coastal areas, there is an increase in loss of life and property in successive storms in countries like India, Bangladesh, Myanmar, etc.

REFERENCES

AghaKouchak, A., Farahmand, A., Melton, F.S., Teixeira, J., Anderson, M.C., Wardlow, B.D. and Hain, C.R. 2015. "Remote sensing of drought: Progress, challenges and opportunities." *Reviews of Geophysics*, 53(2), 452–480.

Aldrich, J. H. and Nelson, F. D. 1984. "Linear probability, logit, and probit models." Sage, Beverly Hills, CA.

Arkin, H., Xu, L.X. and Holmes, K.R. 1994. "Recent developments in modeling heat transfer in blood perfused tissues." *IEEE Transactions on Biomedical Engineering*, 41(2), 97–107.

Box, G.E.P., Jenkins, G.M. and Reinsel, G.C. (1994). "Time series analysis, forecasting and control." Prentice Hall, Englewood Cliffs, NJ.

Central Water Commission, India, "Estimation of Design Flood Peak." Flood Estimation Directorate, Report No. 1/73, New Delhi, 1973.

Çinlar, E. (1975). "Introduction to stochastic processes." Prentice-Hall, New Jersey.

Chow, V. T. 1964. Handbook of Applied Hydrology, McGraw-Hill, New York, NY.

Indian Bureau of Standards, "Guidelines for Fixing Spillway Capacity." IS: 11223–1985.

Joyce, R. and Arkin, P.A. 1997. "Improved estimates of tropical and subtropical precipitation using the GOES precipitation index." *Journal of Atmospheric and Oceanic Technology*, 14(5), 997–1011.

Khushalani, K. B. and M. Khushalani. 1971. Irrigation Practice and Design, Vol. 1, Oxford & IBH, New Delhi.

Kisi, O. and Cimen, M. 2011. "A wavelet-support vector machine conjunction model for monthly streamflow forecasting." *Journal of Hydrology*, 399(1–2), 132–140.

Klir, G. J. and Yuan, B. 1995. "Fuzzy Sets and Fuzzy Logic: Theory and Applications." P. Hall, Ed. Prentice Hall, Englewood Cliffs, NJ.

Menard, S. 1995. "Applied logistic regression analysis." Sage, Thousand Oaks, Ca.

Moreira, E.E., Coelho, C.A., Paulo, A.A. et al. (2008). "SPI-based drought category prediction using loglinear models." *Journal of Hydrology*, 354, 116–130.

Mishra, A.K., Desai, V.R. and Singh, V.P. 2007. "Drought forecasting using a hybrid stochastic and neural network model." *Journal of Hydrologic Engineering*, 12(6), 626–638.

Mishra, A. K. and Singh, V. P. (2010). "A review of drought concepts." *Journal of Hydrology*, 391(1–2), 202–216.

Mohanta, Nihar Ranjan, Niharika Patel, Kamaldeep Beck, Sandeep Samantaray, and Abinash Sahoo. 2020. "Efficiency of river flow prediction in river using wavelet-CANFIS: a case study." In Intelligent Data Engineering and Analytics. 435–443, Springer, Singapore.

Njoku, E.G., Jackson, T.J., Lakshmi, V., Chan, T.K. and Nghiem, S.V. 2003. "Soil moisture retrieval from AMSR-E." *IEEE Transactions on Geoscience and Remote Sensing*, 41(2), 215–229.

Paulo AA, Pereira LS (2007) Prediction of SPI drought class transitions using Markov chains. *Water Resour Manag* 21(10):1813–1827.

Prasad K., Dash, S. K. and Mohanty, U. C. 2010. "A logistic regression approach for monthly rainfall forecasts in meteorological subdivisions of India based on DEMETER retrospective forecasts." International Journal of Climatology, 30, 1577–1588.

Prasad Reddy P. V. G. D. and Sudha K. R. 2011. "Application of Fuzzy Logic Approach to Software Effort Estimation." International Journal of Advanced Computer Science and Applications, 2(5), 87–92.

Rath, A., S. Samantaray, K. S. Bhoi, and P. C. Swain. 2017. Flow forecasting of Hirakud reservoir with ARIMA model. International Conference on Energy, Communication, Data Analytics and Soft Computing (ICECDS), IEEE, 2952–2960.

Sahoo, A., S. Samantaray, and D. K. Ghose. 2019. "Stream flow forecasting in Mahanadi River Basin using artificial neural networks." Procedia Computer Science, 157, 168–174.

Sahoo, A., S. Samantaray, S. Bankuru, and D. K. Ghose. 2020. "Prediction of flood using adaptive neuro-fuzzy inference systems: a case study." In Smart Intelligent Computing and Applications. 733–739, Springer, Singapore.

Samantaray, S., O. Tripathy, A. Sahoo, and D. K. Ghose. 2020. "Rainfall forecasting through ANN and SVM in Bolangir watershed, India." In Smart Intelligent Computing and Applications. 767–774, Springer, Singapore.

Sandya, H.B., Kumar, P.H., Bhudiraja, H. and Rao, S. K. (2013) "Fuzzy rule based feature extraction and classification of time series signal." International Journal of Soft Computing Engineering.

Subramanya, K. 2008. Engineering Hydrology, 3rd Edition, New Delhi, Tata McGraw-Hill

Sandeep Samantaray and Abinash Sahoo. 2020. Estimation of Flood Frequency Using Statistical Method: Mahanadi River Basin, IWA Publishing, India, pp. 189–207.

Sykes, A. O. 1993. "An Introduction to Regression Analysis Coase Sandor Working Paper Series in Law and Economics, 20.

UNO/WMO, "Assessment of Magnitude and Frequency of Flood Flows." United Nations Publications, Water Resources Series No. 30, 1967.

Vapnik, V.N. 2000. "The Nature of Statistical Learning Theory." Second Edition, Springer, 1–324.

Yang, Z., Li, Y., Chen, W. and Zheng, Y. 2012. "Dynamic hand gesture recognition using hidden Markov models." *7th International Conference on Computer Science & Education (ICCSE)*, 360–365.

Zadeh, L. A. (1965). "Fuzzy sets." *Information and Control*, 8, 338–353.

8 Sediment Sampling and Transport

8.1 INTRODUCTION

Many of the earlier civilizations came into being in the fertile valleys of large rivers. Civilizations prospered in the Nile valley in Egypt, along the Tigris and the Euphrates rivers in Mesopotamia, along River Indus in India, and along River Yellow in China. As early as 4000 BCE, people built dams across the rivers to store water, dug canals for navigation purposes, and also carry water to the fields to produce much-needed food. Together with the problems associated with the irrigation works, these earlier civilizations were confronted with the problems or flood control and the Chinese had developed excellent systems of dikes for the protection of inhabited areas against floods. Thus, since the beginning of civilization, the human kind had to face problems related to rivers, and have been able to solve them to best of their capability. In modern times, more complex problems are encountered because, with the population increasing day by day, rivers are being harnessed to a greater extent for multipurpose usage such as water supply, flood control, power generation, navigation, and irrigation for which artificial changes are being made in water courses (Sahoo et al., 2021; Mohanta et al., 2020). These complications have turned out to be problematic because of the fact that in most cases, rivers and other water courses, run through loose materials and water transports some of these materials along with it. The presence of sediment in water also create problems in the working process of pumps and turbines through which the sediment-laden water flows (Ghose and Samantaray, 2018, 2019). Further, a huge amount of money is spent to treat water for removing sediment and making it suitable for domestic as well as industrial consumption. Other problems connected with the training and harnessing of rivers carrying sediment also engage the attention of a large number of engineers. In fact, if the water courses were to flow through non-erodible material, many of the above problems would not exist. Human understanding of the environment and the awareness of the dangers of environmental pollution have increased enormously in the last few decades. While a lot of attention has naturally been placed on chemical, bacterial, and thermal pollution, environmentalists have also studied the influence of high sediment concentration on aquatic animal and plant life.

8.2 SEDIMENT SAMPLING

8.2.1 Bed Load Transport

When average shear stress on an alluvial channel bed surpasses the critical tractive stress of the bed material statistically, the particles on bed may start to travel in the flow direction in accordance with generally accepted hypotheses. The particles move in diverse ways based on flow conditions, ratio or the densities of fluid and sediment, and size of the sediment. One mode of movement of sediment particles is by rolling or sliding along the bed. Such movement of the sediment is usually discontinuous; the particle may roll or slide for some time or remain stationary for a while and again start rolling or sliding (Garde, 1968). Sediment transported in this way is known as contact load. A second method of sediment movement is by bouncing or hopping along the bed whereby, for some time, the particle loses contact with the bed. Material transported in this way is known as saltation load. Saltation is a significant method of transport in case of no cohesive materials having reasonably high fall velocities, like sand in air and, to a smaller degree, gravel in water (Toffaleti 1969). The third mode of transport is in a state of suspension. In this case, the particles are supported by the turbulent fluctuations. Material supported in this way and transported by the flow is known as suspended load. Subcommittee on Sediment Terminology of American Geophysical Union has described different loads as stated below:

Material slid or rolled along the bed in significantly continuous connection with the bed is known as the contact load.
Material bouncing along the bed, or moving indirectly or directly by effect of bouncing elements is called saltation load.
In a fluid, material moving in suspension, being kept in suspension by turbulent variations, is called suspended load.

For a particular ratio of mass densities of the sediment and the fluid, the modes of transport generally depend on the average shear stress on the bed. For relatively low shear stresses, the material is transported almost entirely as contact load. Some material is transported as saltation load at slightly higher shear stresses, if such a type of motion can occur in significant amounts for the given value of ρ_s/ρ_f. With an additional increase in shear stress, a part of the material is transported in a state of suspension.

8.2.2 Bed Load Equations

Du Boys in 1879 developed the first bed load equation. Since then, a number of equations have been projected for predicting rate of bed load transport. Some of these are completely empirical in nature; some arc obtained from dimensional considerations and the others are based on a semi-theoretical approach. These equations have been discussed in the ensuing sections of this chapter.

The first to propose a bed load relation was Du Boys and the form of the equation proposed by him has been used subsequently by many researchers. Assuming that movement of bed material occurs in a series of layers parallel to bed, velocity of each layer varies from maximum at top layer on bed surface to zero at the bottommost layer at some depth. The flow model conceived by Du Boys is shown in Figure 8.1. Let N be the number of layers in motion and Δh the thickness of each layer. The lowest layer is assumed to be the one with zero velocity. If ΔV is the velocity of the second layer from the bottom, the velocity of the surface layer will be $(N-1)\Delta V$, when a linear variation of velocity is assumed. Hence, the bed load transport q_b will be

$$q_B = \gamma_s N\Delta h(N-1)\frac{\Delta V}{2} \tag{8.1}$$

Since the lowest layer is at rest, the resisting force at this elevation must equal the tractive force on the bed, i.e.

$$\tau_0 = (\gamma_s - \gamma_f)N\Delta h \tan\phi \tag{8.2}$$

where ϕ is the angle of repose for the bed material.

The value of N can be obtained by assuming that a single layer is moving under the critical condition:

$$\tau_{0c} = (\gamma_s - \gamma_f)\Delta h \tan\phi \tag{8.3}$$

$$N = \tau_0 / \tau_{0c} \tag{8.4}$$

Therefore,

$$q_B = \frac{\gamma_s \Delta h \Delta V \tau_o (\tau_0 - \tau_{0c})}{2\tau_{0c}^2} \tag{8.5}$$

$$\text{where } A = \frac{\gamma_s \Delta h \Delta V}{2\tau_{0c}^2} \tag{8.6}$$

Equation 8.5 is Du Boys' bed load equation.

8.2.2.1 Empirical Bed Load Equation

Several empirical equations or a form similar to the Du Boys' equation were proposed by some of the early investigators working after Du Boys. These, along with some of the recent empirical equations, are discussed in this section.

In the equations listed herein, values of the parameters A, m, etc. are usually given as functions of sediment size. Since these equations are dimensional, these values are applicable only for the system of units used by the particular investigator. A majority of these empirical equations do not include explicitly the effect of bed forms on the transport rate. However, equations proposed by Chang and US WES include the roughness coefficient of the bed and hence are expected to be valid for various bed configurations. As will be seen from some of the rational equations discussed later, the bed load rate decreases with increase in Manning's n, all other parameters being kept constant. But it may be noticed that Chang's equation indicates an increase in qB with an increase in n, which does not seem logical.

The most commonly used empirical equation is the one by Meyer-Peter and Muller. Prior to this study Meyer-Peter, Favre, and Einstein had proposed the following equation for bed load transport of uniform material with different relative densities:

$$\frac{q_B}{\gamma_s^{2/3}\left(\gamma_s-\gamma_f\right)^{1/9}}=2.16\frac{\left(UR_b\right)^{2/3}}{d\left(\gamma_s-\gamma_f\right)}-20.70 \tag{8.7}$$

R_b = (m)
d = (m)
U = (m/s)
γ_s = (tonnes/m^3)
γ_f = (tonnes/m^3)
q_B = bed load transport rate in dry weight (tonnes per second per meter width)

The foregoing equation has been modified by Meyer-Peter and Muller.

8.2.3 Saltation

It has been mentioned earlier that under certain conditions, the moving particles lose contact with the bed for some time and again hit the bed, thus moving in a series of bounces. Material transported in this manner is known as the satiation load. Probably McGee was the first to use this term, though later Bagnold (1956), Gilbert, and Danel, Durand and Condolios have described this method of transport in detail.

8.3 SUSPENDED LOAD TRANSPORT

The material moves as contact load or saltation load at low values of average shear stress on an alluvial channel bed, and the stream will only have clear water flow. However, as discussed earlier, with more rise in shear stress, some bed particles are carried away to the main flow and thus loses connection with the bed. These particles move with a velocity nearly equal to the velocity of flow and they comprise the suspended load.

Transport of suspended load is a progressive stage of bed load transportation. Thus, in instance of uniform sediment at low shear stresses, only bed load transport is expected, whereas at high shear stresses, both suspended load and bed load transport would take place. For non-uniform sediment, the finer size of bed material may predominantly travel in suspension, whereas the coarser parts of the bed material may mostly move as bed load, if they at all move.

Observations in natural streams and laboratory flumes have revealed that in a vertical, there is a decrease in suspended load concentration with increase in distance from bed. The suspended sediment concentration (SSC) can be articulated in several means:

i. *Absolute volume of solids per unit volume of water-sediment mixture:* First dry weight of sediment in a unit volume of mixture is determined in this method. Then divide this by specific weight of sediment which gives the absolute volume of sediment per unit mixture volume. Also, this can be expressed in terms of percentage by volume.
ii. *Dry weight of solids per unit volume of mixture:* Usually this is expressed as gram/litre, kN/cubic meter, or lb/cubic foot.
iii. *Dry weight of solids per unit weight of mixture:* Typically it is articulated in parts per million (ppm). One percent (%) is equal to 10,000 ppm.
iv. Dry weight of solids per unit weight of pure water equals the sample volume.

To signify the SSC, first and third methods are generally utilized among the above-mentioned methods.

8.3.1 Mechanism of Suspension

One of the most intersecting problems in mechanics of suspension is the study of the exact method by which sediment particles resting on the bed are carried in suspension. When lift on a particle is more than its submerged weight, the particle travels upwards into the flow. On the other hand, it is also supposed by some that the turbulent fluctuations near the boundary are responsible for entrapment of the sediment particle in flow; it is known that for rigid plane bed, the vertical turbulent fluctuation must reduce to zero at the bed. However, the alluvial bed is porous, it can permit vertical turbulent fluctuation of applicable magnitude to occur at the bed.

Laursen (1958) visualized a somewhat different mechanism of sediment entrainment. When a particle is moving either over the surface of the dune or over any small irregularity on the bed, a stage is reached when the particle loses contact with the bed momentarily. In such a case, the gravitational force is small and the flow pattern and the velocity of the particle are such that it can be taken into the main flow. The particle will move into the main flow. Since the amount of material in such an action will depend on

i. the number of particles moving,
ii. size of each particle, and
iii. velocity of each particle,

Laursen concluded that the rate or bed load will govern the rate of suspended load transport.

Sutherland (1967) has thoroughly witnessed the flow when particles are under condition of incipient motion on the bed and also when the particles move in suspension.

The structure of events which leads to entrainment is hypothesized as given below: turbulent flow can be visualized as comprising oval-shaped or rounded eddies. As they approach the bed, these eddies are distorted and the velocity of fluid increases inside the eddy. The laminar sublayer is disrupted by such eddies and they also encroach on surface layer of the particles. As a result, the local shear stress at that spot increases and causes rolling of the particles at the incipient motion condition. Since the eddy is much larger than the particle, many particles move and then come to rest outside the area of influence of the eddy. At the incipient motion condition, the eddies impinge at one spot once in a while and hence sediment movement is intermittent. At high rates of sediment transport, often when eddies encroach onto surface layer of particles and at many places, they apply substantial drag on particles and accelerate them. Certain particles, due to their position or due to their rolling up over neighbouring particles, project beyond the average bed level. In such a location, they are expected to be entrained due to the vertical velocity constituent of fluid inside the eddy. If velocity constituent is sufficiently large and upwardly inclined, the particle will leave the bed. An additional factor that can assist suspension is the lift on the particle. Also, if sediment bed is enclosed with dunes, the characteristics of bed assist the entrainment procedure, the troughs and upstream slopes of the dunes being the most active regions.

It is fairly conclusive from the variation theories for the entrainment of sediment that turbulence in flow is accountable for suspension of particles in the stream. In general, at any point the instantaneous velocity will have three components u, v, w in x, y, z direction in respective manner. If $\bar{u}, \bar{v}, \bar{w}$ represent average values of u, v, w over an adequately large interval of time and u', v', w' are turbulent fluctuation velocities that are entirely arbitrary, the relation can be written as follows:

$$\left.\begin{aligned} u &= \bar{u} + u', \quad \overline{u'} = 0 \\ v &= \bar{v} + v', \quad \overline{v'} = 0 \\ w &= \bar{w} + w', \quad \overline{w'} = 0 \end{aligned}\right\} \tag{8.8}$$

Here, quantities with bars represent mean value. Consider a steady uniform flow in a wide channel with x-axis as general direction of flow, $\bar{v}$, $\bar{w}$ will be zero, since averaged over time, there is no net flow in y as well as z direction. Flow can be considered to be two-dimensional for very wide channel because the conditions of flow do not change in z direction.

Suspended material is directed to two actions. First is the action of downward and upward turbulent velocity constituents v'. Second is the action due to gravity causing settlement of sediment that is heavier than water. It can be assumed that suspended material concentration is constant in vertical direction. The upward and downward fluid flow must be equal since there is no net flow in upward direction. However, settling will assist in downward movement of sediment; therefore, it can be seen that transportation of sediment will be more in downward compared to upward direction. This is true in the beginning, but will lead to greater concentration of sediment at greater distances from the water surface and thus result in formation of concentration gradient, ultimately causing the upward and downward transport of sediment to be such that equilibrium will be maintained. This sediment transfer from one elevation to another is analogous to the momentum transfer. The net upward sediment transport will be proportional to the concentration gradient $\partial C / \partial y$ and hence it can be represented by €, it is the sediment transfer or sediment diffusion coefficient. The net downward sediment transport at any horizontal will be $\omega_0 C$, in which ω_0 is the fall velocity of sediment and C is the sediment concentration. It may be mentioned that C is the time-averaged concentration, but the bar over it has been omitted for convenience. Hence, under equilibrium conditions, the equation

$$\omega_0 C + \varepsilon_x \frac{\partial C}{\partial y} = 0 \tag{8.9}$$

8.3.2 General Equation of Diffusion

Equation 8.2 for equilibrium of sediment can also be obtained from the general diffusion equation in three dimensions. Dobbins (1944), Iwagaki (1953), and McNown have given a general differential equation for the diffusion of foreign particles in a fluid.

Considering an elementary cube of sides δx, δy, δz and assuming ϵ_x, ϵ_y, and ϵ_z as coefficients of sediment diffusion for the diffusion along x-, y-, and z-axes. In addition, let u, v, w be time-averaged velocities in three directions and let the sediment concentration be C. Then inflow and outflow of sediment flux per unit time through different faces will be as shown in Figure 8.1. Terms similar to the ones corresponding to the x and y directions can also be written for the z direction. Then equating the total rate of change of sediment in the volume $\delta x \delta y \delta z$ to the change per unit time due to diffusion, one gets Equation 8.10.

$$\left(\frac{\partial C}{\partial t}\right)\delta x \delta y \delta z = \begin{bmatrix} -\dfrac{\partial}{\partial x}(uC) + \dfrac{\partial}{\partial x}\left(\in_x \dfrac{\partial C}{\partial x}\right) - \dfrac{\partial}{\partial y}\{(v - \omega_0)C\} \\ + \dfrac{\partial}{\partial y}\left(\in_y \dfrac{\partial C}{\partial y}\right) - \dfrac{\partial}{\partial z}(wC) + \dfrac{\partial}{\partial z}\left(\in_z \dfrac{\partial C}{\partial z}\right) \end{bmatrix} \delta x \delta y \delta z \tag{8.10}$$

$$\frac{\partial C}{\partial t} + \frac{\partial}{\partial x}(uC) + \frac{\partial}{\partial y}(vC) + \frac{\partial}{\partial z}(wC) = \frac{\partial}{\partial x}\left(\in_x \frac{\partial C}{\partial x}\right)$$

$$+ \frac{\partial}{\partial y}\left(\in_y \frac{\partial C}{\partial y}\right) + \frac{\partial}{\partial z}\left(\in_z \frac{\partial C}{\partial z}\right) + \omega_0 \frac{\partial C}{\partial y} \tag{8.11}$$

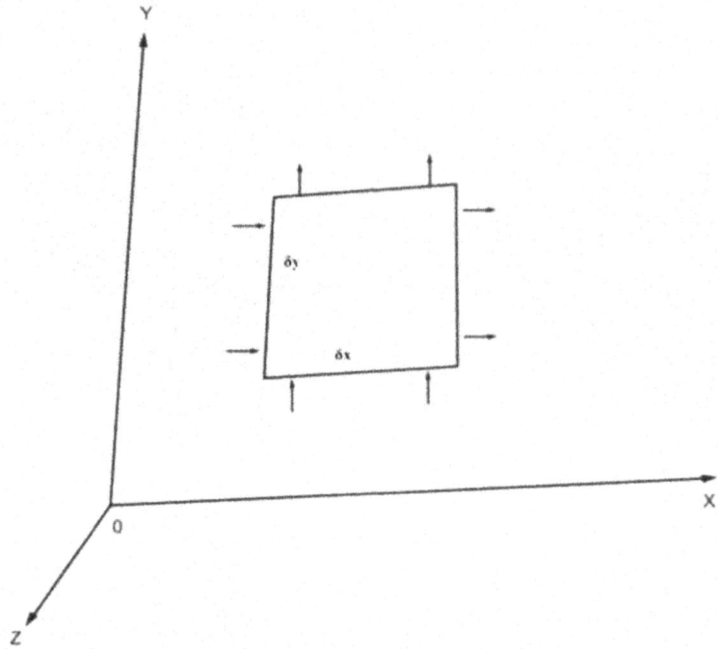

FIGURE 8.1 Inflow and outflow of sediment flux through various faces of a parallel-piped.

The foregoing equation was also given by Hayami. Equation 8.11 can be considerably simplified for the case of steady uniform flow in a very wide channel. In such a case, v and w are zero and all derivatives with respect to x, t, and z vanish. Hence, Equation 8.8 reduces to

$$\frac{\partial}{\partial y}\left(\in_y \frac{\partial C}{\partial y}\right) + \omega_0 \frac{\partial C}{\partial y} = 0 \tag{8.12}$$

Integrating Equation 8.12 with respect to y, we get

$$\in_y \frac{\partial C}{\partial y} + \omega_0 C = C_1(x, z, t) \tag{8.13}$$

where C_1 = constant of integration, which will be either constant or zero for steady, uniform, two-dimensional flow. However, since there is no sediment transfer across the free surface, $C_1 = 0$. Replacing $\in_y$ by $\in_x$ in Equation 8.12, one gets Equation 8.14

$$\omega_0 C + \varepsilon_x \frac{\partial C}{\partial y} = 0 \tag{8.14}$$

The shear stress in turbulent flow can be expressed as

$$\tau = \epsilon_m \frac{\partial}{\partial y}(\rho_f u) \tag{8.15}$$

where

ϵ_m = momentum transfer coefficient
τ = shear stress at a distance y from the boundary

This equation was proposed by Boussinesq Karman who gave the analogy between transfer of mass, heat, and momentum. Therefore, an equation similar to Equation 8.15 can be written for the transfer of suspended sediment:

$$G = \epsilon_s \frac{\partial}{\partial y}(C) \tag{8.16}$$

G = rate of sediment transport per unit area in the upward direction.

8.3.3 Wash Load

Study of suspended load and corresponding bed materials of different streams for their size analyses has revealed that the suspended load can be distributed into two parts based on size of material in suspension in relation to size investigation of bed material. One portion of suspended load is collection of sediment sizes found in plenty in the bed. Next portion of the load is collection of fine sizes which are not accessible in considerable quantities in bed. This latter part of the suspended load is called wash load.

According to Geophysical Union, wash load is the part of sediment load of a stream which is composed of particle size smaller than those found in appreciable quantities in the shifting portions of the stream bed.

Probably Vetter was the first to recognize the division between the two parts of the suspended load mentioned above. While analysing the data from the Enoree River (USA), Einstein et al. found that the bed material consisted of 10% (or less) of material smaller than 0.35 mm in diameter. Analysis of the suspended load data revealed that the amount of material coarser than 0.35 mm in size had a definite relationship with the discharge. However, no such relationship existed between the amount of suspended load smaller than 0.35 mm in size and the water discharge.

The above illustration, which is typical of many streams, shows that the amount of wash load carried by a stream may not be related to the discharge. Wash load originates from caving of the banks of the main stream or its tributaries, erosion of the gullies, and sheet erosion. This material washes through the reach without appreciable deposition. One of the reasons why the amount of wash load carried by the stream cannot be predicted is that this amount is dependent on the availability of the material in the watershed. On the other hand, the quantity of bed material load transported depends on the hydraulic conditions and the characteristics of the bed material.

The definition of wash load will indicate that it is not plausible to specify the size limit for wash load. In case of sandy streams having flat slopes, it may be within the range of clay to silt (Stevens 1936). In contrast, for mountain streams having steep slopes, it may be in coarse and fine sand range. Einstein recommends that restrictive size for wash load may be chosen arbitrarily from mechanical study of bed material, as that particle size of which 10% of bed material is finer.

While studying the effect of a large standard deviation of the bed material on the transport rate, Einstein and Chien have discussed the concept of wash load. They contradict the common belief that wash load is not deposited in the channel. Their experiments indicated that this fine material does temporarily settle on bed. Thus, the bed material both load and settle out on the bed. For wash load, however, a small amount in the bed is related to very large quantities of this material in the flow. In the case of bed material load, a fairly large fraction of bed material load, a fairly large fraction of the material in the bed is related to large quantities of this material in motion. Now as the flow conditions change, there is sufficient amount of coarse material available for transport underneath the bed layer. For finer material (wash load) however, the change of material due to change of flow will be limited to the surface layer only. Working from these ideas and after experimentation, they have found that if the active surface layer is taken as the bed material. Einstein's bed load function can be applied for the determination of wash load as well as bed material load. Yet, as Einstein and Chein put it, 'this does not imply, yet in a river reach there exists a definite relationship between flow for wash load and rate of transport, as in case of bed material load'.

8.3.4 Non-Equilibrium Transport of Suspended Sediment

The preceding discussion in this chapter was devoted to fully developed flows in which equilibrium was maintained between sediment inflow and sediment outflow for any reach. However, in many situations when clear water flows over an erodible bed, one would like to know how the concentration profiles develop until the concentration distribution reaches the equilibrium profile. This problem has been investigated by Kalinske, Apmann, and Runner; Mei, Kerssens, and Van Rijn; Hjclmfelt and Lcnau; and Yalin and Finlayson. The important concepts brought out in these papers are presented below.

If one assumes that

i. $v = w = 0$,
ii. Flow is uniform, i.e. flow depth is independent of x,
iii. There is no diffusion in z direction, or $\in_x = 0$
iv. Fall velocity ω_0 is constant
v. $\partial^2 C / \partial x^2 \ll \partial^2 C / \partial y^2$, hence can be neglected
vi. Unsteady concentration term $\partial C / \partial t$ is relatively small compared to other term, hence it can be neglected.

Equation 8.10 becomes

$$u\frac{\partial C}{\partial x}=\frac{\partial}{\partial y}\left(\in_y\frac{\partial C}{\partial y}\right)+\omega_0\frac{\partial C}{\partial y} \tag{8.17}$$

It is further assumed that $\in_y$ and u do not vary with y,

$$U\frac{\partial C}{\partial x}=\in\frac{\partial^2 C}{\partial y^2}+\omega_0\frac{\partial C}{\partial y} \tag{8.18}$$

$$\frac{\partial^2 C_*}{\partial Y^2}+\bar{Z}\frac{\partial C_*}{dY}=\bar{R}\frac{\partial C_*}{dX} \tag{8.19}$$

where $C_*=\frac{C}{C_0}$, $Y=\frac{y}{D}$, $X=\frac{x}{D}$, $\bar{R}=\frac{UD}{\bar{\in}}$

$$\bar{Z}=\frac{\omega_0 D}{\bar{\in}}$$

C_0 = reference concentration. The boundary conditions are

$$\left.\begin{aligned} C_* &= 0 \ at \ X=0 \\ C_* &= 1 \ at \ X>0,\ Y=0 \\ \frac{\partial C_*}{\partial Y}+ZC_* &= 0 \ at \ X>0,\ Y=1 \end{aligned}\right\} \tag{8.20}$$

8.3.4.1 Total Load Transport

For studying many aspects of alluvial streams, a knowledge regarding rate of total sediment transport for given flow, fluid, and sediment characteristics is essential (Yang, 1973). Engineers often forget the fact that alluvial streams not only carry water but also sediment and that sediment transport rate is closely linked with the stability or the streams. Alluvial channels must be so designed that they convey definite discharges of water and sediment. It can be said otherwise that the rate of total load transport must be treated as a variable having an impact on design of the channel. In the same way, understanding of total sediment transport rate is necessary for estimating rate of silting in reservoirs. Problems related to degradation and aggradation, river training, etc., also require understanding about total load transport for their solution.

A study of the total load transport is the next logical step. The sum of suspended load, bed load, and wash load is known as the total load. For experiments conducted in flume (excluding special cases), wash load does not exist and total load would be the bed material load. In contrast, wash load is invariably present in natural streams and total load is sum of wash load and bed material load. Moreover, it is complicated in separating out wash load from total load measured. Therefore, due to no difference in nature of field and laboratory data as mentioned above, it is challenging to unite total load data collected in field and in laboratory. However, a majority of total load relationships are mainly based on river data and flume data where wash load is estimated and omitted from measured total load; as such to yield the bed material load, these relationships can be expected.

8.4 MICROSCOPIC METHODS

The two commonly used methods under this category are those by Einstein (1950) and Colby and Hembree (1955). The latter is a modification of Einstein's method and usually goes under the name modified Einstein procedure. These methods have been described in detail below.

8.4.1 Einstein's Method

Einstein has related the sediment concentration at $2d$, i.e. C_{2d}, to the rate of bed load transport. The knowledge of the value of C_{2d} enables reconstruction of the curve of concentration distribution in the vertical. Later it was shown how the (concentration × velocity) curve can be integrated. With this information, it is now possible to describe Einstein's method of total load computation. In short, Einstein's method involves the computation of bed load and suspended load for a given size range and addition of these to get the total load corresponding to that size range. The process is repeated for other size ranges and the loads are added.

The use of Einstein's relation for reference concentration C_{2d} in terms of the bed load transport rate and subsequent integration of (concentration × velocity) curve yields the following relationship:

$$i_s q_s = i_B q_B \left(PI_1 + I_2 \right) \tag{8.21}$$

$P = 2.3\log\left(30.2D/\Delta\right)$, where $\Delta = d_{65}/x$

i_s = fraction of suspended load in a given size range
X = correction factor introduced by Einstein and Barbarossa

The total load in a given size range will be given by

$$i_r q_r = i_B q_B \left(PI_1 + I_2 + 1 \right) \tag{8.22}$$

where i_r = fraction of total load.

8.4.2 Macroscopic Method

Several methods of macroscopic nature using a single representative size of bed material have been proposed for the determination of the total bed transport. The salient feature of these investigations and the final relationship are discussed below.

8.4.3 Laursen's Method

He considered the following parameters to be important in the study of total sediment transport:

$\bar{C}$ = total load concentration
τ_0' = grain shear stress
τ_{0c} = critical shear stress

$$\frac{\bar{C}}{\left(\frac{d}{D}\right)^{\frac{7}{6}}\left[\left(\tau_0'/\tau_{0c}-1\right)\right]} = f\left(\frac{u^*}{\omega_0}\right) \tag{8.23}$$

τ_0' is calculated by combining Manning equation with Strickler equation

$$\tau_0' = U^2 d^{\frac{1}{3}} \Big/ 30D^{\frac{1}{3}} \tag{8.24}$$

Critical shear stress (τ_{0c}) is obtained from shield curve.

He used primarily flume data to determine the relation between two parameters in Equation 8.36.

8.4.3.1 Garde and Albertson's Equation

In their discussion to Laursen's paper, Garde and Albertson suggested that the total load concentration can be expressed as

$$\overline{C_v} = f\left(\frac{u_* D}{\nu}, \frac{d}{D}, \eta\right) \tag{8.25}$$

where η is the function of sediment size.

This approach has been further refined by Garde and Dattatri (1963) by expressing η as a function of $\frac{\omega_0 d}{\nu}$. With this, the functional relationship in Equation 8.37 was expressed as

$$\frac{u_* D}{\nu}, \frac{1}{(C_v^{1/3})} = f\left(\frac{D}{d}\frac{1}{\eta}\right)^{1.5} \tag{8.26}$$

and $\eta = 3.3\left(\frac{\omega_0 d}{\nu}\right)^{-0.683}$

where $\overline{C_v}$ is the concentration of total load.

8.4.4 Some Approximate Methods

If some suspended load data are available for a stream, an approximate estimate or the total load can be made by adding to the measured suspended load an estimated unmeasured load (i.e. bed load plus the measured suspended toad). Lane and Borland (1951) cite the following classification of Maddock (Table 8.1), in which the percentage of unmeasured load is related to the concentration of suspended load, type of bed material, and the texture of suspended material.

It will be easily realized that this is only an approximate method and should be used when no other data are available or when time does not permit detailed analysis. Colby (1957) has studied the variation of unmeasured sediment load of several streams in the United States and has found that the unmeasured sediment load q_{sum} increases with the mean velocity of flow. The relationship obtained between these two parameters is

$$q_{sum} = 353.6U^{3.1} \tag{8.27}$$

where q_{sum} is expressed in kN/day/m and U in m/s.

It is obvious that the unmeasured load will also depend on the ratio of the unsampled depth to the total depth and on ocher hydraulic and sediment characteristics of the stream.

TABLE 8.1
Maddock's Classification for Determining Unmeasured Bed Load

Type of Material	Concentration of Measured Suspended Sediment (ppm)	Texture of Suspended Material	Unmeasured Load (%)
Gravel, rock	>7500	≤25% of sand	2–8
Sand	>7500	Bed material	5–15
Gravel, rock	1000–7500	25% of sand /clay	5–12
Sand	1000–7500	Bed channel	10–35
Gravel, rock	<1000	Few amount of sand	5–12
Sand	<1000	Bed material	25–150

8.5 SEDIMENT YIELD FROM CATCHMENT

The soil cover on the catchment is detached by the impact of raindrops and by runoff and is transported from the catchment and brought into the stream. It is estimated that the surface of the earth is eroded in this way at an average rate of 30 mm per thousand years. Naturally this erosion rate will vary from year to year and from region to region. Erosion can be classified into sheet erosion (erosion from land surface due to impact of raindrops and its transportation), gully erosion (erosion due to widening and deepening of gullies), and channel erosion. Landslides, mud flows, and rock falls may also contribute to the sediment entering the streams. The factors affecting the erosion rate include the intensity and duration of rainfall, temperature, soil characteristics, slope, and vegetal cover. Some of these factors are not completely independent of others. For example, the type and amount of vegetation would depend on rainfall, soil characteristics, and temperature. The mechanics of sheet erosion has been studied by Mutchler and Young, Wischmeier and Smith, and others and empirical equations for the rate of sheet erosion have been obtained. However, these equations are valid for small areas and have limited utility as far as erosion from large catchments is concerned. The universal soil loss equation used by Soil Conservation Service in the United States predicts soil erosion rates from cultivated lands. The equation is

$$E = RICLSCP \tag{8.28}$$

E = soil loss in tons per year per acre of area
R = rainfall
K = soil stability factor
L = slope length factor
S = slope steepness factor
C = crop management factor
P = erosion control practice factor

This equation has been used either to find erosion rates from cultivated lands or to determine the change in erosion rate due to changes made on the land in terms of crop pattern, irrigation practice, and erosion control. Morgan and Vanoni have given details of determination of the various factors. Vanoni also gives equations for erosion from gullies.

In the case of large catchments, one defines the annual sediment yield at a gauging station on the stream, which can be expressed in terms of the apparent volume V_S or the absolute volume V_{SA} of sediment passing the gauging station each year. The important parameters affecting V_S or V_{SA} are the catchment area A in km^2, the slope of the catchment S, the mean annual rainfall P, the mean annual runoff Q, the drainage density D_d, and the erodibility factor F_c.

The drainage density is defined as the total length of the channels divided by the catchment area. The erodibility factor can be defined if one knows the type of vegetation in the catchment. Miraki studied the data from 32 Indian catchments.

Based on a study or the types of vegetal cover in these, as obtained from the Irrigation and Forest Atlas for India, he classified the vegetation into different categories and considered erodibility factors for reserved and protected forest, unclassed forest, arable area, scrub and grass, and waste area are 0.2, 0.4, 0.6, 0.8, and 1, respectively.

Weighted average of the catchment is obtained by

$$F_c = \frac{0.20A_1 + 0.40A_2 + 0.60A_3 + 0.80A_4 + A_5}{A_1 + A_2 + A_3 + A_4 + A_5}$$

Miraki used regression analysis and developed the following equation for soil yield:

$$V_{SA} = 1.182 \times 10^{-6} A^{1.026} P^{1.259} Q^{0.287} S^{0.075} D_d^{0.398} F_c^{2.422} \quad (8.29)$$

$$V_{SA} = 1.067 \times 10^{-6} A^{1.292} P^{1.384} S^{0.129} D_d^{0.397} F_c^{2.510} \quad (8.30)$$

$$V_{SA} = 2.410 \times 10^{-6} A^{1.1540} P^{1.071} S^{0.060} F_c^{1.898} \quad (8.31)$$

$$V_{SA} = 4.169 \times 10^{-6} A^{0.841} P^{0.139} Q^{0.312} \quad (8.32)$$

Depending on the data available, any one of these equations can be used to determine V_{SA}.

8.6 CONCLUSION

Knowledge regarding transport processes, sediment types, and circulation patterns accessible today can be efficiently employed for designing, operation, and installation of new amenities that will have minimum maintenance dredging costs or minimum sedimentation rates. Use of existing information for modifying prevailing conveniences for reducing maintenance costs or sedimentation rates is limited to cost of relocating facilities or adjusting installed works. Transforming constructing enclosures and dredged cuts have effectively reduced sedimentation rates in prevailing amenities. Additional field testing of newly developed techniques to reduce sedimentation rates is necessary for evaluating long-term and large-scale effects (Samantaray and Ghose, 2018, 2019, 2020; Samantaray and Sahoo, 2020; Samantaray et al., 2020a, b). Testing is also necessary for evaluating environmental effects of the approaches.

REFERENCES

Bagnold, R. A. 1956. "The flow of cohesionless grains in fluids." *Philosophical Transactions of the Royal Society of London A*, 249 (264), 235–297.

Colby, B. R. and C. H. Hembree. 1955. Comparison of Total Sediment Discharge, Niobrara River Near Cody, Nebraska, USGS Water Supply Paper 1357.

Colby, B. R. 1957. "Relationship of unmeasured sediment discharge to mean velocity." *Eos: Transactions American Geophysical Union*, 38 (5), 708–717.

Dobbins, W. E. 1944. "Effect of turbulence on sedimentation." *Transactions of the ASCE*, 109, 629–653.

Einstein, H. A. 1950. "Bed Load Function for Sediment Transportation in Open Channel Flows." USDA Technical Bulletin No. 1026.

Iwagaki, Y. 1953. "Theory of Flow in Open Channels. Congress of Modern Hydraulics." Chap. 1, Japan Society of Civil Engineers.

Garde, R. J. and J. Dattari. 1963. "Investigations of the total sediment discharge of alluvial streams. University of Roorkee Research Journal, VI (II).

Garde, R. J. 1968. "Analysis of distorted river models with movable beds." *Water and Energy International*, 25 (4), 421–431.

Ghose, D. K. and S. Samantaray. 2018. "Modelling sediment concentration using back propagation neural network and regression coupled with genetic algorithm." *Procedia Computer Science*, 125, 85–92.

Ghose, D. K. and S. Samantaray. 2019. "Sedimentation process and its assessment through integrated sensor networks and machine learning process." In Computational Intelligence in Sensor Networks. 473–488, Springer, Berlin.

Lane, E. W. and W. M. Borland. 1951. "Estimating bed load." *Eos: Transactions American Geophysical Union*, 32 (1), 121–123.

Laursen, E. M. 1958. "Total sediment load of streams." *Journal of the Hydraulics Division*, 84 (1), 1–36.

Mohanta, Nihar Ranjan, Paresh Biswal, Senapati Suman Kumari, Sandeep Samantaray, and Abinash Sahoo. 2020. "Estimation of sediment load using adaptive neuro-fuzzy inference system at Indus River Basin, India." In Intelligent Data Engineering and Analytics. 427–434, Springer, Singapore.

Sahoo, Abinash Ajit Barik, Sandeep Samantaray, Dillip K Ghose. 2020. "Prediction of sedimentation in a watershed using RNN and SVM." Communication Software and Networks. 701–708, Springer, Singapore.

Samantaray, S. and D. K. Ghose. 2018. "Evaluation of suspended sediment concentration using descent neural networks." *Procedia Computer Science*, 132, 1824–1831.

Samantaray, S. and D. K. Ghose. 2019. "Sediment assessment for a watershed in arid region via neural networks." *Sādhanā*, 44 (10), 219.

Samantaray, S. and D. K. Ghose. 2020. "Assessment of suspended sediment load with neural networks in arid watershed." Journal of the Institution of Engineers (India): Series A, 1–10.

Samantaray, S. and A. Sahoo. 2020. "Assessment of sediment concentration through RBNN and SVM-FFA in arid watershed, India." In Smart Intelligent Computing and Applications. 701–709, Springer, Singapore.

Samantaray, S., A. Sahoo, and D. K. Ghose. 2020a. "Prediction of sedimentation in an arid watershed using BPNN and ANFIS." In ICT Analysis and Applications. 295–302, Springer, Singapore.

Samantaray, S., A. Sahoo, and D. K. Ghose. 2020b. "Assessment of sediment load concentration using SVM, SVM-FFA and PSR-SVM-FFA in arid watershed, India: a case study." *KSCE Journal of Civil Engineering*, 24 (6), 1944–1957.

Stevens, J. C. 1936. “The silt problem.” Transactions of the ASCE, 101, 207–250.

Sutherland, A. J. 1967. “Proposed mechanism of sediment entertainment by turbulent flows.” *Journal of Geophysical Research*, 72 (24), 6183–6194.

Toffaleti, F. B. 1969. “Definitive computations of sand discharge in rivers.” *Journal of the Hydraulics Division*, 95 (1), 225–248.

Yang, C. T. 1973. “Incipient motion and sediment transport.” *Journal of the Hydraulics Division*, 99 (10), 1679–1704.

9 Runoff

9.1 INTRODUCTION

In hydrology, runoff is the excess water from rain (precipitation) coming from land surface over the earth's surface to main channel. It includes water travelling over land surface and through channels for reaching a stream and interflow as well (Sun and Peng, 2019; Gao et al., 2020; Goudarzi et al., 2020; Samantaray and Sahoo, 2020a; Samantaray and Sahoo, 2020b). Interflow refers to water which penetrates the surface of soil and travels by gravitational force towards a stream (above central groundwater level at all times) and ultimately drains to the channel. Runoff also includes groundwater flow or subsurface flow (Chow, 1964). It is a part of infiltrated water that infiltrates downwards to ground and laterally flows for emerging in depressions and rivers and joins surface flow. Total runoff is equal to total precipitation minus losses due to storage (as in temporary ponds), evapotranspiration (loss to atmosphere from plant leaves and soil surfaces), and other abstractions.

Water is pulled by forces of gravity through land surface, refilling surface water and groundwater as it moves into a river, stream, or watershed or infiltrates into an aquifer. It originates from unabsorbed water from rainfall, irrigation, snowmelt, or any other sources, containing a substantial part of water cycle along with water supply as soon as it drains into a catchment. Moreover, contribution of runoff towards erosion is a key aspect carving out gorges, canyons, and related type of land.

The intensity, distribution, and type of precipitation affect runoff, including slope, topography of land over which it travels, and vegetation cover. Climate change has a serious effect on runoff patterns. More precipitation falling as rain and a smaller amount as snow, the melting of snow is faster as it is smashed by rain. Also, warmer temperature alters the runoff patterns.

Winter provides the prospect of purposefully spreading runoff onto fields for percolating into the aquifer. But receiving water into those aquifers is stalled by availability of floodwater and limited capability of transport arrangement for moving water wherever it is necessary. Dynamics of runoff has heavily changed due to urbanization by affecting the land itself (Maidment 1993). Growth in infrastructure decreases volume of penetrable land into which water can absorb, initiating more storm water to pass over surface, requiring drainage system for preventing flood events. Increased runoff has attributed to connection between wildfires and subsequent flooding which is not able to absorb into burned earth. In addition, more pollutants like pesticides can be familiarized as runoff flows over polluted regions.

9.1.1 Runoff Is a Mixture of Interflow, Surface Runoff, and Base Flow

Surface runoff: Overland flow and saturation excess is the cause of surface runoff. Urban sources like pavement or roofs are examples of overland flow, whereas melted snow or precipitation is of saturation excess which basically could not be absorbed into ground (Samantaray et al., 2020; Samantaray and Ghose, 2020a). Gravitational force drives the surface runoff downhill from both of these sources.

Storm interflow: Generally interflow approaches after a heavy amount of rainfall. It travels horizontally above the water table but below the surface, in a region called 'aeration zone' or 'vadose zone'.

Base flow: The direct outflow from groundwater to surface water that brings with it any kind of chemicals which groundwater has collected for thousands of years travelling below the earth's surface.

The term 'runoff' is usually used for distinguishing the water flowing off the earth's surface during and soon after rain, from extended term of groundwater flow to rivers. This difference is accomplished by analysing perennial rivers flow data and streams in humid climate. However, in several water harvesting and agro-hydrological circumstances, there is no contribution from groundwater and entire flow is taken as runoff. Almost certainly this is a situation in semi-arid and arid climate.

Accumulation of runoff data is specific to purpose and location of study, both in the way regarding kind of required information and the way in which it can be obtained in best possible manner (Patra, 2000). In all climates, occurrence of runoff is less common than occurrence of rainfall. In low rainfall areas, where agricultural, water harvesting, and hydrological projects are generally situated, up to 10 runoff events occur for every season. This adds uncooperative information that faces difficulties with equipment and installation is not encountered till the occurrence of runoff. If above-mentioned complications are not quickly resolved, then large numbers of dataset get simply misplaced. Furthermore, both experimental and equipment designs must manage a great range of peak flows and runoff volume. Thus, cautious planning and a rapid answer to unanticipated circumstances are very significant in accumulating precise and comprehensive information.

9.2 FACTORS AFFECTING RUNOFF

There are various factors which affect runoff. Some of the factors are discussed below (Reddy, 2001).

9.2.1 Rainfall

Runoff is directly affected by quantity of rainfall. As per expectation, if more rainfall falls on ground, it will in turn create more runoff. Snowmelt is a similar case of rainfall (Ghose and Samantaray, 2019; Samantaray and Sahoo, 2020c). If there is huge amount of melting of snow within short period, it produces huge quantity of runoff.

9.2.2 Permeability

The quantity of surface runoff occurrence is affected by the capability of ground surface in absorbing water. It can be noticed that water sinks into the sand almost immediately after water is poured onto sand. In contrast, water will not sink but run off to the gutter or a ditch if poured on street. Ground absorb less quantity of water which results in more surface runoff, i.e. inversely proportional to each other. This phenomenon is termed as permeability. It indicates surface with high absorption ability has high permeability.

9.2.3 Vegetation

Vegetation requires water for surviving, and root system of a plant is designed for absorbing water from soil. Hence, runoff is less in high vegetation areas as plants use water instead of flowing off ground surface.

9.2.4 Slope

Similarly, slope is very significant to runoff quantity generated from surface. A steeper surface allows runoff to flow down the slope faster, whereas a flat surface will allow time for water to absorb.

9.3 EFFECTS OF RUNOFF

Though different aspects like amount of rainfall, vegetation, etc. has an effect on runoff, excessive of anything affecting factor can have a bad impact on environment also. Few examples involve pollution and erosion.

9.3.1 Pollution

A substantial part of precipitation absorbed into soils in wooded watersheds (infiltration) is put in storage as groundwater and is gradually discharged to streams via springs and seeps. Flood in such more natural surroundings is not much of importance as some quantity of runoff is absorbed into the ground during a storm, thereby reducing the runoff amount into a stream in the course of a storm. Due to urbanization in watersheds, considerable amount of vegetation is substituted by impervious surfaces, thereby decreasing area where infiltration to groundwater can take place. As a result, more storm water runoff takes place. This runoff must be collected by wide drainage systems combining ditches, curbs, and storm sewers for carrying storm water runoff straight to streams.

Runoff from agronomic land (and even our own yards) can transmit excessive nutrients, viz. phosphorus and nitrogen, into lakes, streams, and groundwater supplies. These excessive nutrients have the ability of degrading quality of water.

9.3.2 Erosion

During flow of runoff, it collects a lot of things that come its way, transports them, and when water slows down drops them off someplace down the stream. It can be noticed that things float in streams or rivers. Progressing water is a tough force which is capable of moving these things. Huge quantity of runoff implies flash flood is able to wash away houses, car, etc., whereas small runoff moves items like leaves and pebbles.

9.4 RUNOFF ESTIMATION

Step 1: These are vital monitors for taking decisions regarding which runoff measurement system to utilize, either continuous or volumetric. Subsequent to making the decisions, these estimations must be utilized for determining capacity and size (flow volume and peak flow) of the apparatus.

Step 2: If runoff is not measured, then computations must be done for estimating design conditions of channels, bunds, etc., which are to be employed in field layouts and mechanism of water harvesting.

9.4.1 Estimation of Flow (Theoretical)

An investigation has been carried out for developing hydrological models for predicting runoff volumes and peak flows. However, majority of them are not suitable for overall usage. Occasionally, they are very complicated; however, most often they are restricted by hydrological conditions and geographical localities inside which data were composed. Several models are based on regression analysis and outside their individual specific conditions, their value is tough to assess.

Three models which are suitable for estimating runoff volumes and five models that can be utilized for predicting peak flows are presented here. These models are appropriate for usage with a broad range of watershed conditions and sizes. However, certain drawbacks are associated with these estimation methods, i.e. they can be comparatively erroneous because of making very simple assumptions. They demand accessibility of certain prime data like rainfall and physical characteristics of catchment. Yet, these models have been utilized successfully for some time in different types of environment and work on the basis of measurements from watersheds having lots of physical feature.

Peak flow determines design specification of structures, for example dams, bunds, bridges, and channels. Also, it determines control section capacity of flow by measurement schemes and transfer conduits and assortment of pipes of volumetric collection basins. Before design of these systems can be finished, some peak flow estimation must be made.

Design peak flow is connected to certain return period, it may be maximum flow at 5, 10, 25, etc. years, and specifications of design are a balance between prevention of structural failure and economic cost. On field bunds, where no serious damage is involved, a minimum return period (5–10 years) may be utilized. For agricultural purposes, 10-year-return period is generally utilized. In cases where loss of life or serious damage is included, then design for larger return period, maybe say 50 or 100 years, is essential. Return period is very much necessary, based on which objectives of projects must be decided upon.

9.4.1.1 Rational Method

Estimating peak flow by rational method is a simple illustration of complex procedure where rainfall intensity and amount, catchment size, and conditions along with human activities determine the amount of runoff; however, it is

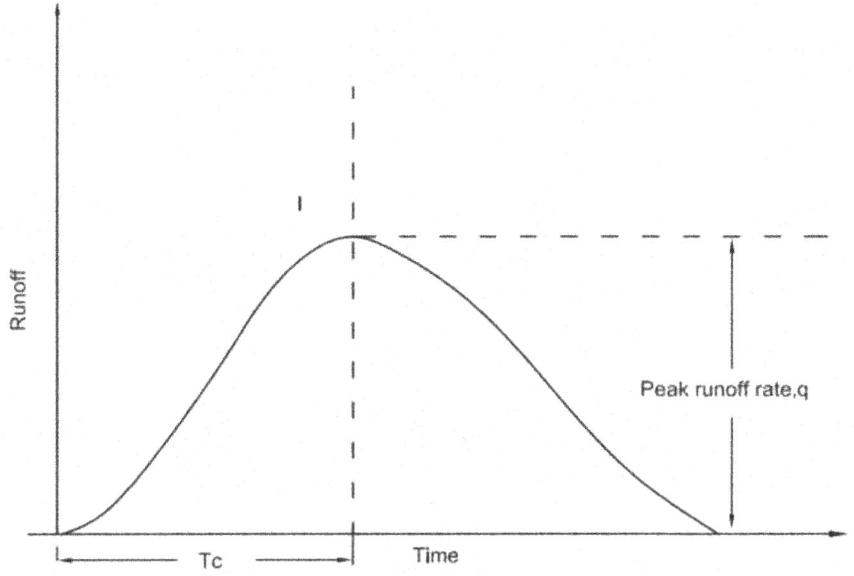

FIGURE 9.1 Hydrograph (rational method).

appropriate where there is limited consequence of structural failure (Mutreja, 1990). Usually this method is limited to small watersheds (area <800 ha) and works on the basis of assumptions made by rainfall/runoff hydrograph as shown in Figure 9.1.

Peak flows is calculated by the following equation:

$$Q = 0.0028CiA \tag{9.1}$$

where

Q = peak flow (m^3 / s) C = coefficient of runoff
i = maximum intensity of rainfall
A = watershed area (hectares)

9.4.1.1.1 Value of Coefficient C

Ratio of peak runoff rate to rainfall intensity determines value of *C*. It is dimensionless and represents amount of rainfall which turns out to be runoff. To a large extent, it is determined by catchment characteristics. Department of US-Soil Conservation Service has allowed effect of a lot of such conditions to be articulated in different *C* values. Table 9.1 provides *C* values for average annual precipitation (700–1000 mm) for the United States and for subtropical region (Central South Africa) with average annual rainfall (less than 400 mm and more than 1000 mm) is presented in Table 9.2. Various hydrological conditions in accordance to different groups of soil are also taken into account.

Malawi values of *C* are presented in Table 9.2.

TABLE 9.1
Coefficient *C* for USA

	Coefficient C for Rainfall Rates 25,000 and 200 mm/h											
	A			B			C			D		
Cover and Condition	**25**	**100**	**200**	**25**	**100**	**200**	**25**	**100**	**200**	**25**	**100**	**200**
Row crop poor	0.56	58	59	63	65	66	69	71	62	71	73	74
Row crop good	0.40	0.48	0.53	0.47	0.56	0.62	0.51	0.61	0.68	0.54	0.64	0.72
Small grain poor	0.33	0.33	0.33	0.38	0.38	0.38	0.42	0.42	.042	0.44	0.44	0.44
Small grain good	0.15	0.18	0.18	0.18	0.18	0.22	0.20	0.23	0.24	0.21	0.24	0.26
Meadow rotation good	0.25	0.29	0.29	0.29	0.36	0.39	0.33	0.41	0.44	0.34	0.42	0.46
Pasture not good	0.01	0.11	0.15	0.02	0.17	0.23	0.02	0.21	0.28	0.03	0.22	0.30
Woodland nature good	0.01	0.05	0.07	0.02	0.10	0.15	0.03	0.13	.019	.001	0.14	0.21

Soil groups A. Deep sands, permeable loss, lowest runoff potential
B. Soils less deep than A, moderately low runoff potential
C. Shallow soils, moderately high runoff potential
D. Highest runoff potential, soils nearly impermeable

Source: US SCS *National Engineering Handbook*, Hydrology, USDA ARS

TABLE 9.2
Coefficient *C* for Malawi

	Gentle Slopes		Medium Slopes		Steep Slopes	
	Degree of Protection					
Cover and Condition	**Good**	**Poor**	**Good**	**Poor**	**Good**	**Poor**
Rocky areas and saturated soils		0.56		0.70		0.80
Cultivated land poor crop cover		0.44		0.60		0.72
Poor crop cover and tied ridges		0.42		0.55		0.65
Mature crops good cover		0.36		0.50		0.60
Mature crops with poor cover and immature crops with good cover and tied ridges	0.28		0.40			0.48
Grassland good cover, forest poor cover, and rain <1000 mm/year	0.26		0.35			0.42
Forest moderate cover and rain 1000–1250 mm/year	0.19		0.30			0.47
Mature crops, good cover, tied ridges	0.15		0.20		0.24	
Forest good cover and rain >1250 mm/year	0.14		0.18		0.21	
Poor cover crops	Cotton, tobacco in all circumstances					
Good cover crops	Cereals, root crops, groundnuts, coffee, and tea if well managed					

Source: Land Husbandry Manual, Ministry of Agriculture and Natural Resources, Malawi.

9.4.1.1.2 *Rainfall Intensity, i_r*

In accordance to desired return period for designing the structure under investigation, the value of rainfall intensity utilized in the rational method is chosen. In case of rational method, duration of rainfall intensity is said to be equal to T_c of runoff.

Based on maximum intensity and specific duration of rainfall, specific return period is determined by plotting a graph or set of graphs (for this method equal to T_c). Graphs of such types require available data of many years, as they signify line of best fit using a cluster of data points drawn from a broad range of rainfall and measurement of their intensity. Wide range of data are particularly needed for long intensity-duration periods that are not frequently experienced. Evidently, climatic conditions of geographical regions will differ and also local alterations can be abundant where a country displays a noticeable diversity in topographical form. Unique sets of rainfall intensity graphs should be provided to regions under uniform characteristics of rainfall. Figure 9.2 demonstrates the way through which graphs are plotted.

Return period is widely used for various structures: large farm dams, 50 years; small farm dams and gully control, 20 years; field structures, 5–10 years.

9.4.1.1.3 *Time of Concentration, T_c*

Time taken for a drop of water flowing from farthest point of watershed to reach its outlet is designated as T_c. For estimating T_c, subsequent formula has been established, with example values specified in Table 9.3.

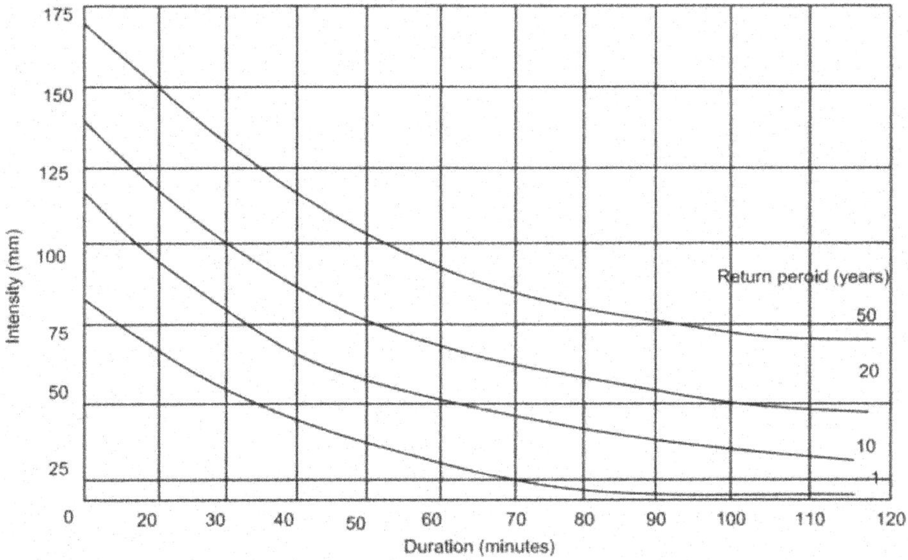

FIGURE 9.2 Intensity, return period, and duration graph (= T_c).

TABLE 9.3
Values for Different Conditions of Catchment

Relief	Soil Infiltration	Vegetation Cover	Surface Storage	Conditions
(40) Steep and rugged slopes >30%	(20) No effective soil with negligible infiltration	(20) No effective cover	(20) Negligible ponds or marshes	Extreme peaks (100)
(30) Hilly land slopes 10–30%	(15) Slow to take water clays, low infiltration	(15) Poor natural cover <10% or clean crops	(15) Low, no ponds, well-defined drainage	High peaks (75)
(20) Rolling slopes 5–10%	(10) Normal deep loam infiltration good	(10) Fair cover grass or wood. Not >50% clean cultivation	(10) Normal lakes, ponds <20% considerable depression storage	Normal peaks (50)
(10) Flat land slopes 0–5%	(5) Deep sand takes up water rapidly	(5) Good to excellent cover 90% grass or wood or equivalent	(5) High surface depression storage, drainage not well-defined	Low peaks (25)

$$T_c = 0.0195L^{0.77}S^{-0.385} \tag{9.2}$$

where

L = maximum length of catchment (m)
S = catchment slope ($\mathrm{mm^{-1}}$) over total length L
T_c, as computed from Equation 9.2

In an alternative way, T_c can be determined: by dividing evaluated flow length by estimated velocity of flow. For estimating flow velocities, Manning's formula can be utilized, even though flow velocity estimation utilizing Manning's formula can be a difficult issue for larger watersheds. This is because of changes in channel size, form, slope, and roughness, which vary to a great extent and where assessment of these features may be a difficult task. The values of T_c for a variety of slope categories, catchment areas, and protection qualities are provided in Figures 9.3 and 9.4. In all circumstances, it is significant for calculating runoff peaks for conditions of catchment that are most likely of producing them, so as to estimate peak flows.

9.4.1.2 Cook's Method

This method was developed by Unified Soil Classification System (USCS), basically providing a more generalized and simpler, but a similar approach for estimating peak flows as that of rational method. Catchment condition and size are given in Table 9.4, which provides the detailing of watershed condition.

Conditions of a catchment are evaluated and numerical values allocated to each are summed up together. For instance, if given conditions are in right column of Table 9.4, an aggregate value of 25 would be determined and peak flow is estimated to be low, whose precise size depends on area of catchment. List of conditions of a

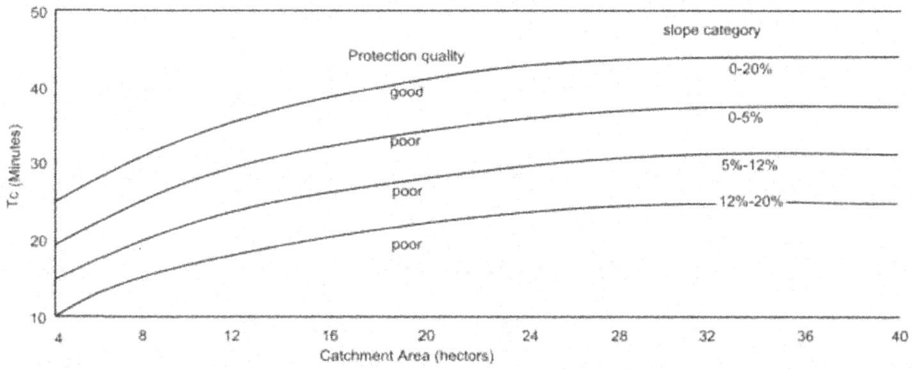

FIGURE 9.3 Time of concentration for catchment areas 0–36 hectares.

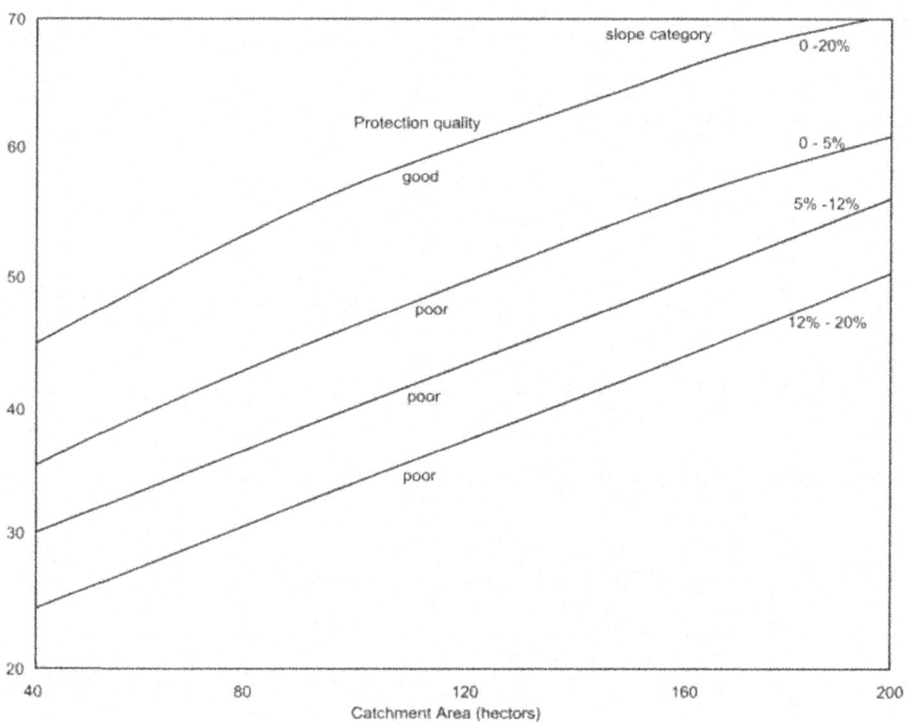

FIGURE 9.4 Time of concentration T_c for catchment areas 40–200 hectares.

Source: Land Husbandry Manual, **Ministry of Agriculture and Natural Resources, Malawi**

TABLE 9.4
CNs for Soils and Catchment Condition, AMC II

Land Use	Treatment	Condition	Soil Group A	B	C	D
Fallow row crops	Contoured	Good	62	71	78	81
	Row	Poor	66	74	80	82
	Straight	Good	67	78	85	89
	Straight	Good	65	75	82	86
	Row	Poor	72	81	88	91
	Row	Poor	70	79	84	88
	Straight		77	86	91	94
Small grain	Straight	Good	59	70	78	81
	Terraced	Poor	63	74	82	85
	Straight	Good	61	73	81	84
	Terraced	Good	63	75	83	87
	Contoured	Poor	65	76	84	88
	Row	Poor	61	72	79	82
Close seeded or rotation meadow	Terraced	Poor	63	73	80	83
	Straight	Good	61	67	76	80
	Terraced	Good	55	69	78	83
	Contoured	Good	58	72	81	85
	Contoured	Poor	64	75	83	85
	Row	Poor	66	77	85	89
Pasture or range	Contoured	Poor	47	67	81	88
	Row	Good	39	61	74	80
	Terraced	Good	6	35	70	79
	Straight	Fair	49	69	79	84
	Contoured	Fair	25	35	70	79
	Row	Poor	68	79	86	89
Permanent meadow	Terraced					
		Good	30	58	71	78
		Poor	45	66	77	83
Woods		Fair	36	60	73	79
	Contoured	Good	25	55	70	77
Farmsteads	Contoured		59	74	82	86
Roads	Contoured		74	84	90	92

Source: US SCS 1964.

specific catchment can most likely be found in various columns, however, relief condition is weighted most heavily. Generally the four columns give conditions which define 'type' of catchments. Drainage conditions and soil type were observed to be particularly significant.

9.4.1.3 TRRL Model

UK Transport and Road Research Laboratory (TRRL) developed a model for overcoming two critical complications linked with data in several developing countries from the work they carried out in East Africa. It refers to development of rainfall/runoff correlations by utilizing large amount of data and prodigality in data is sporadic. USSCS technique was observed not to give satisfactory outcomes for conditions in East Africa.

The concept of 'contributing area' (C_A) is utilized for avoiding usage of a uniform coefficient all over the catchment. Initial rainfall fills up early retention (Y) and runoff at this period is zero. *Incorporation of a lag time (K) is done for accounting routing on bigger catchments.* Total volume of runoff was determined by

$$Q=(P-Y)C_A A10^3 \tag{9.3}$$

where

P = time period of storm rainfall (mm)
Y = primary retention (mm)
C_A = coefficient of contributing area
A = area of catchment $\left(\text{km}^2\right)$

Average flow Q_M is expressed by

$$Q_M = 0.93\left(Q/3600\right)T_B \tag{9.4}$$

T_B is the hydrograph base time (in hours).

9.4.1.3.1 Contributing Area (C_A)

Slope, soil type, catchment wetness, and land use were observed to be most powerful aspects to determine contributing area of catchment. Design value is expressed in terms of

$$C_A = C_s C_w C_L \tag{9.5}$$

where

C_L = land use factor
C_w = catchment wetness factor
C_s = standard value of coefficient of contributing area at field capacity for grassed catchment

9.4.1.3.2 Lag Time (K) It was determined simply for having a relationship with vegetation cover. Simulation studies revealed that T_B could be determined by the following equation:

$$T_B = T_p + 2.3K + T_A \tag{9.6}$$

$$T_A = 0.028L \Big/ Q_M^{0.25} S^{0.5} \tag{9.7}$$

where

- L = length of main stream
- Q_M = average flow during base time
- S = slope of average mainstream
- K = lag time
- T_p = rainfall time

Using trial and error repetition of Equation 9.6, Q_M value can be estimated, with initial value of T_A being zero.

9.4.1.4 USSCS Method

This method is based on relationship between rainfall and runoff for triangular hydrograph demonstrated in Figure 9.5. Significantly, it can be noted that this method is applied for calculating peak flow of a known runoff occurrence or for calculating peak flow for a desired or expected runoff occurrence. It should be designed for a specific discharge. It is not needed to have information regarding rainfall intensity. Expression for peak flow is given by

$$Q = 0.0021qA / T_p \tag{9.8}$$

where

- Q = peak flow
- q = depth of runoff volume
- A = area of watershed (hectares)
- T_p = time to peak, described by

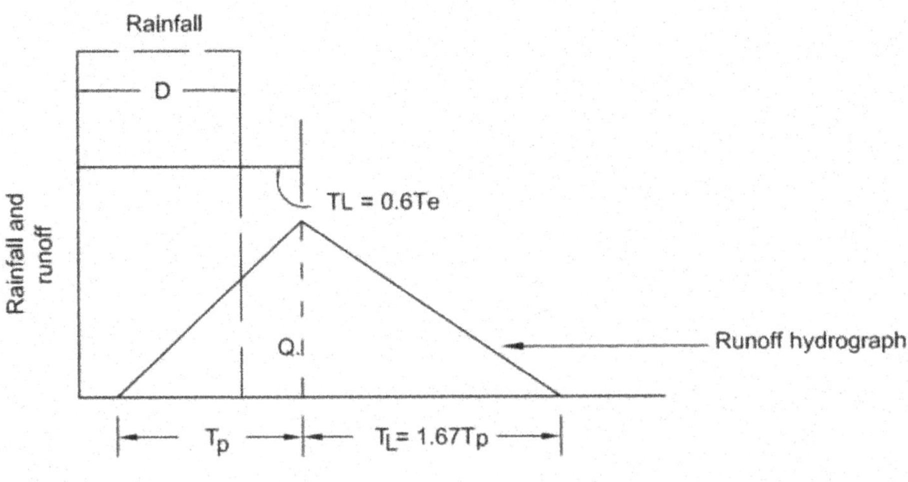

FIGURE 9.5 The USSCS triangular hydrograph.

$$T_p = D/2 + T_L \tag{9.9}$$

where

D = excess rainfall duration

T_L = time of lag, which is an estimate of mean travel time

In an alternative way, time of lag = $0.6 \times T_c$, which is the longest runoff travel time.

9.4.1.5 Izzard's Method

For estimating runoff rates from watersheds which ranges in size from a few hectares to some hundred hectares, previous techniques have been developed. But experiments related to agricultural and hydrological processes make usage of smaller runoff plots frequently, i.e. only tens of square meters in area. Two reasons are behind working of such experiments. Firstly, they are easy in replicating and several such plots can be positioned in a small area for studying a range of watershed characteristics. Secondly, they can be appropriately utilized for looking at intrusions which work on a smaller scale and that are projected to be installed inside borders of separate fields. A method developed in these conditions for estimating runoff from limited channel flow and sheet flow may be more suitable. Izzard conducted a wide range of experiments with flows from different planes over comparatively smaller regions.

$$\text{The flow at equilibrium } q_e = iL / 3.6 \times 10^6 \tag{9.10}$$

where

i = rainfall rate

L = flow surface length

For the purpose of computing peak flow only, there is no need of entering into relations amid other parameters of runoff which permit plotting of overland flow hydrograph and computation of total volume of flow. This is discussed in Section 9.6.

$$\text{Peak flow, } q_p \left(\text{m}^3/\text{s}\right) = 0.97 \left(iL / 3.6 \times 10^6\right) \tag{9.11}$$

i and L are same as mentioned in Equation 9.10.

9.5 EMPIRICAL RELATIONSHIPS FOR DETERMINATION OF RUNOFF

Empirical relationships can be applied to regions for which these are developed. Some popular runoff formulae in use in India are as follows.

9.5.1 Khosla's Formula

Khosla in 1960 developed the following empirical relationship between the runoff and the rainfall in a catchment based on the monthly data (Subramanya, 1994):

$$q_m = P_m - l_m \tag{9.12}$$

where

q_m = monthly runoff, cm (≥ 0)
P_m = monthly rainfall, cm
l_m = monthly runoff loss, cm

If t_m = monthly temperature of the catchment (°C), then for

i. $t_m > 4.5°C, l_m = 0.48t_m$
ii. $t_m \leq 4.5°C, l_m = 2.17$ at 4.5°C
= 1.78 at −1°C
= 1.52 at −6.5°C

9.5.2 Ryves Formula

Ryves formula was reported in 1884. It states that

$$Q_p = C_r r\, A^{2/3} \tag{9.13}$$

where Q_p is the peak discharge rate (m^3/s), A is the drainage basin area (km^2), and C_r is a constant (Ryves).

The Ryves formula is recommended for southern states of India.

9.5.3 Dickens Formula

This formula was developed in 1865. It states that

$$Q_p = C_d A^{3/4} \tag{9.14}$$

where Q_p is the peak discharge rate (m^3/s), C_d is a constant (Dickens) ranging from 6 to 30, and A is the drainage basin area (km^2).

9.6 RUNOFF VOLUMES

Accurate estimation of size of probable runoff volumes is very essential, as they will find whether continuous or volumetric data collection techniques must be utilized. In case of volumetric data collection technique, the estimations will make sure that design of size of collection tank is appropriate. Tanks which are very small will be overfilled and data collection during large runoff events will be lost. Loss of data will be a huge setback since procurement of information regarding huge runoff volumes and probabilities related to them is crucial for purpose of agricultural planning and management. Overdesigning of collection tanks sustains needless expenses and also leads to complications in installing the large tank.

It is essential for estimating size of runoff volumes for water harvesting structures which the catchments are expected to shed. Excess estimation of runoff volumes leads to severe under supplies of additional water, while volume much higher than expected volume results in occurrence of floods and cause physical damage to structures and crops. However, again it is significant to stress that techniques to calculate runoff volumes presented below can give estimations only.

9.6.1 USSCS Method

USSCS method is employed for small agricultural catchments. It has been successfully utilized in other areas, though it was developed from several years of data obtained from the United States (U.S. Soil Conservation Service. 1973). This method is based on relation between amount of rainfall and direct runoff. This relationship is described by a series of curvilinear charts called 'curves'. Every curve signifies relation amid rainfall and runoff for a series of hydrological conditions and every single is provided with a 'curve number' (CN), commencing from 0 to 100. The governing equation depicting relation between rainfall and runoff is

$$Q=(P-0.2S)^2/P+0.85S \tag{9.15}$$

where

Q = direct depth of surface runoff

P = storm rainfall

S = maximum possible difference amid rainfall and runoff, beginning at time the storm originates

Constraint S is basically losses due to infiltration, interception, etc. from runoff.

S is computed by

$$S=(25{,}400/N)-254 \tag{9.16}$$

Where N is the curve number (0–100)

CN 100 adopts entire runoff from rainfall and hence $S=0$ and $P=Q$.

Values of CNs for various agro-hydrological conditions are specified in Tables 9.10 and 9.11. It is to be noted that values of these tables are alienated based on antecedent moisture condition (AMC). It refers to state of soil 'wetness' prior to precipitation. It can be basically assumed that wet soil shed a greater portion of precipitation as runoff compared to dry soil and hence identical soil will be of a higher CN when wet compared to when dry.

Among various local conditions, rate of evapotranspiration must be particularly taken into consideration for assessing whether or not the categorization of AMC I and II ought to be reformed. Taking an example, soil from an area with summer rainfall and growing in summer season may come under category 1, regardless of a preceding 5-day precipitation of 40 mm. In the same way, soils growing in a winter season and same previous rainfall may perhaps come under category III.

Derivation of rainfall parameter P value is a complicated issue during utilization of USSCS method (USSCS 1972). This constraint is generally described as a

TABLE 9.5
Curve Numbers for Catchment Condition and Soils

Conversion Values for Conditions I and III, from Condition II		
Curve Number for Condition II	**Factor for Condition I**	**Factor for Condition III**
100	1.00	1.00
90	0.87	1.07
80	0.79	1.14
70	0.73	1.21
60	0.67	1.30
50	0.62	1.40
40	0.55	1.50
30	0.50	1.67
20	0.45	1.85
10	0.40	2.22

particular return period storm for a known duration. Rainfall P is computed from a comparatively multifaceted linear relation between several return period factors and rainfall duration. These data are available without any difficulty in the United States, which can be obtained from available maps. Even though local differences in relation which define rainfall intensity parameter exists for covering climatic variation, there are severe complications faced during transfer of such type of information to other geographical regions. Such comprehensive data is unlikely to be available in developing countries. Although the data might be available, work required to convert raw data into a sequence of valuable graphs, tables, or maps would be outside the scope of maximum projects where everything that is required is an estimation of runoff.

Alternatively, in countries with only basic meteorological information, long-term daily rainfall is generally available. The frequently used value is the 24-hour rainfall. It is preeminent to employ a simpler rainfall estimate in these circumstances which can be found from a listing of annual maxima. Regression analysis can be used to establish relation between daily and other rainfall period where records are available. When records are accessible, there must be utilization of these records, although they are less acquiescent to erudite treatment.

9.6.2 Izzard's Method

For calculation of peak flow, Izzard method was discussed in relevant section. Also, results can be utilized for calculating runoff volumes, flow hydrographs, and small runoff plot.

$$\text{Based on Izzard's method, time of equilibrium } t_e = 2\ V_e / 60\ q_e \tag{9.17}$$

where t_e is the time when flow is 97% of supply rate and V_e is the equilibrium volume of water in detention. V_e (m^3) can be determined by

$$V_e = kL^{1.33} i^{0.33}/288 \text{ (mm/hr)} \tag{9.18}$$

where L is the strip length (m) and on experimental basis, k can be found by the following equation:

$$k = 2.76 \times 10^{-5} i + c/s^{0.33} \tag{9.19}$$

where s is the surface slope and c as provided.

$$\text{Average depth over strip} = V_e/L = kq_e 0.33 \tag{9.20}$$

The value of C (Surface Retardance Coefficient) are 0.0600 (Dene Bluegrass, Turf), 0.0460 (Closely Clipped Bed), 0.0120 (Concrete) and 0.0075 (Tar and Sand Pavement).

For small rainfall intensities and low slopes, value of c is comparatively significant. Procedures for calculating runoff volume:

- q_e and t_e are computed from Equations 9.10 and 9.14 in respective order.
- With q_e and t_e known, rising limb plot of overland flow hydrograph; plot between q (vol) and t (min), which can be obtained from Figure 9.6.
- Plotting of recession curve of hydrograph utilizing factor B is

$$B = 60 q_e t_a / V_o$$

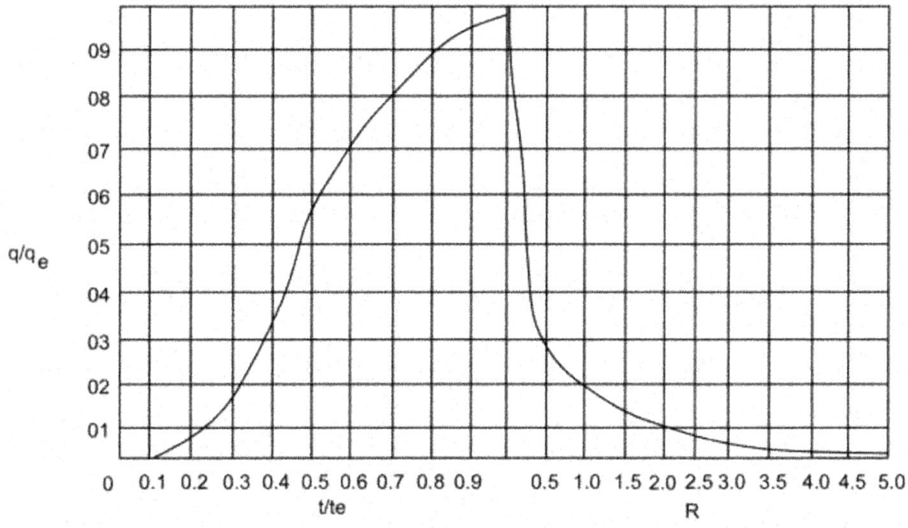

FIGURE 9.6 Dimensionless hydrograph according to Izzard.

where V_o is the detention volume (Equations 9.15 and 9.16), considering $i = 0$ and t_a is any time after rain ends. Runoff volume is taken as area under hydrograph, when hydrograph is drawn.

9.6.3 TRRL Model

Section (9.4.1.3) can be referred to for this method for computing runoff volume.

9.7 CONCLUSION

Runoff is probably the most complex yet important hydrologic process required to be understood by the hydrological engineers or scientists. Tenets of the surface water runoff process presented in this chapter provide the foundation for the study, preservation, and management of this natural resource. Accurate surface runoff estimation techniques suitable for ungauged watersheds are relevant to areas such as India where hydrologic gauging stations are not widely available. The runoff estimation method is not restricted to use for small watersheds (Jimmy et al., 2020; Samantaray et al., 2019, 2020d; BBN Samantaray and Ghose, 2020b). It applies equally well to other large areas if the geographical variations of storm rainfall and soil-cover complex are taken into account. Estimation of runoff in a watershed is a prerequisite for design of hydraulic structures, reservoir operation, and for soil erosion control measures. Water resource planning and management is important and critical issue in arid and semi-arid regions.

REFERENCES

Chow, V. T. 1964. Handbook of Applied Hydrology, McGraw-Hill, New York.

Gao, S, Y. Huang, S. Zhang, J. Han, G. Wang, M. Zhang, and Q. Lin. 2020. "Short-term runoff prediction with GRU and LSTM networks without requiring time step optimization during sample generation." *Journal of Hydrology*, 589, 25188.

Ghose, D. K. and S. Samantaray. 2019. "Estimating runoff using feed-forward neural networks in scarce rainfall region." In Smart Intelligent Computing and Applications. 53–64, Springer, Singapore.

Goudarzi, F. M., Sarraf, A., and Ahmadi, H. 2020. "Prediction of runoff within Maharlu basin for future 60 years using RCP scenarios." *Arabian Journal of Geosciences*, 13 (14), 1–17.

Jimmy, S. R., A. Sahoo, Sandeep Samantaray, and Dillip K. Ghose. 2020. "Prophecy of runoff in a river basin using various neural networks." *In Communication Software and Networks*. 709–718, Springer.

Land Husbandry Manual, Ministry of Agriculture and Natural Resources, Malawi.

Maidment, D. R. (ed.) 1993. Handbook on Hydrology, McGraw-Hill, New York.

Mutreja, K. N. 1990. Applied Hydrology, Tata Mc-Graw-Hill, New Delhi.

Patra, K. C. 2000. Hydrology and Water Resource Engineering, CRC Press, Boca Raton, FL.

Reddy, P. J. 2001. A Textbook of Hydrology, Laxmi Publication, New Delhi.

Samantaray, S. and A. Sahoo. 2020a. "Appraisal of runoff through BPNN, RNN, and RBFN in Tentulikhunti watershed: a case study." Advances in Intelligent Systems and Computing, 1014. doi:10.1007/978-981-13-9920-6_26.

Samantaray, S. and A. Sahoo. 2020b. "Prediction of runoff using BPNN, FFBPNN, CFBPNN algorithm in arid watershed: a case study." *International Journal of Knowledge-based and Intelligent Engineering Systems*, 24 (3), 243–251.

Samantaray, S. and A. Sahoo. 2020c. "Estimation of runoff through BPNN and SVM in Agalpur watershed." Advances in Intelligent Systems and Computing, 1014. doi:10.1007/978-981-13-9920-6_27.

Samantaray, Sandeep, A. Sahoo, N. R. Mohanta, P. Biswal, and U. K. Das. 2020d. "Runoff prediction using hybrid neural networks in semi-arid watershed, India: a case study." In Communication Software and Networks. 729–736, Springer, Singapore.

Samantaray, S., A. Sahoo, and D. K. Ghose. 2019. "Assessment of runoff via precipitation using neural networks: watershed modelling for developing environment in arid region." *Pertanika Journal of Science & Technology*, 27 (4), 2245–2263.

Samantaray, S. and D. K. Ghose. 2020a. "Modelling runoff in an arid watershed through integrated support vector machine." *H_2Open Journal*, 3 (1), 256–275.

Samantaray, S. and D. K. Ghose. 2020b. "Modelling runoff in a river basin, India: an integration for developing un-gauged catchment." *International Journal of Hydrology Science and Technology*, 10 (3), 248–266.

Subramanya, K. 1994. Engineering Hydrology, Tata McGraw-Hill, New Delhi.

Sun, G. and F. Peng. 2019. "Evaluation of future runoff variations in the north–south transect of eastern China: effects of CMIP5 models outputs uncertainty." *Journal of Water & Climate Change*, 11 (4), 1355–1369.

US SCS. 1972. Hydrology: National Engineering Handbook, Section 4, USDA ARS, Washington, DC.

U.S. Soil Conservation Service, 1964, Hydrology. Section 4, SCS National Engineering Handbook: Washington, D.C., 30.

U.S. Soil Conservation Service. 1973. "A method for estimating volume and rate of runoff in small watersheds," Technical Paper 149, rev., Washington, DC.

10 Application of Artificial Intelligence for Prediction of Ground Water Fluctuation

10.1 INTRODUCTION

A significant freshwater source all around the world is groundwater (GW) (Li et al., 2013). In entire ecosystem, GW plays an important part in arid and semi-arid surroundings. Usage of water in a sustainable way for industrial, agricultural, and wildlife purposes is very essential because of its ever-increasing demand (Ghazavi et al., 2012). In different parts of the world as well as in India, depletion of GW is a growing issue of concern for scientists and engineers. Moreover, groundwater level (GWL) prediction in any concerned area is enormously important for GW resources management. But GWL prediction is very multifaceted and vastly non-linear in nature as it is dependent on various intricate features such as rainfall, temperature, etc. Hence, it is vital for developing efficient models to predict GWL precisely (Verma and Singh, 2013). Several models, for example numerical GW models, non-linear experiential models, and artificial intelligence models, are being utilized for predicting GWL (Emamgholizadeh et al., 2014; Sun and Xu, 2010; Sun et al., 2006). In hydrological study area, artificial neural network (ANN) model is developed and utilized for predicating problems usually non-linear in nature, namely rainfall (Nastos et al., 2014), sediment concentration (Afan et al., 2015), streamflow (He et al., 2014), provincial flood studies (Latt et al., 2015), etc. Machine learning approaches are widely used for prediction of various watershed-affecting parameters (Jimmy et al., 2020; Samantaray et al., 2020; Samantaray et al., 2020c; Samantaray and Sahoo, 2020d; Sridharam et al., 2020).

Üneş et al. (2017) applied autoregressive (AR) and support vector machine (SVM) to predict GWL fluctuations taking GWL data from previous years of Hatay Amik Plain, Kumlu region, Turkey. Accuracy of proposed models were investigated and found that AR model showed better results compared to SVM in GWL predictions. Gong et al. (2016) explored applicability of ANN, ANFIS, and SVM to predict GWL for two wells near Okeechobee Lake in Florida, considering interaction amid surface water and GW. Results demonstrated that all proposed models can be applied for GWL prediction and indicated that SVM model and ANFIS model were more precise compared ANN. Huang et al. (2017) utilized integration of particle swarm optimization and SVM (chaotic PSO-SVM), linear PSO-SVM, and chaotic BPNN models for predicting GWLs of Huayuan landslide on daily basis and GWLs of

Baijiabao landslide on weekly and monthly basis in Three Gorges Reservoir Area, China. Results demonstrated that chaotic PSO-SVM model predicted GWL more accurately than other proposed models for considered test data. Devarajan and Sindhu (2015) compared effectiveness of numerical model considering MODFLOW and empirical model taking RBFN to forecast GWLs of Athiyannoor Block Panchayath of Trivandrum district, Kerala. Based on performance indices of both models, it was observed that RBFNN model is better compared MODFLOW in forecasting GWL on weekly basis. Tapak et al. (2014) evaluated performance of SVM to predict GWL in Hamadan-Bahar Plain, Iran, and compared the outcomes with those of classic time series models. Results demonstrated that SVM outperformed classical model and was found to be a reliable and efficient tool to model and predict GWL fluctuations of the proposed study area. Alizamir et al. (2018) investigated ability of extreme learning method (ELM) and compared with those of ANN, RBFN, and autoregressive moving average to model GWL fluctuations of Shamil-Ashekara Plain, situated in Hormozgan Region of Iran. Results revealed that ELM showed better results and showed better accuracy in forecasting GWL changes than other proposed models. Chang et al. (2013) applied BPNN and RBFN models for forecasting GWL fluctuations in Hetao District of Mongolia. Results suggested that both models showed reasonably accurate results in forecasting GWL but RBFN was observed to be simpler, converged input faster, and produced more steady results.

Ebrahimi and Rajaee utilized simple ANN, MLP, and SVM models and hybridized them with wavelet analysis for simulating one month ahead GWL of Qom region, Iran, and compared all the proposed models based on different statistical parameters. Study showed that implementation of wavelet transformation with simple data-driven techniques improved prediction accuracy and outperformed the simple models. They used various ANN approaches to predict groundwater fluctuation at various gauged watersheds (Ghose and Samantaray, 2019; Samantaray et al., 2019; Samantaray et al., 2020a, d). Ying et al. (2014) evaluated the applicability of integrated time series, ARIMA, RBFN models to predict GWL of Jilin Province, China, on the basis of different performance criteria. Results revealed that all three projected models produced GWL accurately but in terms of reliability, RBFN outperformed the other models for desired study area. Nourani and Mousavi (2016) applied hybrid ANFIS-RBF and ANFIS-ANN models for spatiotemporal modelling of GWL fluctuations of Miandoab plain, northwest of Iran. Results revealed that accurateness of ANFIS-RBF model is more consistent in comparison to ANN-RBF model in all steps. Cortes and Vapnik (1995) urbanized the concept of SVM. Major advantage of SVM is that it beholds strength of ANN and also incapacitates some major complications related to ANN. In hydrology field of research, SVM has substantiated to be a favourable modelling tool. Sattari et al. (2018) used SVM and M5 decision tree models to predict GWL in Ardebil plain, Iran. Results revealed that both proposed models performed well in predicting GWL in proposed study area. However, outcomes obtained from M5 decision tree are slightly more accurate, applied effortlessly, and were easier for interpreting compared to SVM. Suryanarayana et al. (2014) applied integrated wavelet and SVM modelling for predicting monthly GWL fluctuations at different gauge stations in Visakhapatnam city of Andhra Pradesh and compared the hybrid model with SVM, ANN, and

ARIMA models. Study indicated that proposed hybrid model produced more accurate results to predict GWL in proposed study in comparison to other proposed models. Mirzavand et al. (2015) investigated applicability of ANFIS and SVM models to estimate monthly GWL fluctuation in the Kashan plain, Isfahan region, Iran. Results indicated that both models were successfully applied for estimating GWL and also found that ANFIS performed slightly better than SVM model.

Nie et al. (2017) employed RBFN and SVM for simulating GWL variations of Da'an region located in Jilin Area, in eastern China. Results revealed that SVM model possessed greater capability to simulate and predict in accordance to four statistical parameters. Zhou et al. (2017) proposed integrated wavelet-SVM (WSVM) to forecast GW depth of 10 boreholes in Mengcheng Province, China, and made a comparison with regular SVM, ANN, and wavelet-ANN models. Results indicated that WSVM model produced most exact and consistent GW depth prediction than that of other proposed models. Yan and Ma (2016) proposed a combined ARIMA and RBFN model to predict GWL fluctuations for two observation wells in city of Xi'an, China, on monthly basis. Result showed that proposed hybrid model is effective and feasible in GWL prediction and produced higher accuracy rate than simple ARIMA and RBFN models. Yoon et al. (2011) urbanized ANN and SVM models to predict GWL variations of two boreholes at a seaside aquifer in Korea. Based on performance indices, SVM model show similar or better prediction soutput compared to ANN and it is also observed that SVM's ability to generalize is superior than ANN. Yoon et al. (2016) utilized a weighted error function method for improving enactment of ANN and SVM to predict GWL of different gauging stations of South Korea for a long term in response to precipitation. Assessment based on different evaluating criteria revealed that SVM performed superiorly as compared to ANN in this case study. Zhang et al. (2017) compared applicability of grey self-memory model (GSM), RBFN, and ANFIS models in forecasting GWL at five locations of Jilin City in northeast China. Outcomes indicated that all models performed satisfactorily in proposed area lacking hydro-meteorological data and also revealed that ANFIS performed slightly better than the other two at all the sites. Zhao et al. (2011) employed SVM to forecast GWL and a comparison of its results was done with other prediction methods using real GWL data. Results demonstrated feasibility of applying SVM to predict GWL and proved that SVR is applicable and performed well for GW data analysis. Gocić et al. (2015) explored the applicability of genetic programming (GP), SVM-FFA, ANN, and W-SVM for forecasting reference evapotranspiration (ET_0) of Serbia, Balkan Peninsula. Result indicated that W-SVM gives best performance compared to other models in predicting ET_0. Shamshirband et al. (2016a) proposed hybrid SVM-FFA model for estimating mean horizontal universal solar radiation. Outcomes revealed that urbanized SVM-FFA method provided satisfactory estimates with considerably higher accuracy than other studied methods.

Ghorbani et al. (2018) employed ANN, SVM, and SVM-FFA models to improve river flow prediction of River Zarrineh located in Northeast of Iran. Results indicated that proposed hybrid model improved the prediction results to a greater extent with high level of accuracy. Al-Shammari et al. (2016) used SVM-FFA hybrid model to develop short-term multistep ahead heat load prediction models for customers linked to heating system of entire district. Investigational outcomes showed that proposed hybrid model proved to be an efficient model with conviction for predictive approaches in district heating systems. Ghorbani et al. (2017) developed a hybrid SVM-FFA for predicting soil field capacity and permanent wilting point of soil outlines situated in East Azerbaijan regions, Northwestern Iran, and paralleled with ANN and SVM models. Comparative outcomes revealed that proposed hybrid model produced better prediction results than SVM and ANN models. Adewumi and Akinyelu (2016) applied SVM-FFA model for developing an improved phishing e-mail classifier proficient of precisely perceiving new phishing patterns as they happen. Results revealed that SVM-FFA is a reliable optimization technique for building a robust classifier. Ch et al. (2014) proposed novel hybrid SVM-FFA model for forecasting malaria incidences in Jodhpur and Bikaner district of Rajasthan where spread of malaria is unhinged and compared with simple ANN, SVM, and ARMA models. Results indicated that developed hybrid model provided more precise forecasting compared to the other traditional techniques. Shamshirband et al. (2015) developed and compared performance of four hybrid models – SVM-FFA, W-SVM, SVM-RBF, and SVM-PSO – to model energy constraints of DI diesel engine. Outcomes showed that W-SVM method was more competent in energetic modelling of engine compared to other methods. Mladenović et al. (2016) presented SVM-FFA model for predicting thermal ease of visitors on the basis of carbon dioxide emission at an open metropolitan area and compared with simple ANN, SVM, and GP. Simulation results revealed that the hybrid model predicted carbon dioxide emission constructively and provided most accurate predictions. Roy et al. (2016) developed SVM-FFA model to predict significant wave height and compared with ANN and GP models. Based on different statistical indices, obtained results showed that SVM-FFA model performed better than traditional methods and acted as an alternative method to estimate wave height.

The objective of this study is to predict GWL in Kalahandi watersheds using new hybrid machine learning (SVM-FFA) approaches. Also, the SVM-FFA model is compared with the RBFN and SVM approaches for performance evaluation based on different statistical parameters for proposed watershed.

10.2 STUDY AREA

Kalahandi district lies between 19.175489–20.454517°N latitudes and 82.617767–83.794874°E longitudes occupying southwestern part of Odisha (Figure 10.1). It is bordered by Balangir and Nuapada districts in North, by Nabarangpur, Koraput, and Rayagada districts in South, and by Rayagada, Kandhamal, and Boudh districts in East. Climate in Kalahandi district is of extreme kind. It is very dry excluding the monsoon period. Maximum temperature here is about more than 45°C, while minimum temperature is 4°C. Kalahandi experiences average annual precipitation of 1378.20 mm. Monsoon starts late in June and usually continues till September. It is principally an agriculture-based economy. Dharamgarh and Narla subdivisions are considered in present study for GWL prediction.

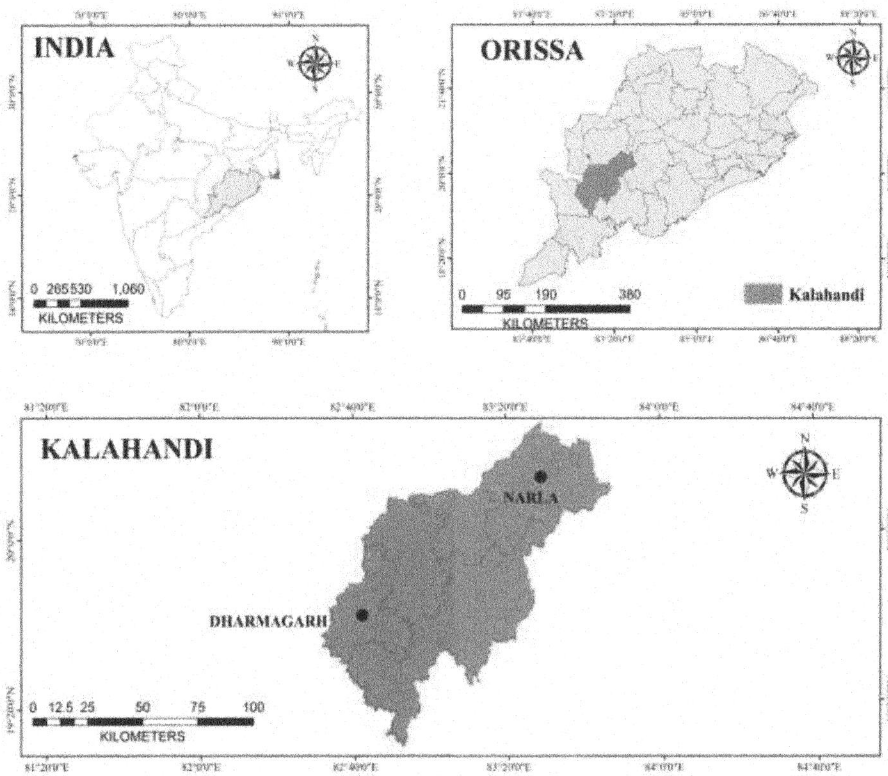

FIGURE 10.1 Proposed watershed.

10.3 METHODOLOGY

10.3.1 RBFN

Moody and Darken proposed RBFN in late 1980s. Since then, it is being extensively utilized to classify and approximate different linear and non-linear functions (Chu et al., 2013). RBFN possesses a resilient biological training and has the capability for approximating any random non-linear function (Schilling et al., 2001). Another advantage of this network is that it comprises optimum estimate point. Fundamentally, RBFN consists of many modest and vastly interrelated simulated neurons which are systematized to several layers, i.e. input, hidden, and output layers as presented in Figure 10.2 (Samantaray and Sahoo, 2020c; Sahoo et al., 2020; Samantaray and Ghose, 2019, 2020a,b).

A single output RBFN with K hidden layer neurons is articulated as

$$y_k = \sum_{i=1}^{m} \omega_{ik} R_i(x) + \theta_k \tag{10.1}$$

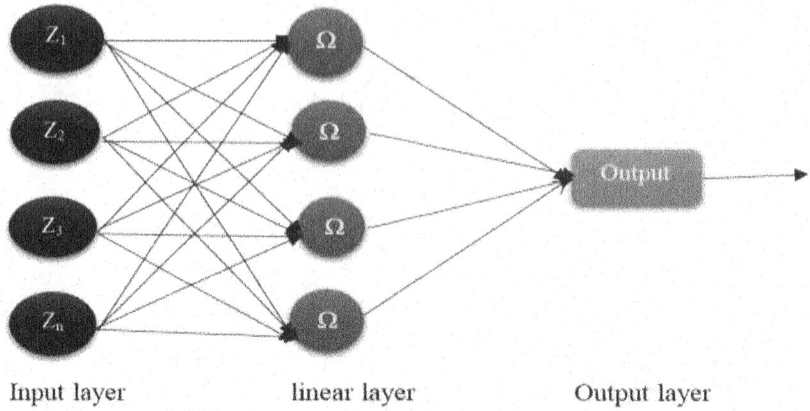

FIGURE 10.2 Architecture of RBFN.

where y_k is the kth output node on output layer; ω_{ik} is the weight connection amid ith hidden node and kth output node; and θ_k is the threshold value of kth output node.

Gaussian function is the most common utilized function in hidden layer, as shown in Equation 10.1:

$$R_i(x) = e\left[-\frac{\|x - c_i\|}{2\sigma_i^2}\right], \quad i = 1, 2, \ldots, m \tag{10.2}$$

where x is the input vector with n dimension; c_i is the ith RBF centre; σ_i is the RBF spread in ith hidden node, indicating radiated distance from RBF centre; m is the number of hidden nodes; and $\|x - c_i\|$ is the radiated distance from X to RBF centre. Quantity of hidden neurons have a big influence on the output of the RBF.

10.3.2 SVM

Among many soft computing techniques, SVM is one of the most recently used techniques applied to many research areas, for example computational field, hydrological field, and in the field of environment studies (Asefa et al., 2006; Sun, 2013). Essentially SVM is used in recognizing various patterns, prediction, forecasts, regression analysis, and different classification. Evolution of SVM theory was proposed by (Vapnik, 2000). While customary approaches aim at minimizing error in local training nodes, SVM focuses to minimize upper bound for simplification error (Vapnik and Vapnik, 1998). Concept of SVM is based on the principle of structural risk minimization, which is a major advantage over traditional soft computing techniques. Moreover, it provides a unique solution because of its convex nature. SVM consists of a high-dimensional kernel function space discreetly containing non-linear transformation where the input data is mapped (Samantaray and Sahoo, 2020b; Samantaray et al., 2020). Different equations related to SVM in accordance to Vapnik's theory is presented in Equations 10.3–10.5. Considering n points in a dataset specified by $\{x_i, d_i\}_i^n$, SVM approximate functions specified in Equations 10.3 and 10.4:

$$f(x) = w\varphi(x) + b \tag{10.3}$$

$$R_{SVM_s}(C) = \frac{1}{2}\|w\|^2 + C\frac{1}{n}\sum_{i=1}^{n} L(x_i, d_i) \tag{10.4}$$

where x_i is the input space vector, d_i is the target value, $\varphi(x)$ is the high-dimension feature space to map input x, w is the normal vector, b is the scalar vector, $C\frac{1}{n}\sum_{i=1}^{n} L(x_i, d_i)$ signifies empirical error, and w and b are linear function $f(x)$ constant.

Constraints w and b are calculated utilizing Equation 10.4. ξi^* and ξi are positive slack variables signifying lower and upper additional deviance.

$$R_{SVM_s}(w, \xi^*) = \frac{1}{2}\|w\|^2 + C\frac{1}{n}\sum_{i=1}^{n} L(\xi_i + \xi_i^*) \tag{10.5}$$

$$\text{Subject to} \quad \begin{cases} d_i - w\varphi(x_i) + b_i \le \varepsilon + \xi_i \\ w\varphi(x_i) - d_i + b_i \le \varepsilon + \xi_i^* \\ \xi_i \xi_i^* \ge 0,\ i = 1, \ldots, l, \end{cases}$$

where $\frac{1}{2}\|w\|^2$ is the regularizing term, ε is the loss function equating to approximate accurateness of training data point, C is the error penalty factor, and l is the training dataset size. A typical architecture of proposed SVM is shown in Figure 10.3.

Determining data correlation using non-linear mapping method is the main aim of SVMs. Kernel techniques permit functions in a higher-dimensional and inherent feature space without computing coordinates of data in corresponding space. Results attained from high-dimensional feature space associates with original low-dimensional data from input space.

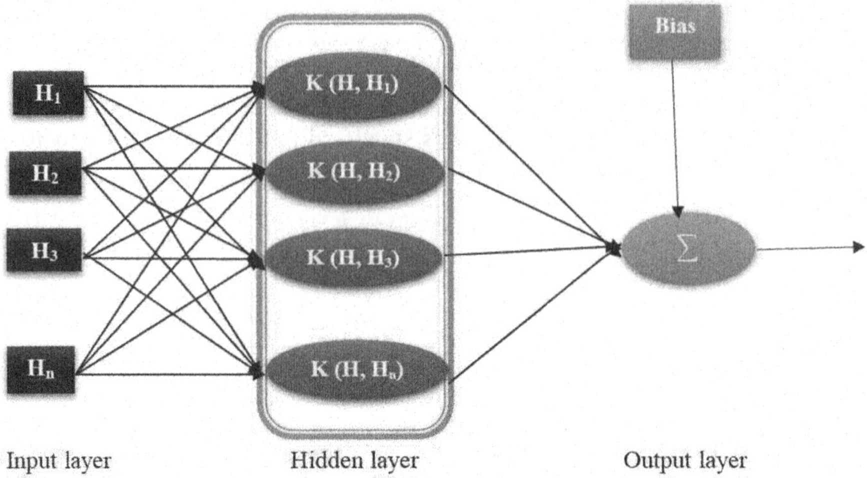

FIGURE 10.3 Architecture of SVM.

10.3.3 SVM-FFA

Nature has inspired many researchers in developing several optimization algorithms to solve traditional problems (Yang and Deb, 2014; Assareh et al., 2010). Among these developed algorithms, ant colony optimization, PSO, cuckoo search (CS), and GA are a few examples. Fundamental concept of such algorithms are based on existence and assortment of fittest species in nature. Most recent biological motivated optimization algorithm was developed by Yang (2009) known as FFA. It was founded based on social movement of fireflies in their natural setting. FFA shows some interactive pattern followed by the fireflies such as their flashing characteristics. A firefly attracts other fireflies and victims utilizing its bio-luminous quality. Luminance yielded by the firefly with high intensity is trailed by the other fireflies following certain path. Research studies recommend that FFA is robust and effective compared to other biologically motivated algorithms (Yang, 2013; Pal et al., 2012). Formulation of attraction and light intensity difference are major shortcomings of FFA. For maximizing the objective function, there is a need to refine the design specifications (Samantaray and Sahoo, 2020a; Samantaray et al., 2020b). The objective function has to be proportional to light intensity generated by a firefly. Light intensity with changing distance in Gaussian form is described as follows:

$$I = I_0 e^{-\gamma r^2} \tag{10.6}$$

where e is the exponential function, I is the light intensity at a distance of r in reference to firefly, I_0 is the intensity of light at distance $r = 0$ from reference firefly, and γ is the light absorption coefficient.

Attractiveness of a firefly trails a light intensity proportion when detected by other fireflies. Attraction x at a distance equal to r from reference firefly is defined as

$$\omega(r) = \omega_0 e^{-\gamma r^2} \tag{10.7}$$

where ω_0 is the attractiveness at $r = 0$ and ω is the light absorption coefficient.

The distance amid any two fireflies i and j at x_i and x_j, correspondingly, is well-defined as the Cartesian distance:

$$r_{ij} = \|x_i - j\| = \sqrt{\sum_{k=1}^{n} x_{i,k} - x_{j,k}} \tag{10.8}$$

where n denotes dimensionality of problem, $x_{i,k}$ is the kth constituent of spatial coordinate x_i of ith constituent.

10.3.4 Data Collection and Model Performance

Altogether, water table depth data of 20 years (2000–2019) were collected from India Meteorological Department (IMD), Groundwater Survey and Investigation (GWSI), Kalahandi district, Orissa. Collected data are utilized to develop models for studying effects of GW fluctuation. The data from 2000 to 2015 (80% of dataset) are utilized to train and data from 2016 to 2019 (20%) are utilized to test the network. Monthly data are transformed from daily data that helps in training and testing proposed model. Input and output data are arranged in such a manner that every data falls inside a quantified series before training. This procedure is known as normalization, with normalized values confined to 0–1 range. Normalization equation which is utilized to scale data is

$$M_t = \frac{M - M_{\min}}{M_{\max} - M_{\min}} \tag{10.9}$$

where M_t = transformed data series, M = actual data set, $M_{\min}$ = minimum of actual dataset, $M_{\max}$ = maximum of actual dataset. Subsequent GWL arrangements are employed as input:

Input	RBFN	SVM	SVM-FFA
i. H_{t-1}	RBFN:1	SVM:1	SVM-FFA:1
ii. H_{t-1}, H_{t-2}	RBFN:2	SVM:2	SVM-FFA:2
iii. H_{t-1}, H_{t-2}, H_{t-3}	RBFN:3	SVM:3	SVM-FFA:3
iv. H_{t-1}, H_{t-2}, H_{t-3}, H_{t-4}	RBFN:4	SVM:4	SVM-FFA:4
v. H_{t-1}, H_{t-2}, H_{t-3}, H_{t-4}, H_{t-5}	RBFN:5	SVM:5	SVM-FFA:5

where

H_{t-1} : one month lag GWL
H_{t-2} : two month lag GWL
H_{t-3} : three month lag GWL
H_{t-4} : four month lag GWL
H_{t-5} : five month lag GWL

10.3.5 Evaluating Criteria

RMSE, MSE, and R^2 are the evaluating standards to determine the best model. To select the perfect model for desired area of study, the criteria followed are MSE and RMSE must be minimum and R^2 must be maximum.

$$(R^2) = 1 - \frac{(\sum_{i=1}^{N} V_{comp} - \bar{V}_{comp})^2}{(\sum_{i=1}^{N} V_{obs} - \bar{V}_{obs})^2} \tag{10.10}$$

$$\text{MSE} = \frac{1}{n}\sum_{j=1}^{n}\left(V_{comp} - V_{obs}\right)^2 \tag{10.11}$$

$$\text{RMSE} = \frac{\sum_{i=1}^{N}(V_{comp} - \bar{V}_{comp})(V_{obs} - \bar{V}_{obs})}{\sqrt{\sum_{i=1}^{N}(V_{comp} - \bar{V}_{comp})^2 (V_{obs} - \bar{V}_{obs})^2}} \tag{10.12}$$

where

V_{comp} = predicted data
V_{obs} = observed data
$\bar{V}_{comp}$ = mean predicted data
$\bar{V}_{obs}$ = mean observed data

10.4 RESULTS AND DISCUSSION

The basic equation used for predicting the GWL of Kalahandi watershed based upon the equation $Q_t = f(H_{t-1}, H_{t-2}, H_{t-3}, H_{t-4}, H_{t-5})$. Results of RBFN are shown in Table 10.1 with the several spread values considered for simulation. Here spread values are taken within 0–1 range, i.e. 0.1, 0.2, 0.3, 0.4, 0.5, 0.6, 0.7, 0.8, and 0.9 preferably to predict GWL from substantial input parameters

TABLE 10.1
Outcomes of RBFN

Station	Input	Architecture	MSE		RMSE		R^2	
			Training	Testing	Training	Testing	Training	Testing
Dharmgarh	RBFN:1	4-0.2-1	0.00908	0.00854	0.07775	0.03926	0.91257	0.89021
	RBFN:2	4-0.8-1	0.00882	0.00842	0.07613	0.0382	0.91289	0.89079
	RBFN:3	4-0.4-1	0.00841	0.00799	0.07377	0.03382	0.91736	0.89411
	RBFN:4	4-0.6-1	0.00826	0.00779	0.07291	0.03249	0.92025	0.89693
	RBFN:5	**4-0.5-1**	**0.00753**	**0.00693**	**0.07224**	**0.03073**	**0.9258**	**0.90424**
Narla	RBFN:1	4-0.5-1	0.00879	0.00834	0.07599	0.03769	0.91389	0.89157
	RBFN:2	4-0.3-1	0.00865	0.00823	0.07487	0.03661	0.91462	0.89259
	RBFN:3	4-0.2-1	0.00836	0.00794	0.07352	0.03363	0.91894	0.89489
	RBFN:4	4-0.1-1	0.00791	0.00757	0.07265	0.03119	0.92171	0.89791
	RBFN:5	**4-0.8-1**	**0.00769**	**0.00681**	**0.07215**	**0.02313**	**0.92855**	**0.90562**

to map output. It is observed that with 0.6 spread value, RBFN illustrates best performance with 4-0.6-1 architecture, which possess MSE training and testing 0.00753 and 0.00693, RMSE training and testing 0.07224 and 0.03073, and R^2 training and testing 0.9258 and 0.90424, respectively, for Dharmgarh watershed. Similarly, for Narla watersheds, the prominent value of R^2 is 0.92855, 0.90562 during training and testing phases.

The graphs with best values for runoff using RBFN with various spread values at Dharmgarh, Narla, during training and testing phases are shown in Figures 10.4 and 10.5, respectively.

Performances of SVM model with different input for two projected gauging sites are shown in Table 10.2. Result found that the best values of performance with R^2 value are 0.95807 and 0.95854 during training phases for Dharmgarh and Narla watersheds, respectively. Correspondingly, during testing phases, the best values of performance with R^2 value are 0.93513 and 0.93741 for Dharmgarh and Narla watersheds, respectively. The performance evaluation measures considered in this study are described below for both testing and training phases. Actual versus predicted graph at Dharmgarh and Narla watersheds using SVM algorithms for training and testing phases are presented in Figures 10.6 and 10.7, respectively.

Among five models, SVM-FFA:5 model indicates best result for training and testing period having R^2 value 0.98719 and 0.96635, respectively, with H_{t-1}, H_{t-2}, H_{t-3}, H_{t-4}, H_{t-5} as input parameters for Dharmgarh gauging site. Similarly, for Narla station, SVM-FFA:5 produced best R^2 value for both periods among five simulations. For training phase, Narla paramount value of R^2 is 0.98807. Outcomes for SVM-FFA model according to RMSE, MSE, and R^2 in training and testing phases are specified in Table 10.3. Actual versus simulated graph at Dharmgarh and Narla

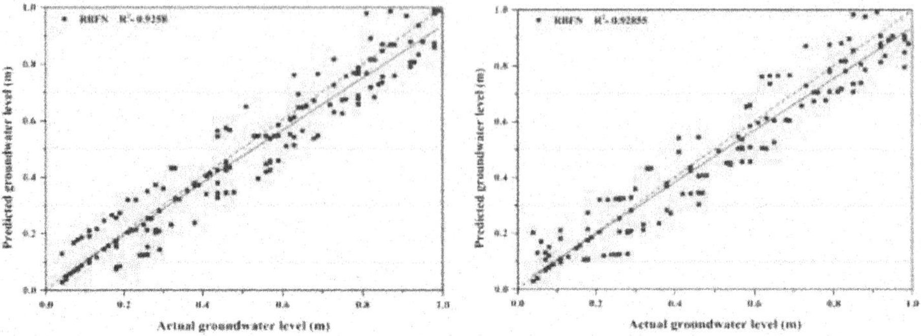

(a) (b)

FIGURE 10.4 Actual versus predicted GWL using RBFN algorithm at (a) Dharmgarh and (b) Narla watersheds during training phase.

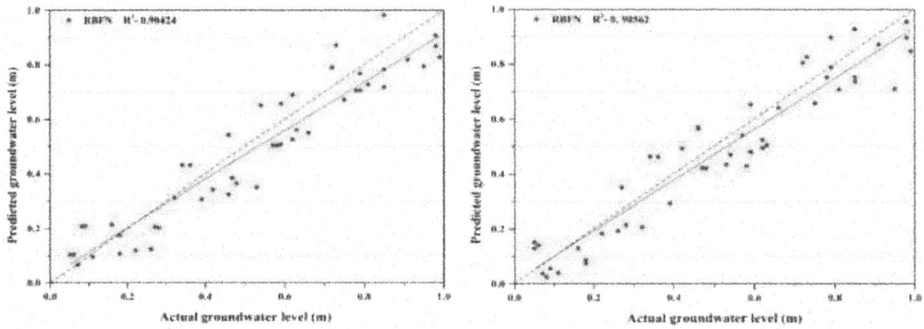

(a) (b)

FIGURE 10.5 Actual versus predicted GWL using RBFN algorithm at (a) Dharmgarh and (b) Narla watersheds during testing phase.

TABLE 10.2
Results of SVM

Station	Input	MSE		RMSE		R^2	
		Training	Testing	Training	Testing	Training	Testing
Dharmgarh	SVM:1	0.00561	0.00536	0.06832	0.02221	0.93995	0.92385
	SVM:2	0.00559	0.00535	0.06816	0.02199	0.94025	0.92411
	SVM:3	0.00545	0.00515	0.06765	0.02163	0.94128	0.92689
	SVM:4	0.00518	0.00456	0.06622	0.02146	0.94221	0.92978
	SVM:5	**0.00485**	**0.00407**	**0.06105**	**0.02123**	**0.95807**	**0.93513**
Narla	SVM:1	0.00562	0.00534	0.0681	0.02192	0.94026	0.92427
	SVM:2	0.00547	0.00525	0.06807	0.0219	0.94033	0.92438
	SVM:3	0.00523	0.00502	0.06712	0.02156	0.94153	0.92711
	SVM:4	0.00502	0.00419	0.06485	0.02133	0.94261	0.93014
	SVM:5	**0.00492**	**0.00406**	**0.06002**	**0.02023**	**0.95854**	**0.93741**

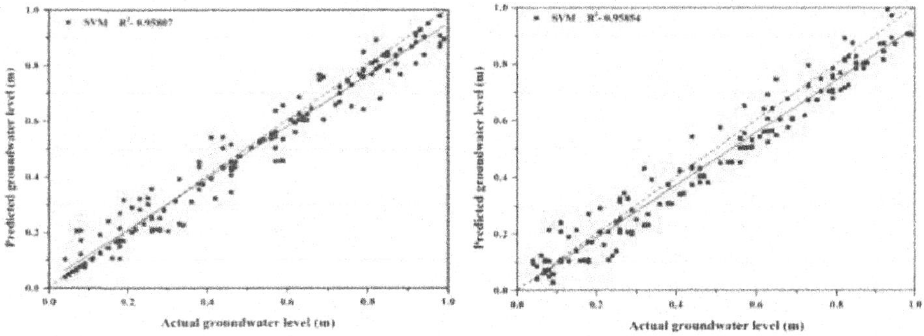

(a) (b)

FIGURE 10.6 Actual versus predicted GWL using SVM algorithm at (a) Dharmgarh and (b) Narla watersheds during training phase.

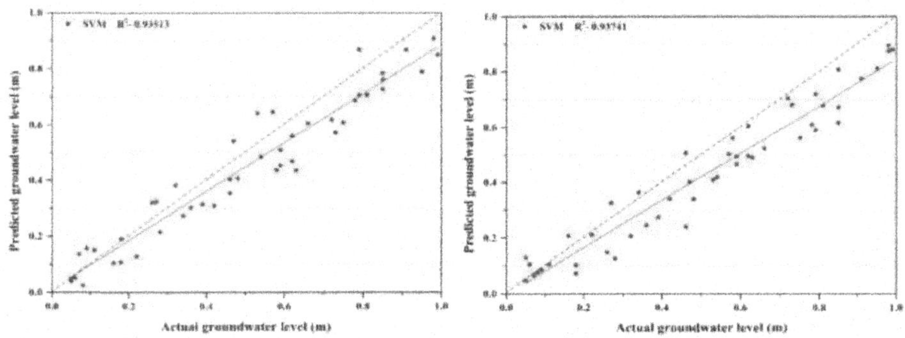

(a) (b)

FIGURE 10.7 Actual versus predicted GWL using SVM algorithm at (a) Dharmgarh and (b) Narla watersheds during testing phase.

TABLE 10.3
Results of SVM-FFA

Station	Input	MSE		RMSE		R^2	
		Training	Testing	Training	Testing	Training	Testing
Dharmgarh	SVM-FFA:1	0.0042	0.00256	0.05729	0.01794	0.97614	0.95449
	SVM-FFA:2	0.00399	0.00242	0.05622	0.01668	0.97775	0.95464
	SVM-FFA:3	0.00371	0.00225	0.05421	0.01414	0.98004	0.95746
	SVM-FFA:4	0.00362	0.0021	0.05284	0.01297	0.98132	0.95947
	SVM-FFA:5	**0.0032**	**0.00187**	**0.04996**	**0.01058**	**0.98719**	**0.96635**
Narla	SVM-FFA:1	0.00387	0.00242	0.05579	0.01626	0.97821	0.95589
	SVM-FFA:2	0.00383	0.00241	0.05562	0.01599	0.97898	0.95597
	SVM-FFA:3	0.00368	0.00219	0.05367	0.01368	0.98042	0.95804
	SVM-FFA:4	0.00339	0.00197	0.05115	0.0115	0.98139	0.96021
	SVM-FFA:5	**0.00328**	**0.00183**	**0.04985**	**0.01027**	**0.98807**	**0.96807**

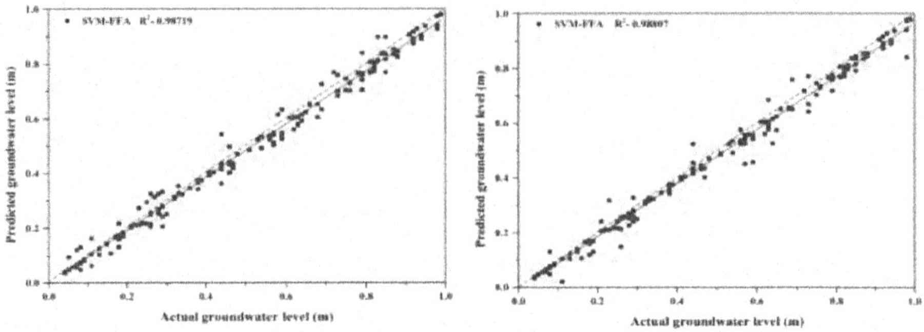

(a) (b)

FIGURE 10.8 Actual versus predicted GWL using SVM-FFA algorithm at (a) Dharmgarh and (b) Narla watersheds during training phase.

stations using SVM-FFA algorithms during training and testing periods are accessible in Figures 10.8 and 10.9, respectively.

10.4.1 Assessment of Actual Versus Predicted GWL at Dharmgarh and Narla During Testing Phase

Variation of actual versus predicted GWL is revealed in Figure 10.10. Results show that predictable peak GWLs are 6.2256 m, 6.4606 m, and 6.6445 m for RBFN, SVM, SVM-FFA, respectively, compared to the actual peak 6.912 m for the watershed Dharmgarh. For station Narla, the estimated peak GWLs are 6.7905 m, 7.0211 m, and 7.2449 m for RBFN, SVM, SVM-FFA, respectively, in contrast to the actual

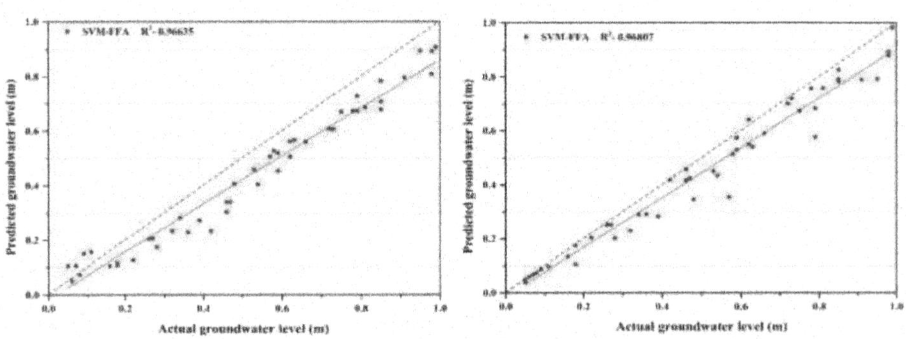

(a) (b)

FIGURE 10.9 Actual versus predicted GWL using SVM-FFA algorithm at (a) Dharmgarh and (b) Narla watersheds during testing phase.

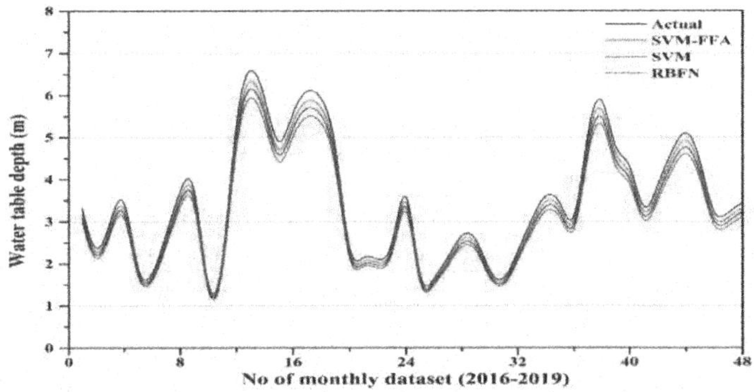

(a)

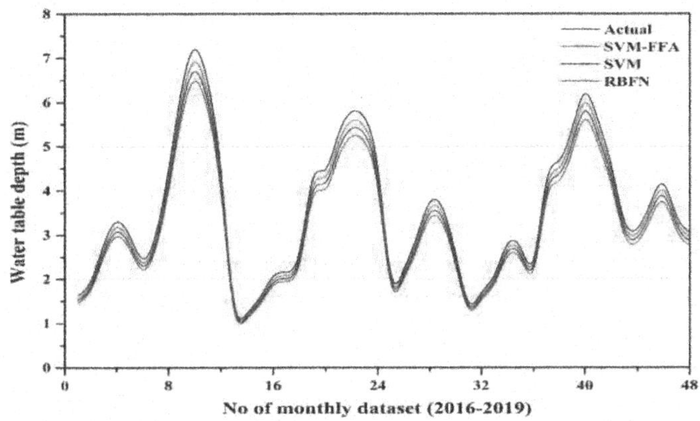

(b)

FIGURE 10.10 Comparison of actual versus predicted GWL at (a) Dharmgarh and (b) Narla station during testing phase.

peak of 7.535 m. This shows the significant potential of GWL and found to be useful for drought-prone region with the predicted water level index.

10.4.2 Comparison of Model Performance for Proposed Station

At Dharmgarh site between three neural network models with evaluation measures RMSE, MSE, and R^2, SVM-FFA performs best with R^2 value 0.98719. Similarly, at Narla, SVM-FFA performs best among three networks with R^2 value 0.98807 for training phases. Among three algorithms, RBFN shows least performance value for all proposed gauge station. The detailed result is compiled in Table 10.4. Since SVM-FFA shows best efficiency, the model will be utilized for predicting GWL of future uses nearest to the watershed and is recommended for irrigation and water resources by the department of Kalahandi.

TABLE 10.4
Comparison of Results for Proposed Watersheds

Station	Techniques	MSE		RMSE		R^2	
		Training	Testing	Training	Testing	Training	Testing
	RBFN	0.00753	0.00693	0.07224	0.03073	0.9258	0.90424
Dharmgarh	SVM	0.00485	0.00407	0.06105	0.02123	0.95807	0.93513
	SVM-FFA	**0.0032**	**0.00187**	**0.04996**	**0.01058**	**0.98719**	**0.96635**
	RBFN	0.00769	0.00681	0.07215	0.02313	0.92855	0.90562
Narla	SVM	0.00492	0.00406	0.06002	0.02023	0.95854	0.93741
	SVM-FFA	**0.00328**	**0.00183**	**0.04985**	**0.01027**	**0.98807**	**0.96807**

10.5 CONCLUSION

Accurate and reliable estimation of GWL is a very significant issue in proper water resources management. In this study, a hybrid model is proposed for overcoming deficiencies of every single model and combine strengths of them. This study proves validity and potential of RBFN, SVM, and SVM-FFA models in prediction of GWL. For studying accuracy of SVM-FFA model, traditional models such as RBFN and SVM are developed. Results reveal that all three developed models have a good fitting accuracy during training period. But RMSE, MAE, and R^2 values show that hybrid model is more proficient to forecast GWL compared to SVM and RBFN model during training and testing period. Results obtained from this study could be used as beneficial plans in choosing suitable modelling methods. Conclusions reveal that SVM-FFA is found to be an effective technique for simulating GWL and investigating model uncertainties with the use of confidence intervals. These can also be useful for facilitating sustainable groundwater management policies. Proposed hybrid model uncovers dynamic variation rule of GWL on the basis of history monitoring data of GWL only. Hence, interaction among various dynamic factors of GWL should be considered in future study.

REFERENCES

Adewumi, Oluyinka Aderemi and Ayobami Andronicus Akinyelu. 2016. "A hybrid firefly and support vector machine classifier for phishing email detection." *Kybernetes*, 45 (6), 977–994.

Afan, Haitham Abdulmohsin, Ahmed El-Shafie, Zaher Mundher Yaseen, Mohammed Majeed Hameed, Wan Hanna Melini Wan Mohtar, and Aini Hussain. 2015. "ANN based sediment prediction model utilizing different input scenarios." *Water Resources Management*, 29 (4), 1231–1245.

Al-Shammari, Eiman Tamah, Afram Keivani, Shahaboddin Shamshirband, Ali Mostafaeipour, Lip Yee, Dalibor Petković, and Sudheer Ch. 2016. "Prediction of heat load in district heating systems by support vector machine with firefly searching algorithm." *Energy*, 95, 266–273.

Alizamir, Meysam, Ozgur Kisi, and Mohammad Zounemat-Kermani. 2018. "Modelling long-term groundwater fluctuations by extreme learning machine using hydro-climatic data." *Hydrological Sciences Journal*, 63 (1), 63–73.

Asefa, Tirusew, Mariush Kemblowski, Mac McKee, and Abedalrazq Khalil. 2006. "Multi-time scale stream flow predictions: the support vector machines approach." *Journal of Hydrology*, 318 (1–4), 7–16.

Assareh, E., M. A. Behrang, M. R. Assari, and A. Ghanbarzadeh. 2010. "Application of PSO (particle swarm optimization) and GA (genetic algorithm) techniques on demand estimation of oil in Iran." *Energy*, 35 (12), 5223–5229.

Ch., Sudheer, S. K. Sohani, Deepak Kumar, Anushree Malik, B. R. Chahar, A. K. Nema, Bijaya K. Panigrahi, and R. C. Dhiman. 2014. "A support vector machine-firefly algorithm based forecasting model to determine malaria transmission." *Neurocomputing*, 129, 279–288.

Chang, Xu, Jia Hui, Wang Rong, and Wu Hao. 2013. "Groundwater level prediction based on BP and RBF neural network." *International Journal of Applied Sciences and Engineering Research*, 2 (3), 263–269.

Chu, H. B., W. X. Lu, and L. Zhang. 2013. "Application of artificial neural network in environmental water quality assessment." *Journal of Agricultural Science and Technology*, 15 (2), 343–356.

Cortes, Corinna and Vladimir Vapnik. 1995. "Support-vector networks." *Machine Learning*, 20 (3), 273–297.

Devarajan, K. and G. Sindhu. 2015. "Application of numerical and empirical models for groundwater level forecasting." *International Journal of Research in Engineering and Technology*, 4 (11), 127–123.

Ebrahimi, Hadi and Taher Rajaee. 2017. "Simulation of groundwater level variations using wavelet combined with neural network, linear regression and support vector machine." *Global and Planetary Change*, 148, 181–191.

Emamgholizadeh, Samad, Khadije Moslemi, and Gholamhosein Karami. 2014. "Prediction of the groundwater level of Bastam Plain (Iran) by artificial neural network (ANN) and adaptive neuro-fuzzy inference system (ANFIS)." *Water Resources Management*, 28 (15), 5433–5446.

Ghazavi, R., A. B. Vali, and S. Eslamian. 2012. "Impact of flood spreading on groundwater level variation and groundwater quality in an arid environment." *Water Resources Management*, 26 (6), 1651–1663.

Ghorbani, M. A., R. Khatibi, V. Karimi, Zaher Mundher Yaseen, and M. Zounemat-Kermani. 2018. "Learning from multiple models using artificial intelligence to improve model prediction accuracies: application to river flows." *Water Resources Management*, 32 (13), 4201–4215.

Ghorbani, Mohammad Ali, Shahaboddin Shamshirband, Davoud Zare Haghi, Atefe Azani, Hossein Bonakdari, and Isa Ebtehaj. 2017. "Application of firefly algorithm-based support vector machines for prediction of field capacity and permanent wilting point." *Soil and Tillage Research*, 172, 32–38.

Ghose, Dillip K. and Sandeep Samantaray. 2019. "Integrated sensor networking for estimating ground water potential in scanty rainfall region: challenges and evaluation." In Computational Intelligence in Sensor Networks. 335–352, Springer.

Gocić, Milan, Shervin Motamedi, Shahaboddin Shamshirband, Dalibor Petković, Sudheer Ch, Roslan Hashim, and Muhammad Arif. 2015. "Soft computing approaches for forecasting reference evapotranspiration." *Computers and Electronics in Agriculture*, 113, 164–173.

Gong, Yicheng, Yongxiang Zhang, Shuangshuang Lan, and Huan Wang. 2016. "A comparative study of artificial neural networks, support vector machines and adaptive neuro fuzzy inference system for forecasting groundwater levels near lake Okeechobee, Florida." *Water Resources Management*, 30 (1), 375–391.

He, Zhibin, Xiaohu Wen, Hu Liu, and Jun Du. 2014. "A comparative study of artificial neural network, adaptive neuro fuzzy inference system and support vector machine for forecasting river flow in the semiarid mountain region." *Journal of Hydrology*, 509, 379–386.

Huang, Faming, Jinsong Huang, Shui-Hua Jiang, and Chuangbing Zhou. 2017. "Prediction of groundwater levels using evidence of chaos and support vector machine." *Journal of Hydroinformatics*, 19 (4), 586–606.

Jimmy, S. R., A. Sahoo, Sandeep Samantaray, and Dillip K. Ghose. 2020. "Prophecy of runoff in a river basin using various neural networks." In Communication Software and Networks. 709–718, Springer.

Latt, Zaw, Hartmut Wittenberg, and Brigitte Urban. 2015. "Clustering hydrological homogeneous regions and neural network based index flood estimation for ungauged catchments: an example of the Chindwin river in Myanmar." *Water Resources Management*, 29 (3), 913–928.

Li, Fawen, Ping Feng, Wei Zhang, and Ting Zhang. 2013. "An integrated groundwater management mode based on control indexes of groundwater quantity and level." *Water Resources Management*, 27 (9), 3273–3292.

Mirzavand, Mohammad, Benyamin Khoshnevisan, Shahaboddin Shamshirband, Ozgur Kisi, Rodina Ahmad, and Shatirah Akib. 2015. "Evaluating groundwater level fluctuation by support vector regression and neuro-fuzzy methods: a comparative study." *Natural Hazards*, 102 (3), 1–15.

Mladenović, Igor, Svetlana Sokolov-Mladenović, Milos Milovančević, Dušan Marković, and Nenad Simeunović. 2016. "Management and estimation of thermal comfort, carbon dioxide emission and economic growth by support vector machine." *Renewable and Sustainable Energy Reviews*, 64, 466–476.

Nastos, P. T., A. G. Paliatsos, K. V. Koukouletsos, I. K. Larissi, and K. P. Moustris. 2014. "Artificial neural networks modeling for forecasting the maximum daily total precipitation at athens, Greece." *Atmospheric Research*, 144, 141–150.

Nie, Siyu, Jianmin Bian, Hanli Wan, Xiaoqing Sun, and Bingjing Zhang. 2017. "Simulation and uncertainty analysis for groundwater levels using radial basis function neural network and support vector machine models." *Journal of Water Supply*, 66 (1), 15–24.

Nourani, Vahid and Shahram Mousavi. 2016. "Spatiotemporal groundwater level modeling using hybrid artificial intelligence-meshless method." Journal of Hydrology, 536, 10–25.

Pal, Saibal K., C. S. Rai, and Amrit Pal Singh. 2012. "Comparative study of firefly algorithm and particle swarm optimization for noisy non-linear optimization problems." *International Journal of Intelligent Systems and Applications*, 4 (10), 50.

Roy, Chandrabhushan, Shervin Motamedi, Roslan Hashim, Shahaboddin Shamshirband, and Dalibor Petković. 2016. "A comparative study for estimation of wave height using traditional and hybrid soft-computing methods." *Environmental Earth Sciences*, 75 (7), 590.

Sahoo, A., U. K. Singh, M. H. Kumar, and Sandeep Samantaray. 2020. "Estimation of flood in a river basin through neural networks: a case study." In Communication Software and Networks. 755–763 Springer, Singapore.

Samantaray, S., A. Sahoo, and D. K. Ghose. 2019. "Assessment of groundwater potential using neural network: a case study." In International Conference on Intelligent Computing and Communication. 655–664, Springer.

Samantaray, Sandeep and Dillip K. Ghose. 2019. "Sediment assessment for a watershed in arid region via neural networks." *Sādhanā*, 44 (10), 219.

Samantaray, S. and A. Sahoo. 2020a. Assessment of sediment concentration through RBNN and SVM-FFA in arid watershed, India. Smart Innovation, Systems and Technologies, 159. doi:10.1007/978-981-13-9282-5_67.

Samantaray, S. and A. Sahoo. 2020b. "Estimation of runoff through BPNN and SVM in Agalpur watershed." Advances in Intelligent Systems and Computing, 1014. doi:10.1007/978-981-13-9920-6_27.

Samantaray, S. and A. Sahoo. 2020c. "Appraisal of runoff through BPNN, RNN, and RBFN in Tentulikhunti watershed: a case study." Advances in Intelligent Systems and Computing, 1014. doi:10.1007/978-981-13-9920-6_26.

Samantaray, S. and Sahoo, A. 2020d. "Prediction of runoff using BPNN, FFBPNN, CFBPNN algorithm in arid watershed: a case study." *International Journal of Knowledge-based and Intelligent Engineering Systems*, 24 (3), 243–251.

Samantaray, S., O. Tripathy, A. Sahoo, and D.K. Ghose. 2020. "Rainfall forecasting through ANN and SVM in Bolangir watershed, India." Smart Innovation, Systems and Technologies, 159. doi:10.1007/978-981-13-9282-5_74.

Samantaray, Sandeep and Dillip K. Ghose. 2020a. "Assessment of suspended sediment load with neural networks in arid watershed." *Journal of the Institution of Engineers (India): Series A*, 101, 371–380.

Samantaray, Sandeep and Dillip K. Ghose. 2020b. "Modelling runoff in a river basin, India: an integration for developing un-gauged catchment." *International Journal of Hydrology Science and Technology*, 10 (3), 248–266.

Samantaray, S., A. Sahoo, and D. K. Ghose. 2020a. Infiltration loss affects toward groundwater fluctuation through CANFIS in arid watershed: a case study. Smart Innovation, Systems and Technologies, 159. doi:10.1007/978-981-13-9282-5_76.

Samantaray, Sandeep, Abinash Sahoo, and Dillip K. Ghose. 2020b. "Assessment of sediment load concentration using SVM, SVM-FFA and PSR-SVM-FFA in arid watershed, India: a case study." *KSCE Journal of Civil Engineering*, 24, 1944–1957.

Samantaray, S., A. Sahoo, and D. K. Ghose. 2020c. Prediction of sedimentation in an arid watershed using BPNN and ANFIS. Lecture Notes in Networks and Systems, 93. doi:10.1007/978-981-15-0630-7_29.

Samantaray, Sandeep, A. Sahoo, N. R. Mohanta, P. Biswal, and U. K. Das. 2020d. "Runoff prediction using hybrid neural networks in semi-arid watershed, India: a case study." In Communication Software and Networks. 729–736, Springer, Singapore.

Sattari, Mohammad Taghi, Rasoul Mirabbasi, Reza Shamsi Sushab, and John Abraham. 2018. "Prediction of groundwater level in Ardebil plain using support vector regression and M5 tree model." *Groundwater*, 56 (4). 636–646.

Schilling, Robert J., James J. Carroll, and Ahmad F. Al-Ajlouni. 2001. "Approximation of nonlinear systems with radial basis function neural networks." *IEEE Transactions on Neural Networks*, 12 (1), 1–15.

Shamshirband, Shahaboddin, Kasra Mohammadi, Chong Wen Tong, Mazdak Zamani, Shervin Motamedi, and Sudheer Ch. 2016. "A hybrid SVM-FFA method for prediction of monthly mean global solar radiation." *Theoretical and Applied Climatology*, 125 (1–2), 53–65.

Shamshirband, Shahaboddin, Meisam Tabatabaei, Mortaza Aghbashlo, Lip Yee, and Dalibor Petković. 2016. "Support vector machine-based exergetic modelling of a DI diesel engine running on biodiesel–diesel blends containing expanded polystyrene." *Applied Thermal Engineering*, 94, 727–747.

Sridharam, S., A. Sahoo, Sandeep Samantaray, and Dillip K. Ghose. 2020. "Estimation of water table depth using wavelet-ANFIS: a case study." In Communication Software and Networks. 747–754, Springer, Singapore.

Sun, Shiliang. 2013. "A survey of multi-view machine learning." *Neural Computing and Applications*, 23 (7–8), 2031–2038.

Sun, Shiliang, Changshui Zhang, and Guoqiang Yu. 2006. "A Bayesian network approach to traffic flow forecasting." *IEEE Transactions on Intelligent Transportation Systems*, 7 (1), 124–132.

Sun, Shiliang and Xin Xu. 2010. "Variational inference for infinite mixtures of Gaussian processes with applications to traffic flow prediction." *IEEE Transactions on Intelligent Transportation Systems*, 12 (2), 466–475.

Suryanarayana, Ch., Ch. Sudheer, Vazeer Mahammood, and Bijaya K. Panigrahi. 2014. "An integrated wavelet-support vector machine for groundwater level prediction in Visakhapatnam, India." *Neurocomputing*, 145, 324–335.

Tapak, Lily, Alireza Rahmani, and Abbas Moghimbeigi. 2014. "Prediction the groundwater level of Hamadan-Bahar plain, west of Iran using support vector machines." *Journal of Research in Health Sciences*, 14 (1), 82–87.

Üneş, Fatih, Mustafa Demirci, Yunus Ziya Kaya, Eyüp İspir, and Mustafa Mamak. 2017. "Groundwater level prediction using support vector machines and autoregressive (AR) models." Vilnius Gediminas Technical University Publishing House "Technika".

Vapnik, V. 2000. "The Nature of Statistical Learning Theory." Springer.

Vapnik, V. N. and V. Vapnik, 1998. Statistical Learning Theory, John Wiley & Sons, Inc., New York.

Verma, A. K. and T. N. Singh. 2013. "Prediction of water quality from simple field parameters." *Environmental Earth Sciences*, 69 (3), 821–829.

Yan, Qiao and Cong Ma. 2016. "Application of integrated ARIMA and RBF network for groundwater level forecasting." *Environmental Earth Sciences*, 75 (5), 396.

Yang, Xin-She. 2009. "Firefly algorithms for multimodal optimization." In International Symposium on Stochastic Algorithms, 169–178, Springer.

Yang, Xin-She. 2013. "Multiobjective firefly algorithm for continuous optimization." *Engineering with Computers*, 29 (2), 175–84.

Yang, Xin-She and Suash Deb. 2014. "Cuckoo search: recent advances and applications." *Neural Computing and Applications*, 24 (1), 169–174.

Ying, Zhao, Lu Wenxi, Chu Haibo, and Luo Jiannan. 2014. "Comparison of three forecasting models for groundwater levels: a case study in the semiarid area of West Jilin Province, China." *Journal of Water Supply*, 63 (8), 671–683.

Yoon, Heesung, Yunjung Hyun, Kyoochul Ha, Kang-Kun Lee, and Gyoo-Bum Kim. 2016. "A method to improve the stability and accuracy of ANN-and SVM-based time series models for long-term groundwater level predictions." *Computers & Geosciences*, 90, 144–155.

Yoon, Heesung, Seong-Chun Jun, Yunjung Hyun, Gwang-Ok Bae, and Kang-Kun Lee. 2011. "A comparative study of artificial neural networks and support vector machines for predicting groundwater levels in a coastal aquifer." *Journal of Hydrology*, 396 (1–2), 128–138.

Zhang, Nan, Changlai Xiao, Bo Liu, and Xiujuan Liang. 2017. "Groundwater depth predictions by GSM, RBF, and ANFIS models: a comparative assessment." *Arabian Journal of Geosciences*, 10 (8), 189.

Zhao, Wei Guo, Huan Wang, and Zi Jun Wang. 2011. "Groundwater level forecasting based on support vector machine." *Applied Mechanics and Materials*. 44, 1365–1369.

Zhou, Ting, Faxin Wang, and Zhi Yang. 2017. "Comparative analysis of ANN and SVM models combined with wavelet preprocess for groundwater depth prediction." *Water*, 9 (10), 781.

11 Prediction of Flood Using Hybrid ANFIS-FFA Approaches in Barak River Basin

11.1 INTRODUCTION

Global flood occurrences have time and again raised questions on predicting flood and measures on its control. Precise and well-timed forecast of imminent flood is a very serious issue for diverting and distributing water, preventing drought, human protection, and sustainable ecology. Consistent flood forecast models are very essential in India because most Indian states are affected by flood events, causing major damage to public property and threat to human and animal life. Recently, artificial intelligence methods are more progressively utilized to forecast flood events. Several scholars have efficaciously used ANN to forecast flood at altered time lead (Sudheer et al., 2002; Chang et al., 2007). Similarly, neuro-fuzzy is an additional field of study that is effectively being used to forecast flood events (Nayak et al., 2005; Singh, 2007).

Che et al. (2011) applied BPNN and GA for training FFNN for coping with weighing adjustment problem and compared their performance considering measurement indicators and experimental data. Outcomes showed that BPNN is superior to GA and had faster training speed than GA. Dar et al. (2015) presented applicability of ANN method to develop flood prediction model for a large-size catchment of Jhelum River in J&K, India. Based on statistical parameters, ANN proved to be a potential model in flood discharge prediction with reliability and accuracy in proposed area of study. Shiau and Hsu (2016) applied FFBPNN and radial basis function network (RBFN) linked with different time-lagged stream flow and precipitation inputs for extending short stream flow record at Lilin gauging site situated in Gaoping River basin, Taiwan. Findings revealed that proposed techniques performed suitably for extending daily stream flow records at the selected gauge station. Pandey and Srinivas (2015) investigated the potential of Auto Regressive Integrated Moving Average (ARIMA), FFBPNN, and RBFN for forecasting daily stream flow at Basantpur gauge station of River Mahanadi, Odisha, India. Results indicated that FFBPNN model performed better compared to other proposed models in daily stream flow forecasts. Wei and Hsu (2008) employed FFBPNN for estimating downstream water levels of River Tanshui located in Taiwan. Results from optimized model were compared with historical proceedings and observed that FFBPNN effectively helped to solve problems in controlling flood events. Arulmurugan and Anandakumar (2018) applied FFBPNN and wavelet transform classifier for recognizing and classifying lung nodules. Results indicated that FFBPNN method produced more accuracy in detecting lung cancer at early stage. Syahrullah and Sinaga (2016) explored the usability of FFBPNN for optimizing and predicting potential of motorcycle fuel injection system of gasoline. Investigation showed that proposed model accurately predicted the potential of motorcycle injection system. Mehr et al. (2015) studied usability of FFBPNN, generalized regression NN, and RBF to predict stream flow in poor rain gauging stations considering monthly stream flow data of two succeeding stations on Coruh River, Turkey. Findings of present study revealed that RBFN performed better than FFBPNN and GRNN in proposed study area. Mehr et al. (2014) investigated efficacy of linear genetic programming (LGP), FFBPNN, GRNN, and RBFN techniques for stream flow prediction. Results specified that LGP technique provided more accurate results for successive station monthly stream flow prediction compared to other ANN techniques. Konate et al. (2015) analysed and compared GRNN and FFBP to model porosity on four wells of Zhenjing oilfield data, China. Findings proved that GRNN made more precise and reliable porosity constraint estimation compared to FFBPNN.

Nowadays ANN is a successive tool commonly applied for prediction of various hydrologic parameters in different watersheds (Mohanta et al, 2020a; Samantaray and Ghose, 2019, 2020; Samantaray and Sahoo, 2020d; Samantaray et al., 2019a, 2019b, 2020b 2020c). Mukerji et al. (2009) used ANN, ANGIS, and ANFIS models for forecasting flood at Jamtara gauge station of River Ajay, Jharkhand, India, and comparative performance of proposed models was conducted. Results revealed that ANGIS predicted flood occurrences with extreme accurateness followed by ANFIS and ANN models. Shu and Ouarda (2008) proposed ANFIS technique for providing regional flood estimation of Quebec province, Canada, and contrasted to ANN, NLR, and NLR with regionalization (NLR-R). Results indicated that ANFIS model generalized input parameters with much better capability compared to NLR and NLR-R models. Sehgal et al. (2014) developed wavelet ANFIS-split data and wavelet ANFIS-modified time series models for forecasting river water levels of Kamla and Kosi River basins in India. Proposed models helped in forecasting river water levels precisely and when the two models were compared, WANFIS-SD gave better performance compared to WANFIS-MS for high river stages. Chau et al. (2005) applied hybrid genetic algorithm based ANN model (ANN-GA) and ANFIS model to forecast flood events in a channel reach of River Yangtze, China. Rezaeianzadeh et al. (2014) used ANN, ANFIS, MLR, and MNLR to forecast peak flows on daily basis at exit of Khosrow Shirin catchment, situated in Fars State, Iran. Model performances were evaluated and outcomes revealed that considering area prejudiced rainfall as input for ANNs and MNLR and spatially dispersed rainfall as input for ANFIS and MLR gave more precise prediction performances. Kisi et al. (2012) evaluated accuracy of several data-driven methods, i.e. ANN, SVM, and ANFIS, to forecast daily intermittent stream flows. Gong et al. (2016) applied ANN, SVM, and ANFIS for groundwater level prediction for wells near Lake Okeechobee in Florida, considering interface amid surface water and groundwater. Results demonstrated that ANFIS and SVM models predicted groundwater level more accurately than ANN model and lake level variations were found to be the key driving factor in groundwater level prediction. Dastorani et al. (2010) used data from bordering sites, ANN and ANFIS models to find missing data of selected gauge stations from different parts of Iran. Findings demonstrated that ANFIS presented a superior ability to predict missing flow data. Mehr et al. (2019) developed an amalgam model integrating SVR and FFA for one-month ahead precipitation forecasting at Tabriz and Urmia sites in northwest Iran. Results of hybrid model were compared with that of regular SVR and genetic programming models and found that SVR-FFA model outperformed other models in forecasting rainfall more accurately in the desired stations. Hussain and Khan (2020) explored potential of MLP, SVR, and random forest (RF) to forecast river flow in Hunza region of Pakistan utilizing in situ dataset. Results showed that RF gave best test results compared to MLP and SVR and proved useful to forecast river flow with great accuracy. Shamseldin (2010) explored applicability of ANN to forecast Blue Nile river flows in Sudan. Results showed that ANN has substantial ability to forecast river flow in emerging countries. Yaseen et al. (2018) proposed a novel hybridized model called ANFIS-FFA to forecast monthly precipitation of Pahang River situated in Malaysia. Projected conjoint model is equated with regular ANFIS model, and results demonstrated that the hybrid model is ascertained to be a prudent modelling approach to simulate monthly precipitation in proposed study site. Zhou et al. (2019) applied ANN, SVM, ANFIS, hybridized ANFIS-FFA, and ANFIS-GA for predicting particle size distribution of muck piling after blast at different timescales in two pit mines located in Iran. On the basis of evaluation criteria, both ANFIS-FFA and ANFIS-GA performed agreeably, but ANFIS-GA model slightly gave better prediction results, making it a potential prediction model. Soft computing is a widely used technique for prediction of flood at a gauge station (Sahoo et al., 2019; Sahoo et al., 2020a, b). Shafiei et al. (2020) used ANFIS and hybrid ANFIS-FFA techniques for modelling discharge coefficient of labyrinth weirs and employed Monte Carlo simulations for enhancing capability of proposed models. ANFIS models were compared with that of computational fluid dynamics model and comparison results revealed that ANFIS-FFA conjoint model is significantly more accurate. Riahi-Madvar et al. (2020) proposed an ANFIS model hybridized with FFA for predicting contaminant dispersal coefficient in river beds using Subcategory Assortment by Maximum Distinction. On basis of results, projected amalgam model exhibited substantial developments than standard ANFIS in dispersion coefficient prediction. Tien-Bui et al. (2018) projected three novel amalgam models integrating ANFIS with bees (ANFIS-BA), invasive weed optimization (ANFIS-IWO), and cultural (ANFIS-CA) to map flood-susceptible areas in Haraz catchment, Iran. Results revealed that ANFIS-BA showed superior prediction ability followed by ANFIS-IWO and ANFIS-CA. Termeh et al. (2018) integrated ANFIS with GA, ant colony optimization (ACO), and particle swarm optimization (PSO) for flood vulnerability mapping over Jahrom Township in Fars Province, Iran, and compared their accurateness. ANFIS-PSO performed best as most practical model to produce extremely engrossed flood-prone maps.

The objective of this study is to utilize hybrid ANN modelling technique for flood forecasting at BP ghat and Dholai gauge stations of Barak River flowing through the Barak valley region in Assam, India. A comprehensive study is carried out to compare relative performances of ANN, ANFIS, and ANFIS-FFA models at proposed study area. Findings of present research could be useful for local and national governing bodies to plan for future and develop suitable new infrastructure to protect the lives and property.

11.2 STUDY AREA

Barak River basin in Barak valley is located in southern region of Assam. Barak is second major river in North East India falling amid 24°8′–25°8′N latitudes and 92°15′–93°15′E longitudes. Barak valley is named after River Barak. The river originates from the Manipur Hills named as the 'Mukru River'. After passing through various northeastern states of India, i.e. Manipur, Meghalaya, Mizoram, and Assam, it reaches Bangladesh. Ultimately the river drains into Bay of Bengal at Bangladesh. Of around 900 km total length, it navigates around 532 km in India with 129 km in Assam. Total basin size is 52,000 km^2 of which 41,723 km^2 lies in India. The study area map of the Barak valley is presented in Figure 11.1, depicting the two proposed gauge stations. The valley receives mean annual precipitation of 2500–4000 mm, with 80–85% from mid of April to mid of October.

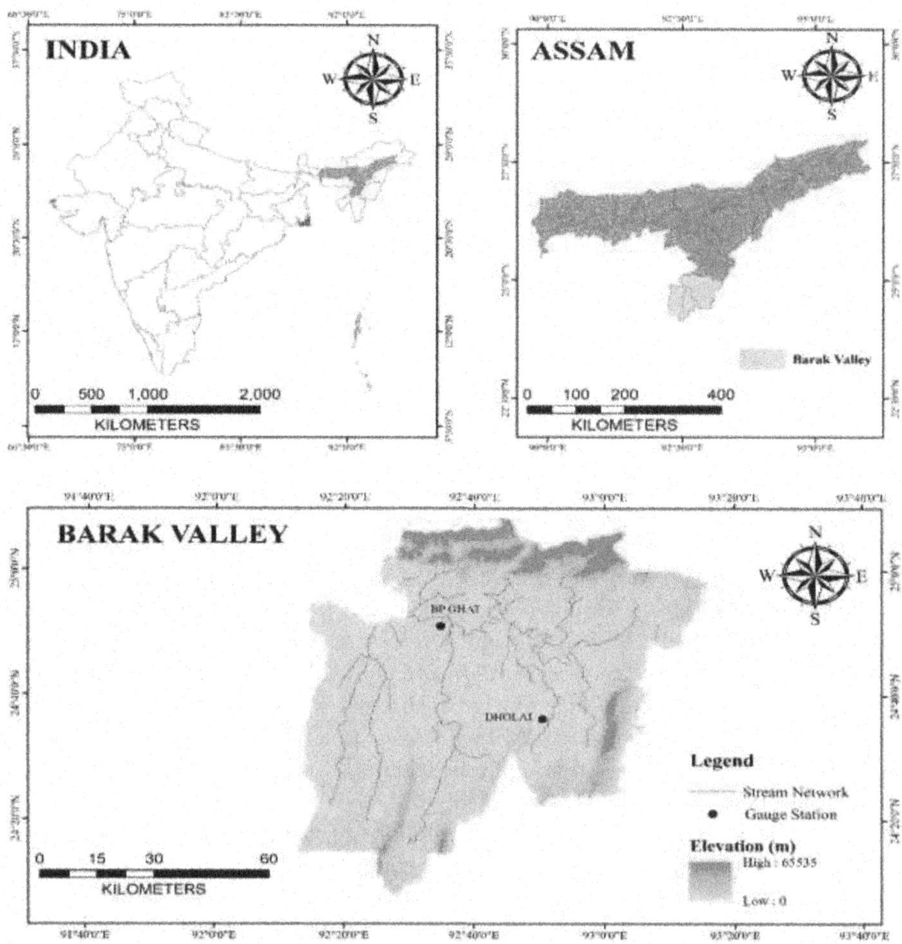

FIGURE 11.1 Study area.

Almost 70–75% of entire inhabitants of Barak valley is dependent on agricultural practices and effect of recurring flood exposes socio-economic development of the area. Assam is highly humid and thus receives heavy precipitation due to its 'tropical monsoon rainforest climate'.

11.3 METHODOLOGY

11.3.1 FFBPNN

FFBPNN is one of the most popular ANN models for engineering applications (Haykin, 2007). A multilayer FFNN comprises input, hidden, and output layers. Artificial nodes are organized in different layers, and every neuron of each layer has links to all neurons in subsequent layer. Related with each link amid these neurons, a weight value is demarcated for representing connected weight. Input signal transmits through network in forward course, layer by layer. For determining weights of connection and connection patterns, BP algorithm is utilized in present research for training the multilayer FFNN (Samantaray and Sahoo, 2020a, c). Typical architecture of FFBPNN is shown in Figure 11.2. BPNN approximates non-linear relationship amid input and output by regulating weight values within. In training procedure, for minimizing errors amid actual output data and target data, BP algorithm utilizes gradient descent technique for modifying arbitrarily nominated weights of nodes. Training procedure halts after minimization of errors or after alternative ending criteria is encountered. After that point, amalgamation of weights is taken as an answer to learning problem.

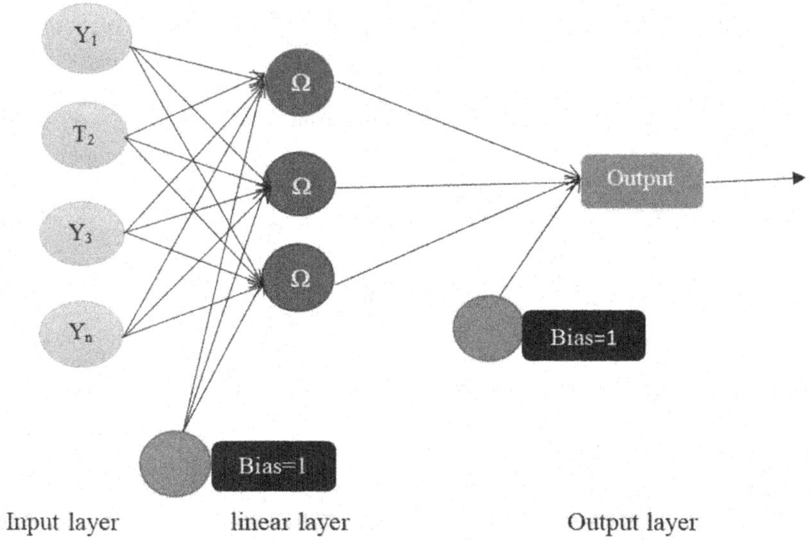

FIGURE 11.2 Architecture of FFBPNN.

BP algorithm is defined in subsequent equations (Haykin, 1999; Elminir et al., 2007). As soon as input vector ($x_1, x_2, \ldots, x_{NI}$) is given to input layer, an output from each neuron of hidden layer is specified by

$$h_j = g\left(\sum_{i=1}^{NI} w_{ji} x_i + b_j\right) \tag{11.1}$$

where $g(.)$ signifies hidden layer activation function, b_j is the bias of j-neuron, w_{ji} is the weight allotted to input i by neuron j, and h_j is the output from neuron j. Network output $\widehat{y_k}$ is specified by

$$\widehat{y_k} = f\left(\sum_{j=1}^{NJ} w_{kj} h_j + b_k\right) \tag{11.2}$$

where $f(\cdot)$ signifies output layer activation function and NJ is the number of neurons in hidden layer. For all entries, computation procedure is repetitive and an output vector $\widehat{y_k}$ is yielded. Training procedure comprises weight modifications for minimizing error amid actual and network's desired output utilizing an iterative process.

11.3.2 ANFIS

In recent years, ANN being a commanding method to model several physical world complications suffers its own inadequacies. In case input data are equivocal or subjected to comparatively high ambiguity, a fuzzy system, namely ANFIS, works as an enhanced choice (Moghaddamnia et al., 2009). Jang (1993) was the first to propose ANFIS technique and used its codes effectively to several real-world difficulties (Lin and Lee, 1995). Present arrangement is a fuzzy Sugeno by forward network architecture. For enabling adaptation, this kind of structures are characteristically urbanized and positioned in outline of a NN model (Gulley and Jang, 1995). Architecture of ANFIS network consisting of binary inputs, single output, and binary rules are shown in Figure 11.3. ANFIS comprises dual inputs x and y and single output and its rule is presented below:

$$\text{If } x \text{ is } A_1 \text{ and } y \text{ is } B_1, \text{ then } f = p_1 x + q_1 y + r_1 \tag{11.3}$$

$$\text{If } x \text{ is } A_2 \text{ and } y \text{ is } B_2, \text{ then } f = p_2 x + q_2 y + r_2 \tag{11.4}$$

A_i and B_i are a set of fuzzy, f_i is the output in fuzzy section indicated by fuzzy regulation, and p_i, q_i, and r_i are project constraints which are found in the course of training procedure. Each node is a fuzzy set of concerned level and any output of any node resembles to associated degree of input variable in the concerned level (Mohanta et al., 2020b; Samantaray et al., 2020a, b). Shape constraints regulate

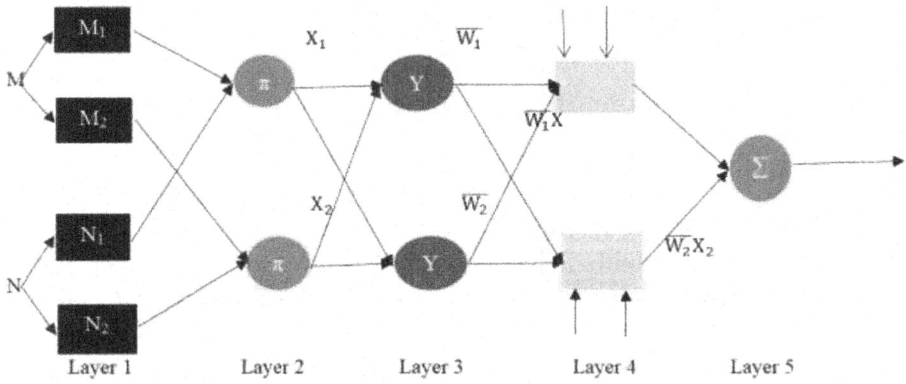

FIGURE 11.3 Architecture of ANFIS.

outline of membership function (MF) of fuzzy set in present level of network (Zadeh, 1965). Grade of certainty of a fuzzy set is signified by MF, ranging from 0 to 1. Final MFs attained subsequently to every regulation's output are alike in form to primary MFs, but vary in values.

11.3.3 FFA

In recent years, nature is observed to be a basis of stimulation for generating numerous influential optimization methods (Fister et al., 2013; Yang & He, 2013). Among these effective methods, FFA is one which influenced considerable admiration in very short period. FFA is motivated by social flashing conducts of fireflies. Drawing probable prey and breeding associates are two major purposes of these flashings. Three elementary assumptions are vital for expansion of FFA characteristics. Firstly, every firefly is unisexual. Second, a brighter one draws less bright firefly and arbitrary activities are deliberated for other fireflies if there is no luminous firefly in comparison to a particular one. Thirdly, problem related to maximization can be abridged considering brightness as an objective function value. In this methodology, light intensity deviation and formulating attractiveness are two major concerns. Equations 11.5 and 11.6 denote strength and desirability features of fireflies as every firefly display its exceptional attractiveness (β) that depicts its appealing ability inside swarm (Yang, 2010):

$$I = I_0 e^{-\gamma r^2} \tag{11.5}$$

$$w(r) = w_0 e^{-\gamma r^2} \tag{11.6}$$

where I and $w(r)$ signify intensity of light and attractiveness, correspondingly at a distance r from firefly. I_0 and w_0 are intensity of light and attractiveness at a distance $r = 0$ from firefly and y is the light immersion coefficient. Distance r amid any two fireflies (i and j) is articulated in subsequent formulation (Yang, 2010):

$$r_{ij} = \|x_i + x_j\| = \sqrt{\sum_{k=1}^{d} (x_{i,k} - x_{j,k})} \tag{11.7}$$

where x_i and x_j are locations of *i* and *j* fireflies in a Cartesian coordinate system.

11.3.3.1 Hybrid Model Based on ANFIS-FFA

FFA algorithm is being utilized more and more for resolving several multifaceted optimization problems (Samantaray et al., 2020d; Samantaray and Sahoo, 2020b). In present study, hybridization of ANFIS and FFA is used for flood prediction. For developing the hybrid model, a preliminary firefly population was arbitrarily produced for initiating primary iteration of ANFIS-FFA arrangement. This process was implemented which will help in mapping each firefly precisely onto ANFIS sets. Intensity of light of the fireflies was very critical feature essential for computing desirability of fireflies and comparing cooperatively. Selection of firefly was done on the basis of their illumination intensity, where fireflies with maximum illumination were selected and others were left to entice each other naturally. The duration of the training process was determined by either achievement of most number of iterations or when a suitable aptness function value was accomplished.

11.3.4 Data Collection and Model Performance

Altogether, 50 years (1970–2019) of discharge data were assembled from India Meteorological Department (IMD) Guwahati, Assam. Collected data are utilized for model development for studying effect of flood forecasting. The data collected from 1970 to 2004 (70% of dataset) are used for training and data from 2005 to 2019 (30%) are used for testing the network. Daily data are converted to monthly data that ultimately assists to train and test the model.

The following flood discharge arrangements are applied as input:

i. F_{t-1}
ii. F_{t-1}, F_{t-2}
iii. F_{t-1}, F_{t-2}, F_{t-3}
iv. F_{t-1}, F_{t-2}, F_{t-3}, F_{t-4}
v. F_{t-1}, F_{t-2}, F_{t-3}, F_{t-4}, F_{t-5}

11.4 RESULTS AND DISCUSSION

Performance of proposed models is assessed during training and testing periods. Table 11.1 indicates performance results of FFBPNN models based on the Tan-sig, Log-sig, Purelin membership function. The best MSE, RMSE, and WI values are found for monthly river flow forecasts with five lag time for FFBPNN algorithm. Result shows that Tan-sig function produces best value of performance for both stations. The WI, RMSE, and MSE values are 0.92715, 0.06943, 0.00623 for training phase and 0.90764, 0.02385, 0.00551 for testing phase, respectively, at Dholai gauge station.

TABLE 11.1
Results of FFBPNN

Station	Membership Function	MSE Training	MSE Testing	RMSE Training	RMSE Testing	WI Training	WI Testing
BP ghat	Purelin	0.00636	0.00566	0.07024	0.02605	0.91387	0.89083
	Log-sig	0.00634	0.00561	0.07011	0.02568	0.91962	0.89803
	Tan-sig	**0.00633**	**0.0056**	**0.07006**	**0.02521**	**0.92481**	**0.9008**
Dholai	Purelin	0.00628	0.00556	0.06994	0.02422	0.91537	0.89278
	Log-sig	0.00624	0.00552	0.06956	0.0241	0.91988	0.89781
	Tan-sig	**0.00623**	**0.00551**	**0.06943**	**0.02385**	**0.92715**	**0.90764**

For ANFIS algorithm, best performance measurement shows while monthly river flow forecasting on 5 months lag time are considered. The MSE, RMSE, and WI values are 0.00577, 0.06852, 0.94223 for training phase and 0.00519, 0.02195, 0.92342 for testing period, respectively for BP ghat gauge station. Correspondingly, for Dholai gauge station, the best values of WI, RMSE, and MSE are 0.94596, 0.06819, 0.00557 for training phase and 0.92691, 0.02158, 0.005 for the testing phase, respectively. Details of model performance indices for both the station are shown in Table 11.2.

Likewise, for ANFIS-FFA, best performance results are found for river flow forecasts on 5 months lag time. Here five membership functions (Tri, Trap, Gbell, Gauss, Pi) are considered for evaluating model performance. Among five functions, Gbell function gives best result for both the gauge stations. For BP ghat and Dholai gauge stations, the prominent values of WI, RMSE, and MSE are 0.97278, 0.05749, 0.00438 and 0.97568, 0.05547, 0.00401 in case of training phase, respectively. Detail performance score for training and testing periods are presented in Table 11.3.

To envisage model performance accurateness on observed data, actual river flow versus simulated river flow by FFBPNN, ANFIS, ANFIS-FFA models are compared and shown in Figures 11.4, 11.5, 11.6, 11.7, respectively. All models illustrate

TABLE 11.2
Results of ANFIS

Station	Membership Function	MSE Training	MSE Testing	RMSE Training	RMSE Testing	WI Training	WI Testing
BP ghat	Tri	0.00587	0.00525	0.06875	0.02236	0.94042	0.92309
	Trap	0.00583	0.00524	0.06863	0.02214	0.94137	0.92394
	Gbell	**0.00577**	**0.00519**	**0.06852**	**0.02195**	**0.94223**	**0.92342**
	Gauss	0.0058	0.00521	0.0686	0.02207	0.9417	0.92443
	Pi	0.00578	0.0052	0.06856	0.02203	0.94208	0.92477
Dholai	Tri	0.00568	0.00513	0.06836	0.02174	0.94406	0.92706
	Trap	0.00565	0.00511	0.0683	0.02166	0.9447	0.9276
	Gbell	**0.00557**	**0.005**	**0.06819**	**0.02158**	**0.94596**	**0.92691**
	Gauss	0.00563	0.00509	0.06827	0.02166	0.94492	0.92784
	Pi	0.0056	0.00504	0.06823	0.02162	0.94534	0.92807

TABLE 11.3
Results of ANFIS-FFA

Station	Membership Function	MSE		RMSE		WI	
		Training	Testing	Training	Testing	Training	Testing
BP ghat	Tri	0.00443	0.00302	0.05859	0.01878	0.96863	0.94841
	Trap	0.00439	0.00248	0.05827	0.01837	0.96915	0.94999
	Gbell	**0.00438**	**0.00241**	**0.05749**	**0.01768**	**0.97278**	**0.95263**
	Gauss	0.00436	0.00246	0.05804	0.01812	0.97051	0.95034
	Pi	0.00434	0.00243	0.05782	0.01778	0.97183	0.95157
Dholai	Tri	0.00417	0.00227	0.05642	0.01642	0.9713	0.95342
	Trap	0.0041	0.00221	0.05599	0.016	0.97284	0.9549
	Gbell	**0.00401**	**0.00224**	**0.05547**	**0.01498**	**0.97568**	**0.95739**
	Gauss	0.00408	0.00229	0.05582	0.01573	0.97307	0.95527
	Pi	0.00405	0.00226	0.05565	0.01526	0.97423	0.95652

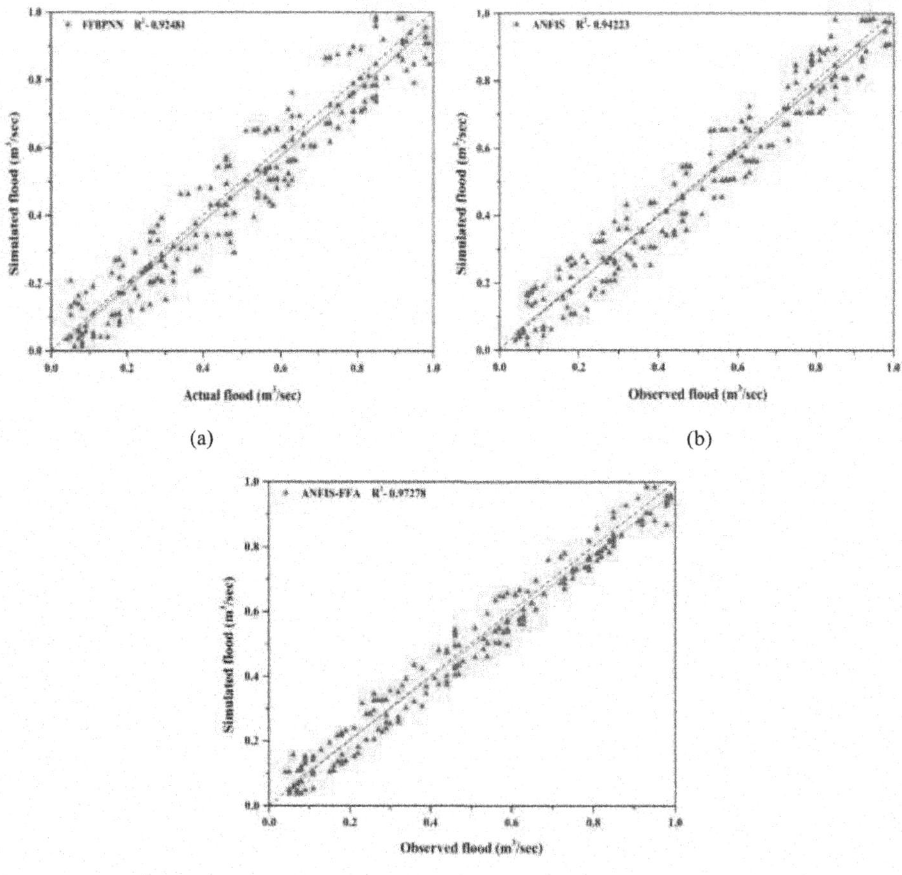

FIGURE 11.4 Observed versus simulated flood at BP ghat gauge station using (a) FFBPNN, (b) ANFIS, and (c) ANFIS-FFA for training phase.

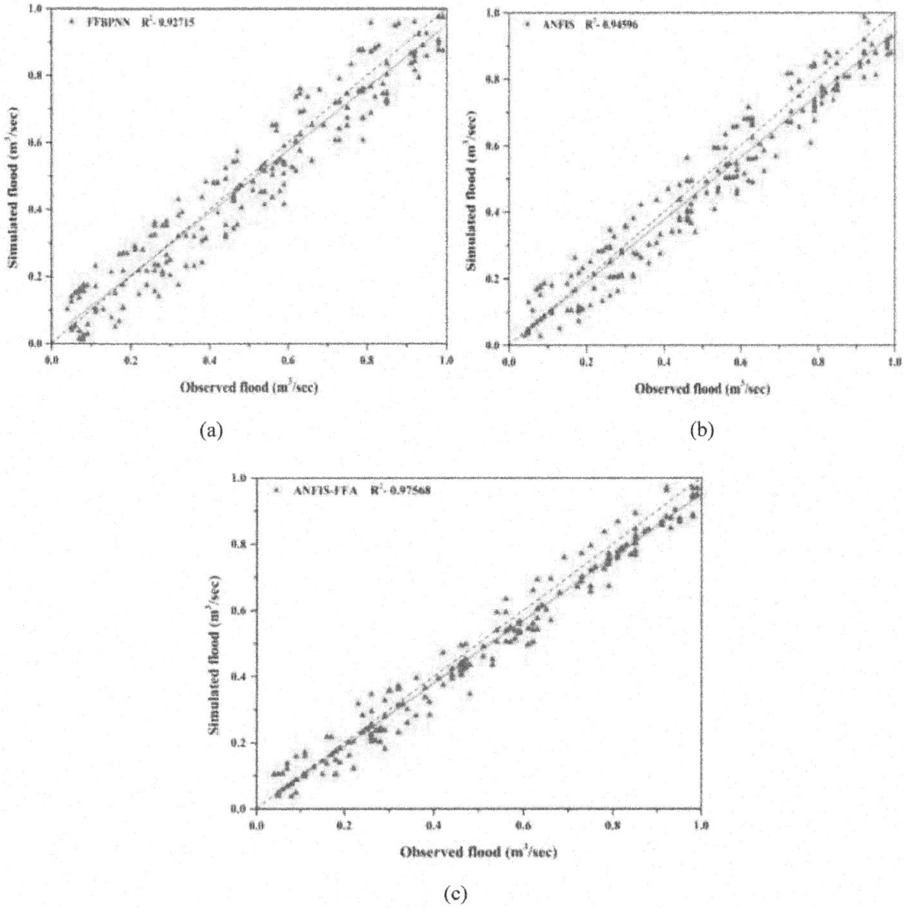

FIGURE 11.5 Observed versus simulated flood at Dholai gauge station using (a) FFBPNN, (b) ANFIS, and (c) ANFIS-FFA for training phase.

generally good agreement with the actual and simulated river flow in study area; however, certain peak flow data are not apprehended very well. Moreover, it is observed that FFBPNN, ANFIS, and ANFIS-FFA model performance during training phase is better than that of testing period performance. Overall, it can be concluded that performance of ANFIS-FFA is better compared to FFBPNN and ANFIS on the basis of defined performance measures.

Large deviance from ideal line demonstrates less substantial prediction accurateness. Scatter plot of ANFIS and ANFIS-FFA models fall nearer to ideal line for actual and predicted values having high coefficient of determinant. On the other hand, FFBPN model illustrates a slight deviance from the ideal line in comparison to other applied models for both training and testing periods. Obtained results evidently indicate that ANFIS-FFA model predicts river flow forecasts more accurately as

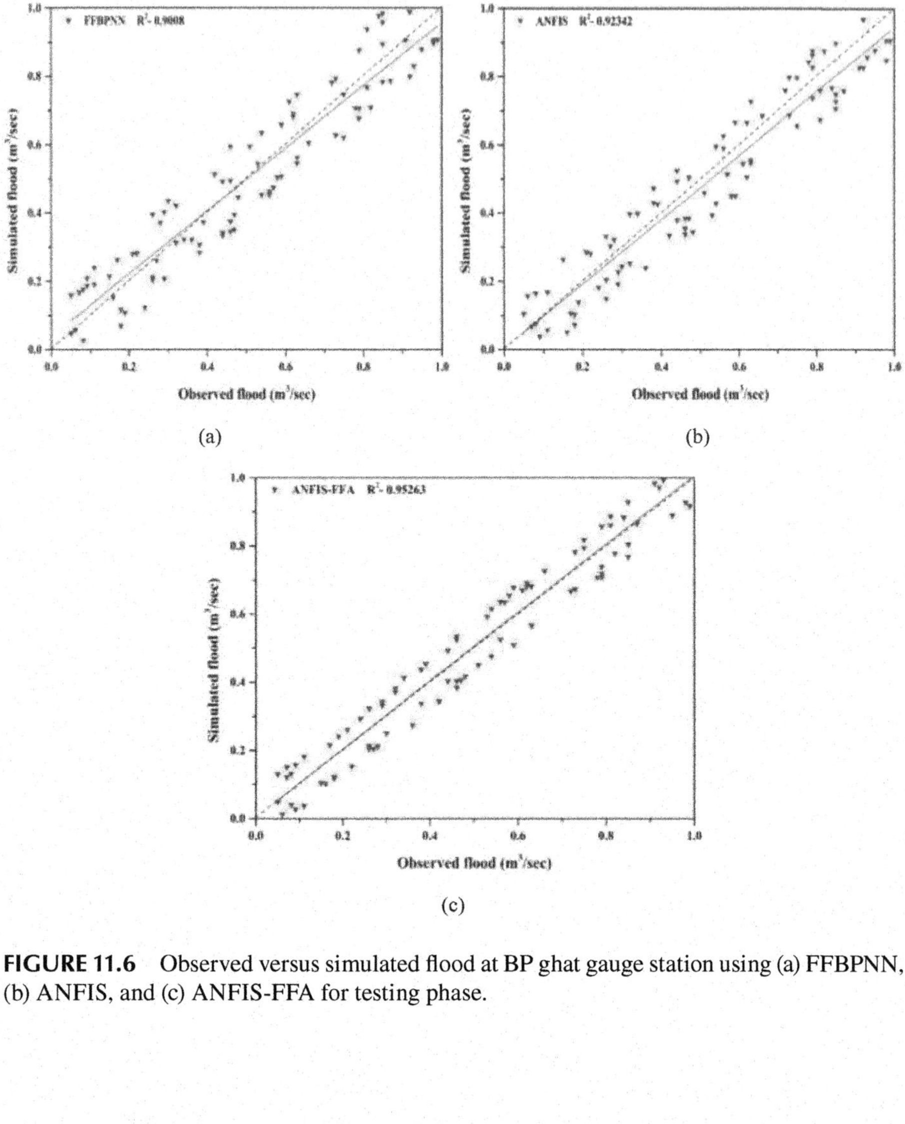

FIGURE 11.6 Observed versus simulated flood at BP ghat gauge station using (a) FFBPNN, (b) ANFIS, and (c) ANFIS-FFA for testing phase.

compared to ANFIS and FFBPNN models. This is because of the fact that ANFIS-FFA model was able to determine non-linear relationship amid input and output values extraordinarily well.

For further analysis of predictive performance, comparison plots are prepared to forecast monthly river flow assessed by FFBPNN, ANFIS, and ANFIS-FFA models for training and testing periods, as shown in Figures 11.8 and 11.9, respectively. Results illustrate that estimated peak flood is 4658.282 m^3/s, 4759.706 m^3/s, and 4913.128 m^3/s for FFBPNN, ANFIS, and ANFIS-FFA, respectively, against actual peak 5148.411 m^3/s for the station BP ghat. Similarly, for Dholai gauging station, tangible flood is 4984.983 m^3/s aligned with predicted

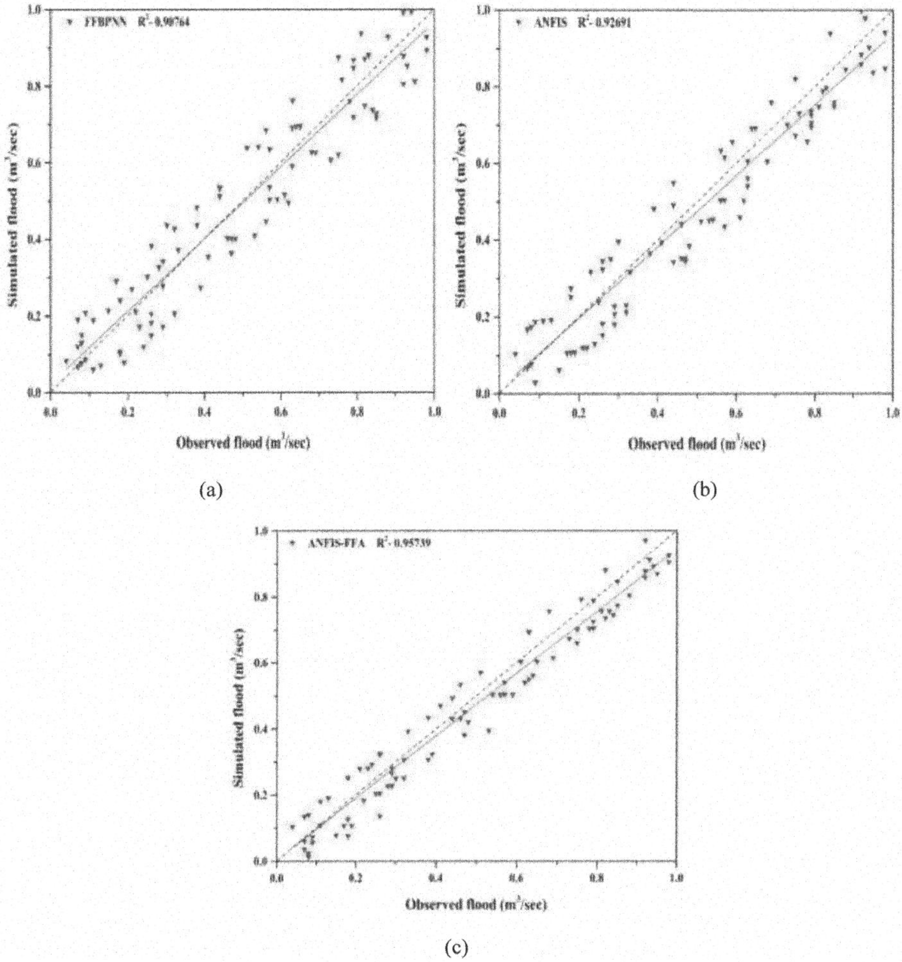

FIGURE 11.7 Observed versus simulated flood at Dholai gauge station using (a) FFBPNN, (b) ANFIS, and (c) ANFIS-FFA for testing phase.

flood 4504.431 m^3/s, 4600.641 m^3/s, and 4751.187 m^3/s for FFBPNN, ANFIS, and ANFIS-FFA, respectively.

11.4.1 Comparison of Preeminent Outcomes

MSE, RMSE, and WI indicators are utilized for evaluating performance of FFBPNN, ANFIS, and ANFIS-FFA models for three gauging sites. Assessment based on performance evaluation measures are quantified in Table 11.4 illustrating efficacy of each model. Assessing flood conditions are vital and hence approaches utilized in this study are important to demonstrate information regarding flood events. It is obvious that ANFIS-FFA model performed well followed by ANFIS and FFBPNN.

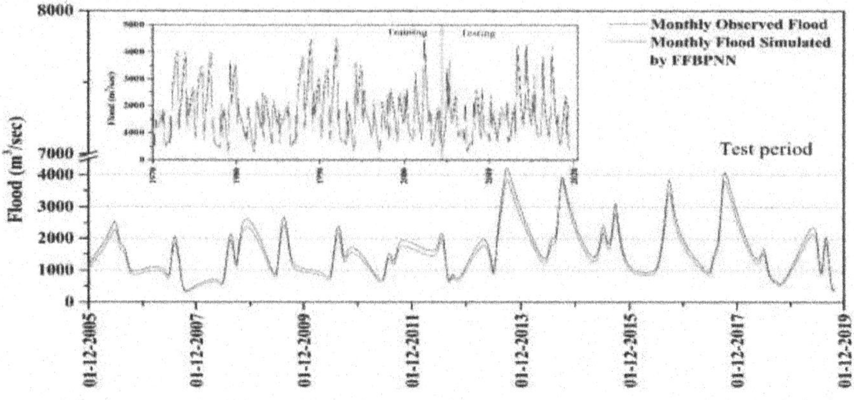

(a)

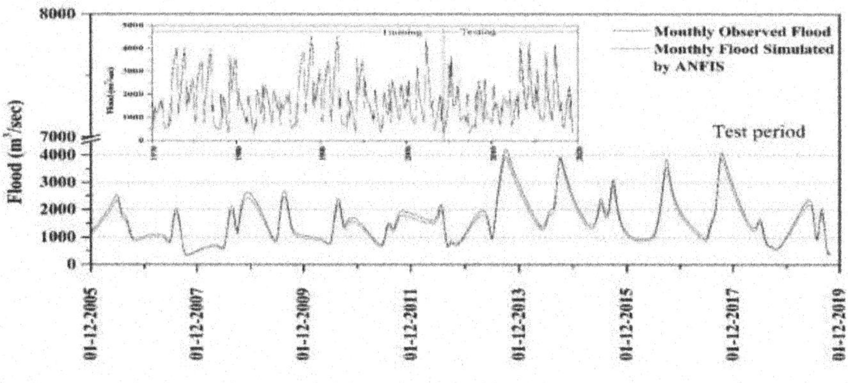

(b)

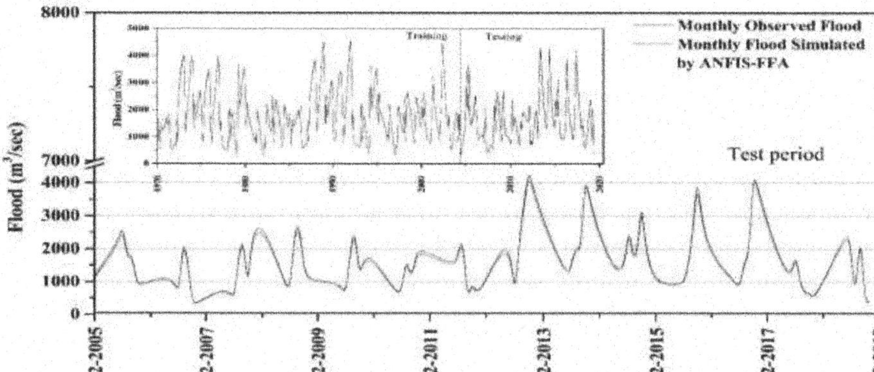

(c)

FIGURE 11.8 Comparison of actual versus predicted flood by (a) FFBPNN, (b) ANFIS, and (c) ANFIS-FFA approaches at BP ghat gauge station.

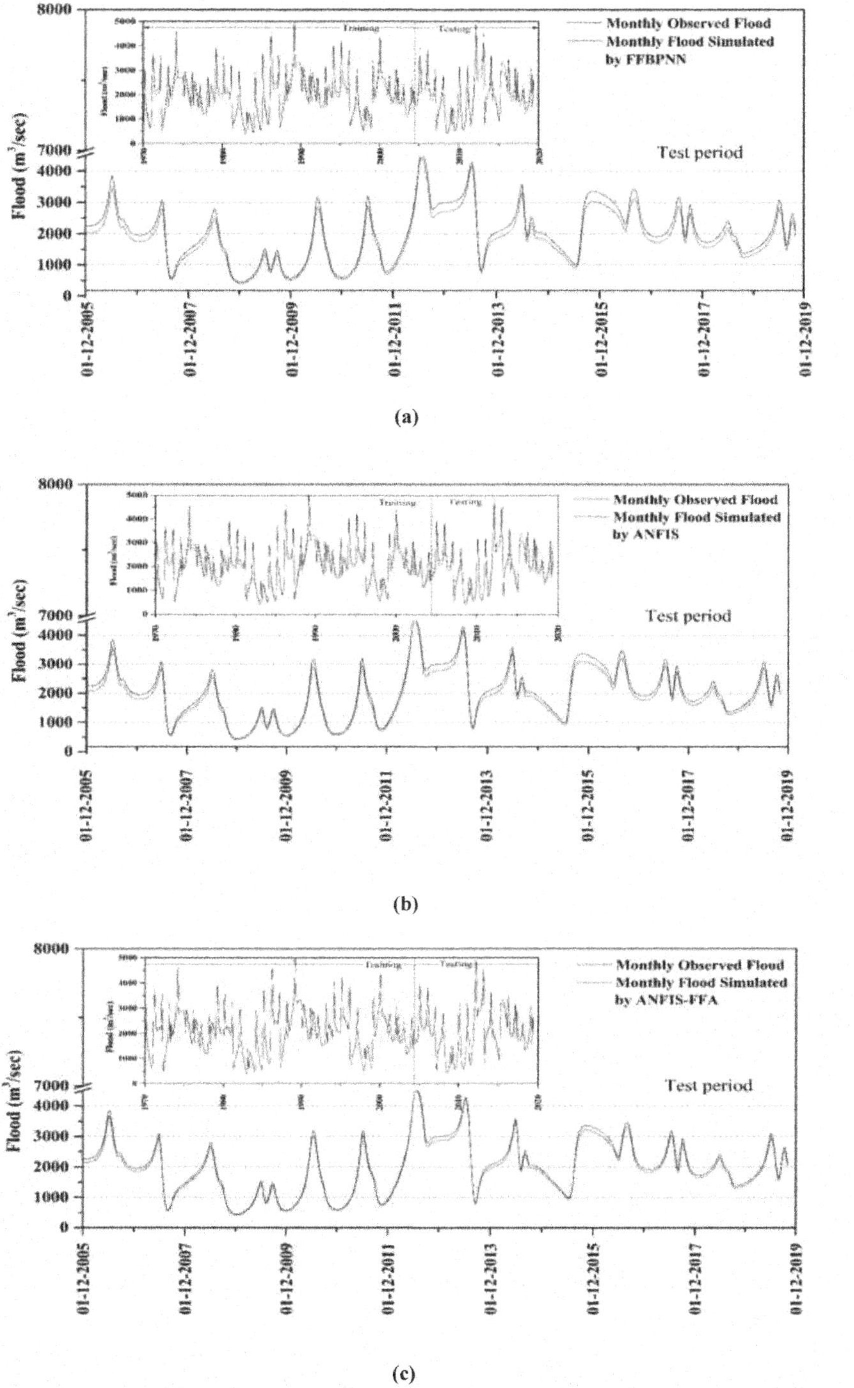

FIGURE 11.9 Comparison of actual versus predicted flood by (a) FFBPNN, (b) ANFIS, and (c) ANFIS-FFA approaches at Dholai gauge station.

TABLE 11.4
Performance Measures of Different Methods for Flood Prediction

		MSE		RMSE		WI	
Station	**Techniques**	**Training**	**Testing**	**Training**	**Testing**	**Training**	**Testing**
BP ghat	FFBPNN	0.00633	0.0056	0.07006	0.02521	0.92481	0.9008
	ANFIS	0.00577	0.00519	0.06852	0.02195	0.94223	0.92342
	ANFIS-FFA	0.00438	0.00241	0.05749	0.01768	0.97278	0.95263
Dholai	FFBPNN	0.00623	0.00551	0.06943	0.02385	0.92715	0.90764
	ANFIS	0.00557	0.005	0.06819	0.02158	0.94596	0.92691
	ANFIS-FFA	0.00401	0.00224	0.05547	0.01498	0.97568	0.95739

11.4.2 Discussion

A direct assessment of three data-driven techniques (FFBPNN, ANFIS, and ANFIS-FFA) is conducted for assessing their performance on the basis of distinct performance evaluation indices. Each technique is examined for river flow observations with different time lag as modelling input for training and testing periods. Generally, model accurateness calculations are conducted on the basis of error deviation amid actual and predicted data, yet a number of investigations utilized different indices for performance evaluation. In present research, MSE, RMSE, and WI are utilized as performance measurement indices for evaluating enactment of proposed models. Based on different evaluation indices, our outcomes reveal that ANFIS-FFA model outperformed ANFIS and FFBPNN models.

11.5 CONCLUSION

Precise flood water level prediction is important as an early cautionary system for public informing them regarding potential incoming flood calamity. ANNs offer fast and malleable ways to generate models for forecasting flood events. This study evaluates the capabilities of FFBPNN, ANFIS, and ANFIS-FFA models for flood prediction. ANFIS-FFA model was obtained by combining two methods ANFIS and FFA. Developed hybrid model was tested employing different input arrangements of monthly discharge data for different time series of BP ghat and Dholai stations located in Barak valley of Assam and were compared to standard ANFIS and SVM models. Based on the evaluation criteria conducted using different statistical parameters, ANFIS-FFA model predicted flood conditions with maximum accuracy than ANFIS and FFBPNN models for identical input dataset. ANFIS and SVM models perform similar in few conditions, although ANFIS produced better prediction results than FFBPNN in most conditions. It can be concluded that FFBPNN, ANFIS, and ANFIS-FFA models can effectively predict flood occurrences happening at proposed area of study.

REFERENCES

Arulmurugan, R. and H. Anandakumar. 2018. "Early detection of lung cancer using wavelet feature descriptor and feed forward back propagation neural networks classifier." In Computational Vision and Bio Inspired Computing. 103–110, Springer.

Chang, Fi-John, Yen-Ming Chiang, and Li-Chiu Chang. 2007. "Multi-step-ahead neural networks for flood forecasting." *Hydrological Sciences Journal*, 52 (1), 114–130.

Chau, K. W., C. L. Wu, and Y. S. Li. 2005. "Comparison of several flood forecasting models in Yangtze river." *Journal of Hydrologic Engineering*, 10 (6), 485–491.

Che, Zhen-Guo, Tzu-An Chiang, and Zhen-Hua Che. 2011. "Feed-forward neural networks training: a comparison between genetic algorithm and back-propagation learning algorithm." *International Journal of Innovative Computing, Information and Control*, 7 (10), 5839–5850.

Dar, A. Q., Ruhhee Tabbassum, and Shahzad Faisal. 2015. "Using artificial neural network for real-time flood prediction in river Jhelum, J&K, India". *Elixir Civil Engineering*, 82, 32710–32713.

Dastorani, Mohammad T., Alireza Moghadamnia, Jamshid Piri, and Miguel Rico-Ramirez. 2010. "Application of ANN and ANFIS models for reconstructing missing flow data." *Environmental Monitoring and Assessment*, 166 (1–4), 421–434.

Elminir, Hamdy K., Yosry A. Azzam, and Farag I. Younes. 2007. "Prediction of hourly and daily diffuse fraction using neural network, as compared to linear regression models." *Energy*, 32 (8), 1513–1523.

Fister, Iztok, Iztok Fister Jr., Xin-She Yang, and Janez Brest. 2013. "A comprehensive review of firefly algorithms." *Swarm and Evolutionary Computation*, 13, 34–46.

Gong, Yicheng, Yongxiang Zhang, Shuangshuang Lan, and Huan Wang. 2016. "A comparative study of artificial neural networks, support vector machines and adaptive neuro fuzzy inference system for forecasting groundwater levels near lake Okeechobee, Florida." *Water Resources Management*, 30 (1), 375–391.

Gulley, Ned and J. S. Roger Jang. 1995. Fuzzy Logic Toolbox for Use with MATLAB, The Mathworks Inc., Natick, MA.

Haykin, S. 1999. Neural Networks: A Comprehensive Foundation, 2nd ed., Prentice-Hall Inc.

Haykin, S. 2007. Neural Networks: A Comprehensive Foundation, 3rd ed., Prentice Hall Inc., Upper Saddle River, NJ.

Hussain, Dostdar and Aftab Ahmed Khan. 2020. "Machine learning techniques for monthly river flow forecasting of Hunza River, Pakistan." Earth Science Informatics, 13, 939–949.

Jang, J-SR. 1993. "ANFIS: adaptive-network-based fuzzy inference system." *IEEE Transactions on Systems, Man, and Cybernetics*, 23 (3), 665–685.

Kisi, Ozgur, Alireza Moghaddam Nia, Mohsen Ghafari Gosheh, Mohammad Reza Jamalizadeh Tajabadi, and Azadeh Ahmadi. 2012. "Intermittent streamflow forecasting by using several data driven techniques." *Water Resources Management*, 26 (2), 457–474.

Konate, Ahmed Amara, Heping Pan, Nasir Khan, and Jie Huai Yang. 2015. "Generalized regression and feed-forward back propagation neural networks in modelling porosity from geophysical well logs." *Journal of Petroleum Exploration and Production Technology*, 5 (2), 157–166.

Lin, C. T. and C. S. G. Lee. 1995. Neural Fuzzy System, Prentice Hall, Englewood Cliffs, NJ.

Mehr, A. Danandeh, E. Kahya, A. Şahin, and M. J. Nazemosadat. 2015. "Successive-station monthly streamflow prediction using different artificial neural network algorithms." *International Journal of Environmental Science and Technology*, 12 (7), 2191–2200.

Mehr, A. Danandeh, Vahid Nourani, V. Karimi Khosrowshahi, and Moahmmad Ali Ghorbani. 2019. "A hybrid support vector regression–firefly model for monthly rainfall forecasting." *International Journal of Environmental Science and Technology*, 16 (1), 335–346.

Mehr, Ali Danandeh, Ercan Kahya, and Cahit Yerdelen. 2014. "Linear genetic programming application for successive-station monthly streamflow prediction." *Computers & Geosciences*, 70, 63–72.

Moghaddamnia, A., M. Ghafari Gousheh, J. Piri, S. Amin, and DJAiWR Han. 2009. "Evaporation estimation using artificial neural networks and adaptive neuro-fuzzy inference system techniques." *Advances in Water Resources*, 32 (1), 88–97.

Mohanta, N. R., N. Patel, K. Beck, S. Samantaray, and Abinash Sahoo. 2020a. "Efficiency of river flow prediction in river using wavelet-CANFIS: a case study." In Intelligent Data Engineering and Analytics. 435–443, Springer, Singapore.

Mohanta, N. R., P. Biswal, S. S. Kumari, S. Samantaray, and Abinash Sahoo. 2020b. "Estimation of sediment load using adaptive neuro-fuzzy inference system at Indus River Basin, India." In Intelligent Data Engineering and Analytics. 427–434, Springer, Singapore.

Mukerji, Aditya, Chandranath Chatterjee, and Narendra Singh Raghuwanshi. 2009. "Flood forecasting using ANN, neuro-fuzzy, and neuro-GA models." *Journal of Hydrologic Engineering*, 14 (6), 647–652.

Nayak, P. C., K. P. Sudheer, D. M. Rangan, and K. S. Ramasastri. 2005. "Short-term flood forecasting with a neurofuzzy model." Water Resources Research. 41 (4).

Pandey, Alok and V. V. Srinivas. 2015. "Use of data driven techniques for short lead time streamflow forecasting in Mahanadi basin." *Aquatic Procedia*, 4, 972–978.

Rezaeianzadeh, M., Hossein Tabari, A. Arabi Yazdi, S. Isik, and L. Kalin. 2014. "Flood flow forecasting using ANN, ANFIS and regression models." *Neural Computing and Applications*, 25 (1), 25–37.

Riahi-Madvar, Hossien, Majid Dehghani, Kulwinder Singh Parmar, Narjes Nabipour, and Shahaboddin Shamshirband. 2020. "Improvements in the explicit estimation of pollutant dispersion coefficient in rivers by subset selection of maximum dissimilarity hybridized with ANFIS-firefly algorithm (FFA)." *IEEE Access*, 8, 60314–60337.

Sahoo, A., S. Samantaray, S. Bankuru, and D. K. Ghose. 2020a. "Prediction of flood using adaptive neuro-fuzzy inference systems: a case study." Smart Innovation, Systems and Technologies, 159. doi:10.1007/978-981-13-9282-5_70.

Sahoo, A., S. Samantaray, and D. K. Ghose. 2019. "Stream flow forecasting in Mahanadi River basin using artificial neural networks." In Procedia Computer Science, 157. doi:10.1016/j.procs.2019.08.154.

Sahoo, A., U. K. Singh, M. H. Kumar, and Sandeep Samantaray. 2020b. "Estimation of flood in a river basin through neural networks: a case study." In Communication Software and Networks. 755–763 Springer, Singapore.

Samantaray, S. and A. Sahoo. 2020a. "Estimation of runoff through BPNN and SVM in Agalpur watershed." Advances in Intelligent Systems and Computing, 1014. doi:10.1007/978-981-13-9920-6_27.

Samantaray, S. and A. Sahoo. 2020b. "Assessment of sediment concentration through RBNN and SVM-FFA in arid watershed, India." Smart Innovation, Systems and Technologies, 159. doi:10.1007/978-981-13-9282-5_67.

Samantaray, S. and A. Sahoo. 2020c. "Appraisal of runoff through BPNN, RNN, and RBFN in Tentulikhunti watershed: a case study." Advances in Intelligent Systems and Computing, 1014. doi:10.1007/978-981-13-9920-6_26.

Samantaray, S. and Sahoo, A. 2020d. "Prediction of runoff using BPNN, FFBPNN, CFBPNN algorithm in arid watershed: a case study." *International Journal of Knowledge-based and Intelligent Engineering Systems*, 24 (3), 243–251.

Samantaray, Sandeep and Dillip K. Ghose. 2019. "Sediment assessment for a watershed in arid region via neural networks." *Sādhanā*, 44 (10), 219.

Samantaray, Sandeep and Dillip K. Ghose. 2020. "Modelling runoff in a river basin, India: an integration for developing un-gauged catchment." *International Journal of Hydrology Science and Technology*, 10 (3), 248–266.

Samantaray, S., A. Sahoo, and D. K. Ghose. 2020a. "Prediction of sedimentation in an arid watershed using BPNN and ANFIS." Lecture Notes in Networks and Systems, 93. doi:10.1007/978-981-15-0630-7_29.

Samantaray, S., A. Sahoo, and D. K. Ghose. 2020b. "Infiltration loss affects toward groundwater fluctuation through CANFIS in arid watershed: a case study." Smart Innovation, Systems and Technologies, 159. doi:10.1007/978-981-13-9282-5_76.

Samantaray, S., O. Tripathy, A. Sahoo, and D. K. Ghose. 2020c. "Rainfall forecasting through ANN and SVM in Bolangir watershed, India." Smart Innovation, Systems and Technologies, 159. doi:10.1007/978-981-13-9282-5_74.

Samantaray, Sandeep, Abinash Sahoo, and Dillip K. Ghose. 2020d. "Assessment of sediment load concentration using SVM, SVM-FFA and PSR-SVM-FFA in arid watershed, India: a case study." *KSCE Journal of Civil Engineering*, 27, 1944–1957.

Samantaray, S., A. Sahoo, and D. K. Ghose. 2019a. "Assessment of runoff via precipitation using neural networks: watershed modelling for developing environment in arid region." *Pertanika Journal of Science and Technology*, 27 (4), 2245–2263.

Samantaray, Sandeep, A. Sahoo, and Dillip K. Ghose. 2019b. "Assessment of groundwater potential using neural network: a case study." In International Conference on Intelligent Computing and Communication. 655–664, Springer.

Sehgal, Vinit, Rajeev Ranjan Sahay, and Chandranath Chatterjee. 2014. "Effect of utilization of discrete wavelet components on flood forecasting performance of wavelet based ANFIS models." *Water Resources Management*, 28 (6), 1733–1749.

Shafiei, Shahabodin, Mohsen Najarchi, and Saeid Shabanlou. 2020. "A novel approach using CFD and neuro-fuzzy-firefly algorithm in predicting labyrinth weir discharge coefficient." *Journal of the Brazilian Society of Mechanical Sciences and Engineering*, 42 (1), 44.

Shamseldin, Asaad Y. 2010. "Artificial neural network model for river flow forecasting in a developing country." *Journal of Hydroinformatics*, 12 (1), 22–35.

Shiau, Jenq-Tzong and Hui-Ting Hsu. 2016. "Suitability of ANN-based daily streamflow extension models: a case study of Gaoping River basin, Taiwan." *Water Resources Management*, 30 (4), 1499–1513.

Shu, C. and T. B. M. J. Ouarda. 2008. "Regional flood frequency analysis at ungauged sites using the adaptive neuro-fuzzy inference system." *Journal of Hydrology*, 349 (1–2), 31–43.

Singh, S. R. 2007. "A robust method of forecasting based on fuzzy time series." *Appl. Math. Comput.*, 188(1), 472–484.

Sudheer, K. P., A. K. Gosain, and K. S. Ramasastri. 2002. "A data-driven algorithm for constructing artificial neural network rainfall-runoff models." *Hydrological Processes*, 16 (6), 1325–1330.

Syahrullah, La Ode Ichlas and Nazaruddin Sinaga. 2016. "Optimization and prediction of motorcycle injection system performance with feed-forward back-propagation method artificial neural network (ANN)." *American Journal of Engineering and Applied Science*, 9 (2), 222–235.

Termeh, Seyed Vahid Razavi, Aiding Kornejady, Hamid Reza Pourghasemi, and Saskia Keesstra. 2018. "Flood susceptibility mapping using novel ensembles of adaptive neuro fuzzy inference system and metaheuristic algorithms." *Science of the Total Environment*, 615, 438–451.

Tien Bui, Dieu, Khabat Khosravi, Shaojun Li, Himan Shahabi, Mahdi Panahi, Vijay P. Singh, Kamran Chapi, Ataollah Shirzadi, Somayeh Panahi, and Wei Chen. 2018. "New hybrids of ANFIS with several optimization algorithms for flood susceptibility modeling." *Water*, 10 (9), 1210.

Wei, Chih-Chiang and Nien-Sheng Hsu. 2008. "Multireservoir flood-control optimization with neural-based linear channel level routing under tidal effects." *Water Resources Management*, 22 (11), 1625–1647.

Yang, Xin-She. 2010. "Firefly algorithm, stochastic test functions and design optimisation." *International Journal of Bio-Inspired Computing*, 2 (2), 78–84.

Yang, Xin-She and Xingshi He. 2013. "Firefly algorithm: recent advances and applications." *International Journal of Swarm Intelligence*, 1 (1), 36–50.

Yaseen, Zaher Mundher, Mazen Ismaeel Ghareb, Isa Ebtehaj, Hossein Bonakdari, Ridwan Siddique, Salim Heddam, Ali A Yusif, and Ravinesh Deo. 2018. "Rainfall pattern forecasting using novel hybrid intelligent model based ANFIS-FFA." *Water Resources Management*, 32 (1), 105–122.

Zadeh, Lotfi A. 1965. "Fuzzy sets." *Information and Control*, 8 (3), 338–353.

Zhou, Jian, Chuanqi Li, Chelang A. Arslan, Mahdi Hasanipanah, and Hassan Bakhshandeh Amnieh. 2019. "Performance evaluation of hybrid FFA-ANFIS and GA-ANFIS models to predict particle size distribution of a muck-pile after blasting." *Engineering with Computers*, 37, 265–274.

12 Prophecy of Sediment Load Using Hybrid AI Approaches at Various Gauge Station in Mahanadi River Basin, India

12.1 INTRODUCTION

River flow transports enormous amount of sediment concentration either in suspended form or as bed load. Yet, sediment load quantification is significant to design and manoeuvre many environmental engineering, water resources, and hydropower projects. Transport of suspended sediment have key impacts on water resources engineering (WRE) as emphasized earlier (Qin et al., 2007; Mohamed and Ouillon, 2007). Approaches to estimate sediment load on the basis of flow properties and sediment have restrictions accredited to generalization of significant constraints and boundary conditions. In circumstances like these as mentioned above, soft computing techniques have been established as a proficient tool to model sediment concentration. ANN is an evolving modelling method that has capability of pattern recognition, self-adaptation, and apprehending non-linear multifaceted behaviour amid input and output constraints. In recent years, ANNs have gained approval in several engineering fields and has also been effectively utilized in WRE to predict different water resources variables (Fernando and Shamseldin, 2009; Kisi et al., 2012), for example runoff (Ju et al., 2009; Evsukoff et al., 2011) and sediment load (Alp and Cigizoglu, 2007; Kisi et al., 2009; Guven and Kisi, 2011). Neural network is a successive tool towards prediction of various climatic indexes at gauge station (Jimmy et al., 2020; Sahoo et al., 2019; Samantaray and Sahoo, 2020c; Samantaray et al., 2019a, 2019b; Samantaray et al., 2020a).

ANNs are well-known for recognizing fundamental behaviour amid variables even if data is raucous and comprise certain errors. These abilities endorse application of ANN to predict sediment concentration in river basins. ANNs generally produce outcomes quicker compared to its physical equivalents and also more precisely, but only in range of values witnessed in data utilized to construct models (Nourani et al., 2009). Melesse et al. (2011) used MLP for predicting daily and weekly SSC at

Mississippi, Missouri, and Rio Grande river systems in the United States and compared the obtained results with autoregressive integrated moving average (ARIMA), multiple non-linear regression (MNLR), and multiple linear regressions (MLR). Based on performance indices, it was observed that MLB produced better and reliable outcomes than other proposed models in all the three rivers. Yadav et al. (2018) evaluated performance of MLP technique for predicting suspended sediment yield at Tikarapara gauging site of River Mahanadi, India. Obtained results were compared with sediment rating curve (SRC) and MLR and observed that MLP performed best with better accuracy followed by MLR and SRC. Soleymani et al. (2016) proposed hybridization of RBF with FFA (RBF-FFA) for predicting water level of Selangor River, Malaysia. Prediction accuracy of proposed hybrid model was compared with support vector machine (SVM) and MLP. Obtained results showed that developed RBF-FFA model provided more accurate and consistent prediction compared to other models. Prediction of SSC in proposed watershed is done through various machine learning approaches (Mohanta et al., 2020; Samantaray and Ghose, 2018; Ghose and Samantaray, 2018; Samantaray and Sahoo, 2020a; Samantaray et al., 2020c). Kumar et al. (2012) used ANN, RBF, fuzzy logic, M5, and REPTree for predicting SSC at Kasol gauge station, located on Sutlej River in north India. Results indicated that M5 model performed well and offered categorical expressions helpful for site engineers compared to other soft computing techniques investigated in this study.

Kumar et al. (2016) employed ANN, RBFN, least-square SVM, classification and regression tree (CART), and M5 to predict SSC at River Kopili in India. Overall performance of models exhibited that all studied models simulated SSC of Kopili River basin adequately. Mustafa and Isa (2014) investigated the applicability of RBF and MLP techniques to predict SSC in Pari River, Perak, Malaysia. Results based on statistical parameters showed that performances of both RBF and MLP models were close to each other. Yet, RBFN revealed certain irregularity during time series data prediction. Memarian and Balasundram (2012) explored the potential of ANN, RBFN, and MLP models to predict SSC at Hulu Langat catchment, Malaysia. Findings from the study indicated that MLP showed a slightly better output than RBFN and ANN models to predict SSC. Malik et al. (2017) applied co-active neuro-fuzzy inference system (CANFIS), MLP, MLR, MNLR, and SRC methods to simulate daily SSC at Tekra gauge station on River Pranhita, Andhra Pradesh, India. Outcomes indicated supremacy of CANFIS model compared to other proposed models in simulating SSC for selected study area. Mohamadi et al. (2020) utilized hybrid neural network techniques integrating shark algorithm (SA) and FFA with ANFIS, MLP, and RBFN and also simple ANFIS, MLP, and RBFN to predict monthly evaporation of Mianeh and Yazd stations located in Iran. Based on evaluation criteria, results proved that developed ANFIS-SA hybrid model was considered as a powerful tool to predict evaporation. Kişi (2004) predicted and estimated SSC utilizing MLP at two gauging sites on River Tongue in Montana, USA, and compared the achieved results with RBF, MLR, and generalized regression NN (GRNN). Based on comparisons, it was found that MLP generally gave better SSC estimation than other NN methods. Olyaie et al. (2017) used MLP, RBF, SVM, and linear genetic programming (LGP) techniques to predict dissolved oxygen in River Delaware situated at Trenton, USA. Comparing estimation accuracies of proposed models demonstrated that SVM successfully developed most precise model for estimating DO followed by LGP and ANN models. Many researchers are now considering hybrid neural network to predict suspended sediment load (SSL) at different gauge station around the world (Samantaray and Ghose, 2019; Samantaray and Ghose, 2020; Samantaray et al., 2020b).

Mustafa et al. (2012) employed MLP with different training algorithms to predict SSC of Pari River at Silibin, Malaysia, and proposed the best algorithm to model SSC. Ghorbani et al. (2018) investigated predictive capability of an amalgam model incorporating optimization algorithm FFA with MLP neural network model (MLP-FFA) to predict water level in Egirdir Lake, Turkey, and compared with simple MLP for different model combinations. Study of results revealed that MLP-FFA model with 4-month lagged amalgamations of water level data performed more precisely than individual MLP4 model. Feyzolahpour et al. (2012) proposed neural differential evolution (NDE) model combining NNs and differential evolution for estimating current suspended sediment of River Givi Chay in northwestern Iran and contrasted with MLP, RBFN, and SRC models. Results indicated that NDE delivered superior performance than MLP, RBFN, and SRC to estimate total sediment load. Rezapour et al. (2010) reviewed the potential of various ANN techniques to estimate SSC which helped researchers and scientists get ideas regarding implementation of such techniques with different input combinations and find suitable model applications. Cobaner et al. (2009) proposed ANFIS for predicting SSC at Mad River Catchment near Arcata, USA, and compared its potential with GRNN, RBFN, MLP, and two types of SRC. Results revealed that ANFIS performed better in comparison to other models for estimating daily SSC. Cigizoglu (2004) studied the applicability of MLP in estimating and forecasting daily SSC at two gauge sites of River Schuylkill in Philadelphia. Based on different statistical parameters, MLP performed relatively better compared to traditional regression models.

Mustafa et al. (2011) compared prediction potential of RBFN and feedforward NN (FFNN) for prediction of suspended sediments discharge at Silibin gauging site of River Pari in Perak, Malaysia. Results showed that RBFN produced slightly better performance than FFNN in predicting suspended sediment discharge. Bouzeria et al. (2017) presented enactments of best training algorithms in MLP NNs to predict suspended sediment transport in Mellah watershed. Model produced suitable outcomes and indicated very good agreement amid predicted and observed data. Alp and Cigizoglu (2007) used RBFN and feedforward backpropagation (FFBP) NN for estimating daily total SSC on Juniata River, Pennsylvania, and compared with MLR in terms of performance criteria. Based on different evaluation criteria, RBFN showed better results compared to FFBPNN and MLR models. Choubin et al. (2018) studied the applicability of CART to estimate SSC for Kareh-Sang River gauge site of Haraz watershed located in northern Iran. Obtained results from CART were compared with MLP, ANFIS, and SVM methods for examining accuracy of proposed model. Outcomes exhibited that CART model performed best in SSC prediction compared to other proposed models at selected area of study. Afan et al. (2014) employed FFNN and RBFN to estimate daily SSC of Rantau Panjang gauging site on River Johor, Malaysia. Assessment of obtained results indicated that FFNN model performed superiorly and more accurately than RBFN model in estimating daily SSC. Khan et al. (2019) employed FFBP, GRNN, and RBF for predicting monthly total sediment loads at Bareilly gauge station of Ramganga River. Tfwala and Wang (2016) explored potential of MLP, time-lagged recurrent networks (TLRN), CANFIS, fully recurrent neural networks (FRNN), and RBFN in improving accuracy of relationship amid streamflow and sediment discharge in the course of storm events in River Shiwen, situated in south Taiwan. Findings suggested that MLP provided most precise estimates of sediment discharge than CANFIS, FRNN, TLRN, and RBFN. Samet et al. (2019) used and examined ANN, ANFIS, and GA methods to forecast total sediment load in Maku dam reservoir, Iran. Results indicated that ANFIS produced best results in forecasting sediment load followed by ANN and GA models. Buyukyildiz and Kumcu (2017) investigated capabilities of SVM, ANNs, and ANFIS models for estimating daily SSC in Ispir Bridge gauge site on Coruh River situated in northeast Turkey. Based on performance indices, SVM performed best compared to ANN and ANFIS in predicting SSC at proposed study area. Emamgholizadeh and Demneh (2019) developed and studied the ability of ANN, gene expression programming (GEP), and ANFIS for estimating the SSC at two gauge sites of Telar River situated in north Iran and compared with traditional SRC model. Results revealed that GEP performed better than ANN and ANFIS in estimating daily SSC. Malik et al. (2019) employed RBFN, self-organizing map (SOM) NN, LSSVM, and multivariate adaptive regression spline (MARS) for daily SSC modelling at three gauging sites situated on Godavari River, Andhra Pradesh, India. Obtained results were compared with SRC models and revealed that all proposed models produced higher accuracy results than SRC with RBFN outperforming other models in proposed area of study. ANN models have demonstrated to be very valuable where physically based models do not perform well such as in case of non-linear modelling applications or multifaceted hydrological processes. Single drawback of ANNs is that these modelling techniques require lengthier time series data for training, testing, and validation.

The objective of present study is to predict SSL using hybrid machine learning approaches. MLP-FFA is introduced here to find the efficacy of model and compare its efficiency with the MLP and RBFN approaches.

12.2 STUDY AREA

Mahanadi is one of the most important rivers flowing through east central India among many other Indian rivers. Its total length is about 858 km from its source till convergence at Bay of Bengal. Of the total length of river basin, 361 km flows through Chhattisgarh and 497 km through Odisha. It emerges at an elevation of around 442 m above mean sea level near Pharsiya village and Nagri town in Raipur district, Chhattisgarh. Mahanadi has a total catchment area of 141,600 km^2. Basin comprises 14 main tributaries. Among these 12 links at upstream, whereas 2 at downstream of Hirakud Dam. Mean yearly precipitation in entire basin varies from 120 to 140 cm. During June–October (monsoon period), it receives approximately 85% of yearly precipitation. In a year, coldest months are December and January with lowest temperature of 11°C; while April and May record maximum temperature of around 40°C. Digital elevation map of River Mahanadi is presented in Figure 12.1.

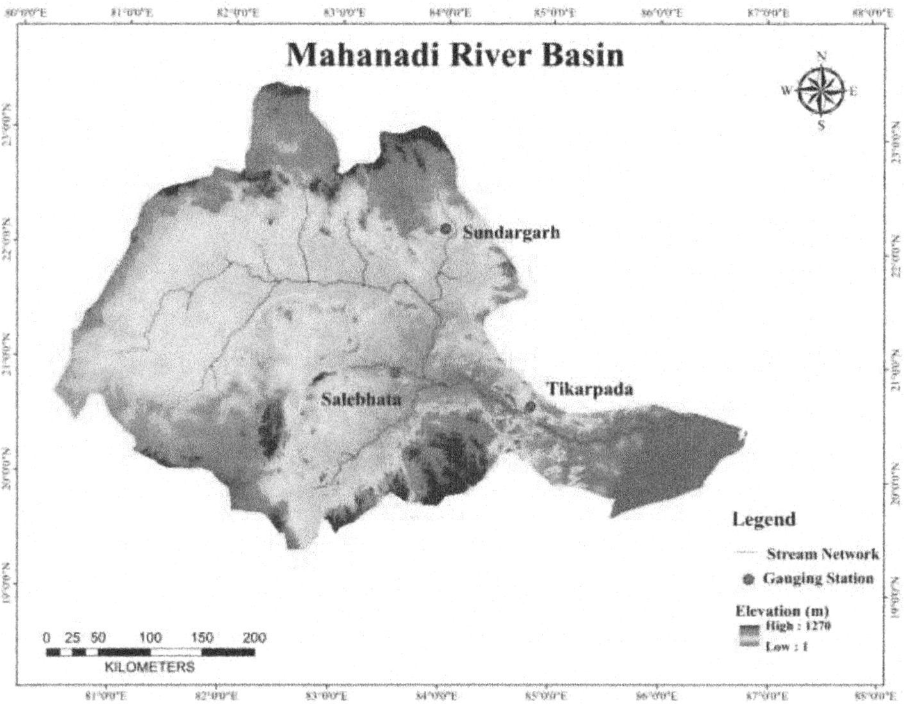

FIGURE 12.1 Proposed gauge station.

12.3 METHODOLOGY

12.3.1 RBFN

Broomhead and Lowe (1988) introduced RBFN to field of neural networks and developed as a variant of ANN. Yet, RBFN are subjective to very old pattern recognition approaches like functional approximation, potential functions, mixture models, clustering, and spline interpolation (Tou and Gonzalez, 1974). RBFNs are entrenched in a two-layer NN, where each unit of hidden layer executes a radially stimulated function. Each node unit of hidden layer signifies a RBF comprising two constraints: a width and a centre. Width regulates range of RBF and how increase in distance from centre to input reduces hidden node activation rapidly. Architecture and flowchart of RBFN is presented in Figure 12.2a and b. Weighted sum of hidden units is implemented in output units. Input to a RBFN is non-linear, whereas output is linear. *k*th output of RBFN can be computed as follows:

$$y_k(x_i) = \sum_{i=1}^{n} w_{ki}\psi_i(x_i) + b_k \tag{12.1}$$

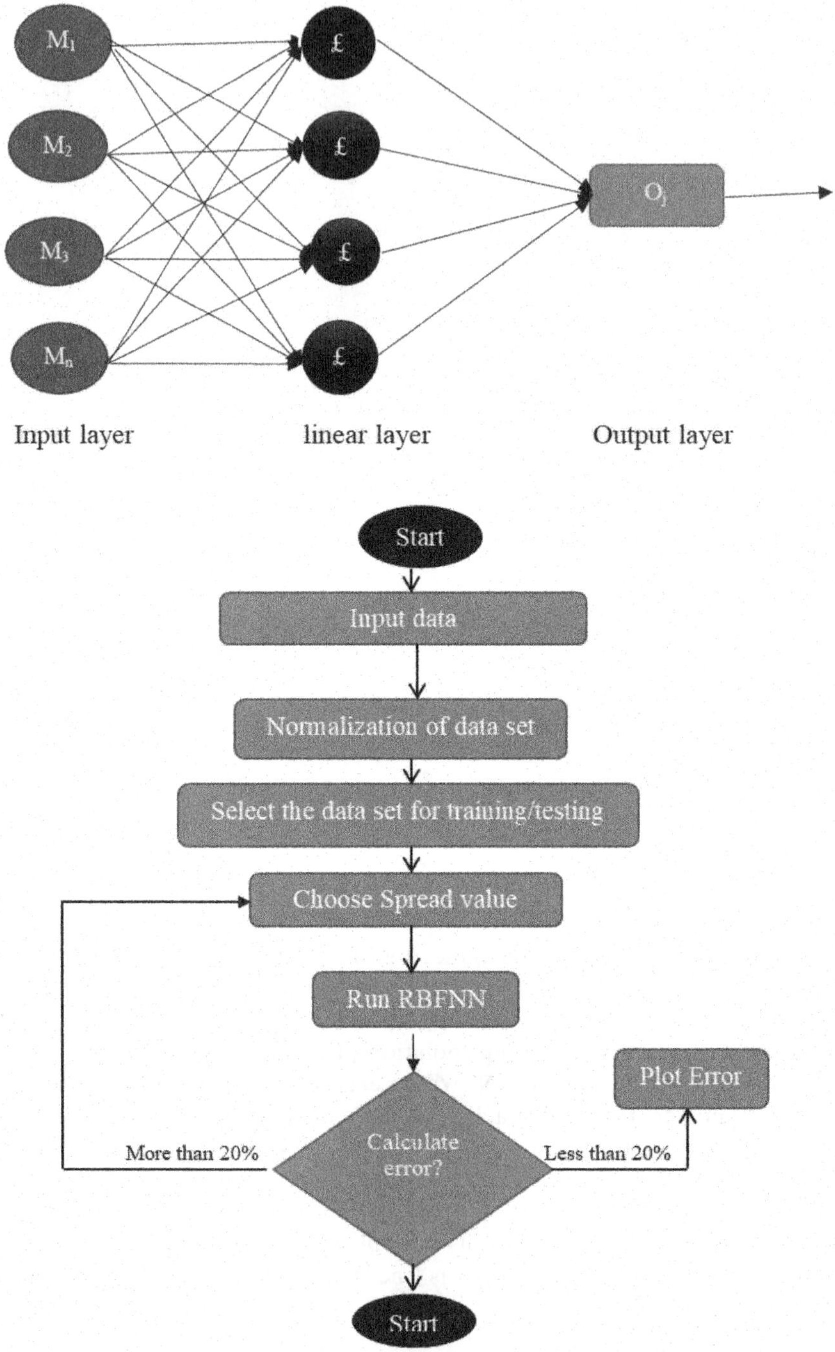

FIGURE 12.2 (a) Architecture of RBFN. (b) Flowchart of RBFN.

where x_i is the input data, n is the number of hidden nodes, $y_k(x_i)$ are output values of kth node, w_{ki} is the linking weight amid ith output node and kth hidden node, $\psi_i(x_i)$ is the RBF of hidden layer, and b_k is the bias constraint of kth node. Gaussian function is generally utilized as radial basis activation function in hidden layer (Ghose and Samantaray, 2019; Sahoo et al., 2020; Samantaray and Sahoo, 2020b).

12.3.2 MLP

MLP is a feedforward ANN consisting of an input, a hidden, and an output layer. Each layer comprises definite quantity of neurons with a stimulation function. Quantity of neurons in input and output layers need to be identical as target and input constraints. Quantity of neurons in hidden layer is obtained by error statistic of NN. Theoretical works have revealed that for approximating any complex non-linear function, a solitary hidden layer is adequate in case of ANNs (Hornik et al., 1989; Cybenko, 1989). Tang et al. (1991) found that a solitary hidden layer is utilized for avoiding increase in intricacy of network. Each network was skilled beneath a number of structures with altered quantity of neurons in hidden layer. Diagram representing MLP network for predicting sediment concentration and related flowchart is depicted in Figure 12.3a and b.

12.3.3 MLP-FFA

Estimation of suspended sediment concentration in this study is based on combination of MLP with FFA as an optimization tool used for training individual MLP. Yang (2010a, b) originally developed nature-inspired FFA technique on the basis of fireflies' movement. Solution of a problematic issue can be anticipated as an agent in this optimization methodology, i.e. firefly that illuminates in proportion with regard to its eminence. Subsequently, each livelier firefly is capable of attracting its associates, irrespective of their gender, making study of pursuit space extra effective (Łukasik and Zak, 2009). As fireflies are enticed to brightness, whole group travels in the direction of brightest firefly that can be conceptually applied in a prediction model for solving optimization approach. Here, attraction of fireflies is directly proportionate to their illumination and brightness is dependent on intensity of agent (Kayarvizhy et al., 2014). A crucial problem for FFA is building an objective function and discrepancy of intensity of light. Principal arrangement of FFA variables preparation are intensity of light $I(r)$, desirability β, and Cartesian space amid any two fireflies i and j at x_i and x_j correspondingly, which can best be formulated as follows (Yang, 2010a, b):

$$I(r) = I_0 \exp\left(-\gamma r^2\right) \tag{12.2}$$

$$\beta(r) = \beta_0 \exp\left(-\gamma r^2\right) \tag{12.3}$$

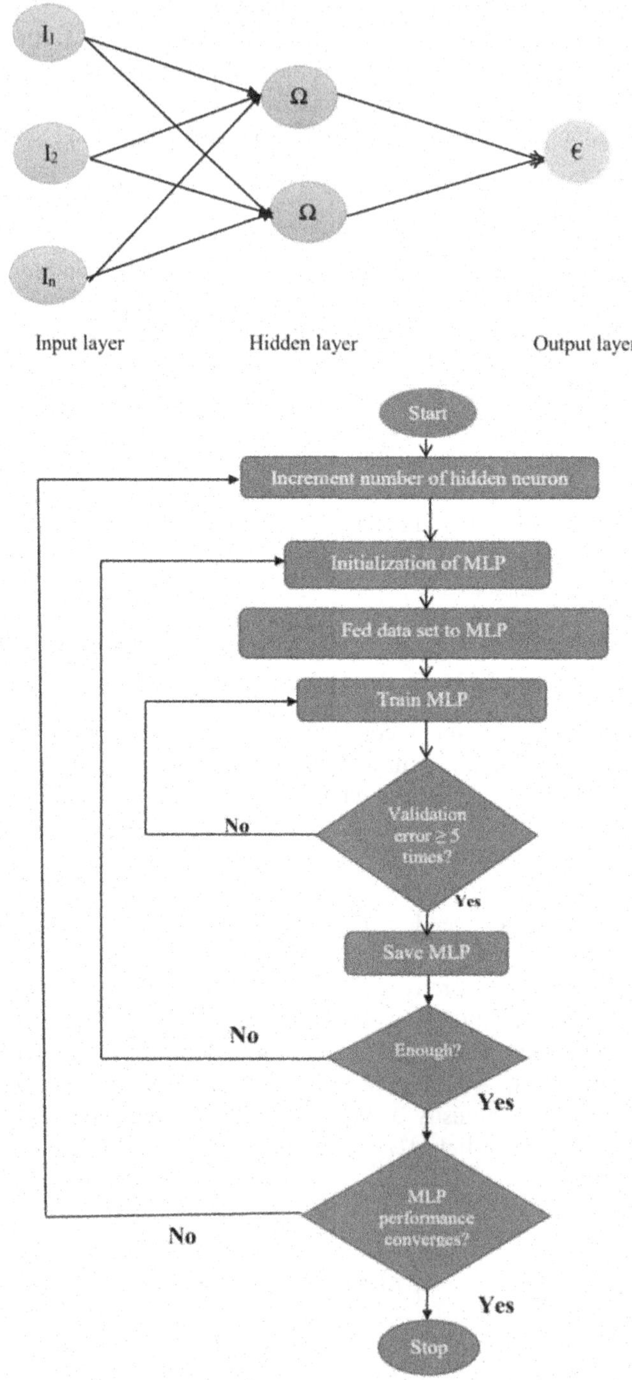

FIGURE 12.3 (a) Configuration of MLP neural network model. (b) Flowchart of MLP neural network model.

$$r_{ij} = x_i + x_j = \sqrt{\sum_{K=1}^{d} (x_{i,k} - x_{j,k})} \tag{12.4}$$

where $x_{i,k}$ is the kth constituent of spatial coordinate x_i of ith firefly, d is the dimensionality of specified problem, γ is the light absorption coefficient, I_0 and $I(r)$ are initial light intensity from a firefly and light intensity at distance r, β_0 and $\beta(r)$ are attractiveness β at a distance $r = 0$ and r.

Subsequent firefly movement signified as i is presented as follows:

$$x_i^{i+1} = x_i + \Delta x_i \tag{12.5}$$

$$\Delta x_i = \beta_0 e^{-\gamma r^2} \left(x_j - x_i\right) + \alpha \epsilon_i \tag{12.6}$$

The first part in Equation 12.5 specifies desirability, while the second part designates randomization procedure, α regulates randomization data positioned amid 0 and 1, and i denotes arbitrary Gaussian distribution quantity (Ch et al., 2014). Figure 12.4

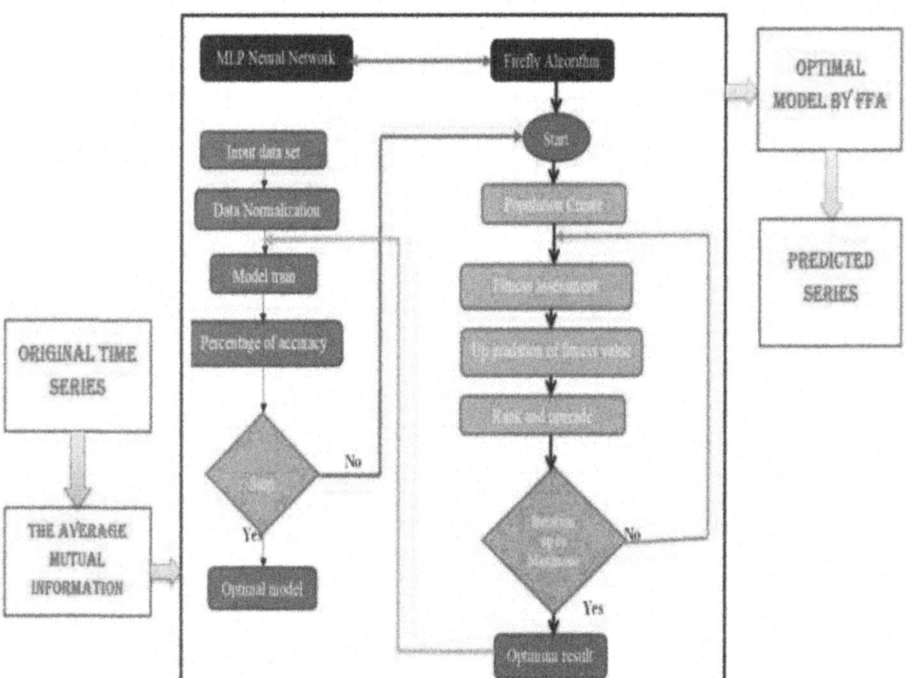

FIGURE 12.4 Flowchart for proposed MLP-FFA model.

indicates procedure used for obtaining optimum weights of MLP, which was combined with FFA.

12.3.4 Data Normalization

Training process of network was hurried up by preprocess of input and target values earlier to training phase. Before employing proposed models, input and output values were standardized in a series of 0 and 1 by utilizing succeeding equation:

$$X_i = \frac{x}{x_{\max}} \tag{12.7}$$

where X_i is the normalized value of certain constraint, x is the measured value of that constraint, and $x_{\max}$ is the maximum value in database for identical constraint, in respective order. Entire data will be in the series of 0 and 1 once the sediment values are normalized. This increases correlation amid sediment values, which will result in improvement of NN model performance.

12.3.5 Data Preparation and Performance Indices

In general, performance of a model is assessed on the basis of a comparison between actual data and calculated output. Predictions of all applied models are evaluated utilizing R^2, RMSE and MAE. Most common statistical performance evaluation parameter used in hydrological modelling is RMSE. A commonly applied measure for forecasting error in time series study is MAE. Size of MAE can be utilized for determining how an urbanized network output fit preferred output. Formulations of RMSE, (R^2), and MAE are specified below:

$$\text{RMSE} = \sqrt{\frac{1}{N}\sum_{t=1}^{n}\left(Q(obser)(t) - Q(simu)(t)\right)^2} \tag{12.8}$$

$$R^2 = \frac{\sum_{t=1}^{n}\left[\left(Q(obser)(t) - \overline{Q(obser)}(t)\right)\left(Q(simu)(t) - \overline{Q(simu)}(t)\right)\right]}{\sqrt{\sum_{t=1}^{n}\left(Q(obser)(t) - \overline{Q(obser)}(t)\right)^2 \sum_{t=1}^{n}\left(Q(simu)(t) - \overline{Q(simu)}(t)\right)^2}} \tag{12.9}$$

$$\text{MAE} = \frac{1}{N}\sum_{t=1}^{n}\left|Q(obser)(t) - Q(simu)(t)\right| \tag{12.10}$$

Data of sediment load for the period 1960–2019 are collected from Indian meteorological department from May to October (monsoon months). Collected data from 1960 to 2001 are applied for training, data from 2002 to 2010 are used for testing, and data from 2011–2019 are used for model validation. Monthly data is converted from daily data that finally assists to train and test the model. The following sediment arrangements are utilized as input:

i. S_{t-1}	RBFN-1	MLP-1	MLP-FFA 1
ii. S_{t-1}, S_{t-2}	RBFN-2	MLP-2	MLP-FFA 2
iii. $S_{t-1}, S_{t-2}, S_{t-3}$	RBFN-3	MLP-3	MLP-FFA 3
iv. $S_{t-1}, S_{t-2}, S_{t-3}, S_{t-4}$	RBFN-4	MLP-4	MLP-FFA 4
v. $S_{t-1}, S_{t-2}, S_{t-3}, S_{t-4}, S_{t-5}$	RBFN-5	MLP-5	MLP-FFA 5

12.4 RESULTS AND DISCUSSION

In this study, sediment data were only considered as inputs for studying various SSL scenarios. Five types of set are trained and tested utilizing RBFN, MLP, and MLP-FFA. The RMSE, R^2, and MAE values for RBFN, MLP, and MLP-FFA are represented in Tables 12.1–12.3. Training and testing results are specified in all three tables which show five input types and each input provide different outcome

TABLE 12.1
Results of RBFN

Station	Input Scenario	MAE		RMSE		R^2	
		Training	Testing	Training	Testing	Training	Testing
	RBFN-1	0.00624	0.00571	0.07016	0.02616	0.91278	0.89188
Sundargarh	RBFN-2	0.00616	0.00563	0.06986	0.02433	0.91498	0.89483
	RBFN-3	0.00605	0.00553	0.06913	0.02358	0.91748	0.89744
	RBFN-4	0.00588	0.00541	0.0689	0.02286	0.91937	0.89964
	RBFN-5	**0.00556**	**0.0052**	**0.06828**	**0.02185**	**0.95397**	**0.93711**
	RBFN-1	0.00622	0.00568	0.07003	0.02579	0.91357	0.89308
Salebhata	RBFN-2	0.00612	0.00559	0.06948	0.02421	0.91579	0.89586
	RBFN-3	0.00598	0.00545	0.06899	0.02327	0.91813	0.89849
	RBFN-4	0.00583	0.00538	0.06882	0.02271	0.91997	0.90078
	RBFN-5	**0.00553**	**0.00517**	**0.06822**	**0.02177**	**0.95461**	**0.93765**
	RBFN-1	0.00621	0.00567	0.06998	0.02532	0.91388	0.89357
Tikarpada	RBFN-2	0.00611	0.00558	0.06935	0.02396	0.91627	0.89625
	RBFN-3	0.00595	0.00544	0.06897	0.02311	0.91849	0.89867
	RBFN-4	0.00582	0.00537	0.06878	0.02266	0.92016	0.90107
	RBFN-5	**0.00551**	**0.00516**	**0.06819**	**0.02177**	**0.95483**	**0.93789**

TABLE 12.2
Results of MLP

Station	Input Scenario	MAE		RMSE		R^2	
		Training	Testing	Training	Testing	Training	Testing
	MLP-1	0.00548	0.00511	0.06815	0.02173	0.95525	0.93812
Sundargarh	MLP-2	0.00527	0.00491	0.06724	0.02141	0.95746	0.94089
	MLP-3	0.00519	0.00448	0.06634	0.02131	0.95841	0.94176
	MLP-4	0.00498	0.00401	0.06248	0.02112	0.96053	0.94411
	MLP-5	**0.00431**	**0.00309**	**0.05851**	**0.01889**	**0.97357**	**0.95146**
	MLP-1	0.00545	0.00507	0.06811	0.02169	0.95548	0.93857
Salebhata	MLP-2	0.00524	0.0049	0.06689	0.02136	0.95787	0.94124
	MLP-3	0.00508	0.00411	0.06497	0.02118	0.95987	0.94318
	MLP-4	0.00496	0.00399	0.06117	0.02108	0.96078	0.94438
	MLP-5	**0.00427**	**0.00255**	**0.05819**	**0.01848**	**0.97406**	**0.95204**
	MLP-1	0.00529	0.00494	0.06777	0.02148	0.95695	0.94027
Tikarpada	MLP-2	0.00521	0.00459	0.06648	0.02134	0.95812	0.94145
	MLP-3	0.00502	0.00402	0.06427	0.02116	0.96026	0.94385
	MLP-4	0.00491	0.00398	0.06014	0.02008	0.96107	0.94465
	MLP-5	**0.00426**	**0.00248**	**0.05741**	**0.01779**	**0.97499**	**0.95288**

subsequent to higher performance standards (Table 12.3). Best set of performance was acquired for five dependent variables (S_{t-1}, S_{t-2}, S_{t-3}, S_{t-4}, S_{t-5}). From others, the performance analysis of RBFN, MLP, and MLP-FFA models for five different input scenarios indicated that aggregating number of input nodes to two, three, and so on, the performance of testing stage of proposed networks increased. In this research, RMSE and MAE were taken as key assessment standards. R^2 gives information regarding linear dependency amid observations and corresponding estimations. Among five models, when I considered input up to five lag months, it showed prominent result for all three proposed watershed.

As described in preceding segments of this study, the best datasets were nominated from each stage. It was problematic in selecting a precise model as there is slight difference in resultant values amid models. But as well-known from previous result, MLP-FFA model has superior estimation than MLP and RBFN models. One significant objective of this research is to make a comparison between three different techniques. Figures 12.5, 12.6, and 12.7 illustrate the scatter plot between actual and predicted sediment concentration at three proposed gauge stations during training, testing, and validation phases. It is clearly observed from figures that peak values for scenario S_{t-1}, S_{t-2}, S_{t-3}, S_{t-4}, and S_{t-5} is more near to fitting line compared to other for all three algorithms.

Linear scale plotting of actual versus predicted sediment concentration of developed models for proposed sites is displayed in Figure 12.8. Outcomes demonstrate that approximated peak SSC is 9,308,754 µg/L, 9,519,093µg/L, and 9,721,458µg/L for RBFN, MLP, and MLP-FFA, respectively, against actual peak 9,968,681µg/L for station Sundargarh. The estimated peak SSC is 9,200,981 µg/L, 9,405,272 µg/L, and

TABLE 12.3
Results of MLP-FFA

Station	Input Scenario	MAE		RMSE		R^2	
		Training	Testing	Training	Testing	Training	Testing
	MLP-FFA 1	0.00405	0.00236	0.05634	0.01653	0.97621	0.95447
Sundargarh	MLP-FFA 2	0.00393	0.00231	0.05557	0.01537	0.97714	0.95557
	MLP-FFA 3	0.00374	0.00211	0.05379	0.01353	0.97906	0.95749
	MLP-FFA 4	0.00361	0.00202	0.05296	0.01282	0.97967	0.95803
	MLP-FFA 5	**0.00334**	**0.00181**	**0.05035**	**0.01068**	**0.98104**	**0.95964**
	MLP-FFA 1	0.00398	0.00234	0.05591	0.01611	0.97675	0.95495
Salebhata	MLP-FFA 2	0.00389	0.00228	0.05539	0.01509	0.97738	0.95589
	MLP-FFA 3	0.00368	0.00208	0.05357	0.01327	0.97924	0.95764
	MLP-FFA 4	0.00345	0.00189	0.05127	0.01145	0.98039	0.95902
	MLP-FFA 5	**0.00329**	**0.00179**	**0.05007**	**0.01043**	**0.98132**	**0.95989**
	MLP-FFA 1	0.00396	0.00233	0.05574	0.01584	0.97698	0.95532
Tikarpada	MLP-FFA 2	0.00377	0.00217	0.05433	0.01399	0.97859	0.95697
	MLP-FFA 3	0.00363	0.00206	0.05321	0.01301	0.97946	0.95789
	MLP-FFA 4	0.00337	0.00184	0.05061	0.01091	0.98088	0.95947
	MLP-FFA 5	**0.00326**	**0.00175**	**0.04997**	**0.01012**	**0.98145**	**0.96008**

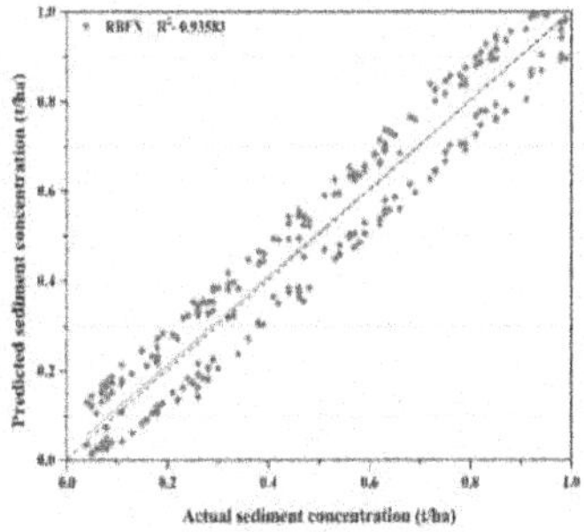

(a. i)

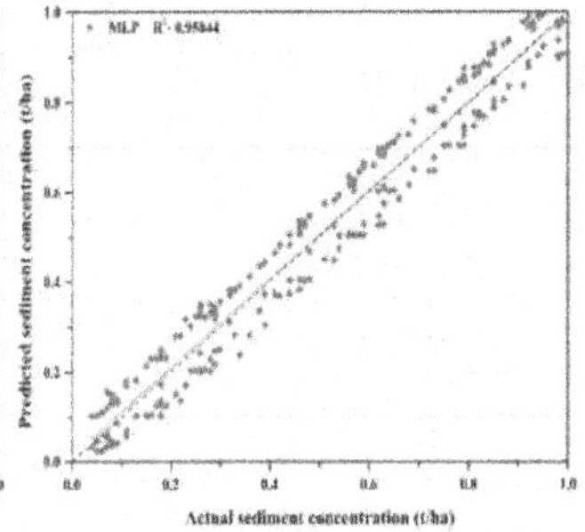

(a. ii)

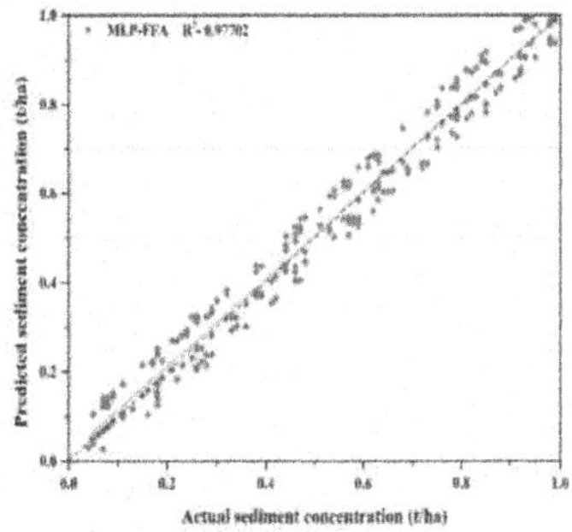

(a. iii)

FIGURE 12.5 Scatter plot for actual and predict sediment using (i) RBFN, (ii) MLP, and (iii) MLP-FFA for (a) Sundergarh, (b) Salebhata, and (c) Tikarpada gauge stations during training phase.

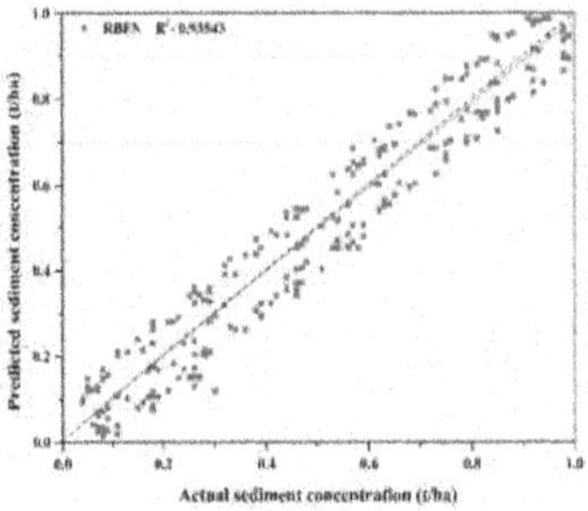

(b. i)

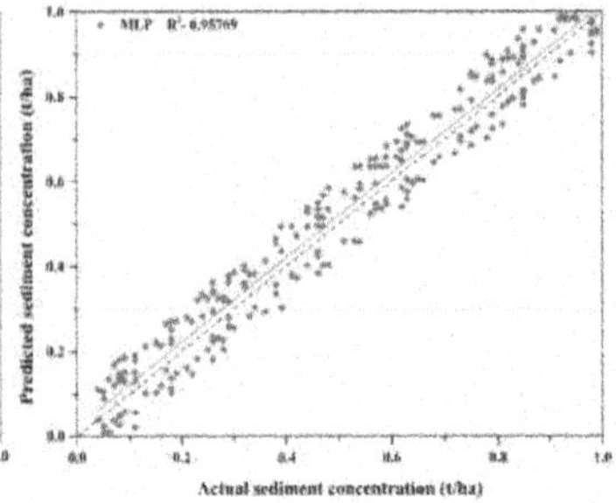

(b. ii)

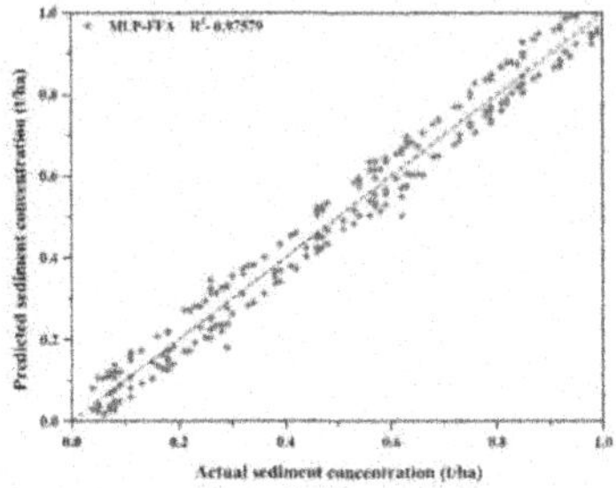

(b. iii)

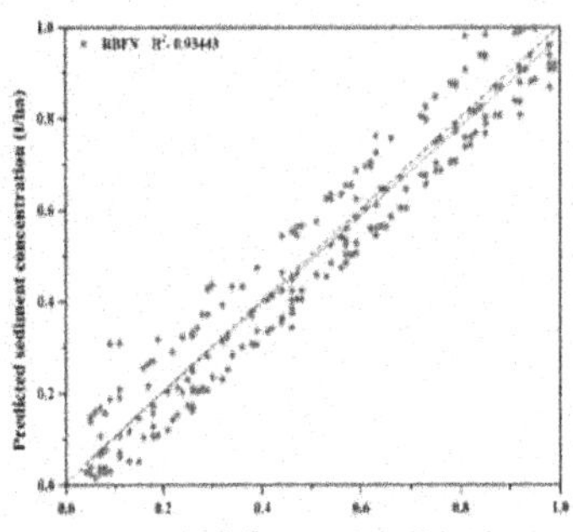

(c. i)

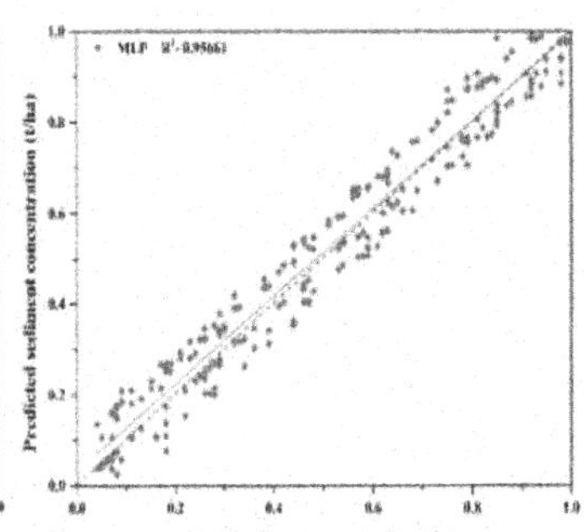

(c. ii)

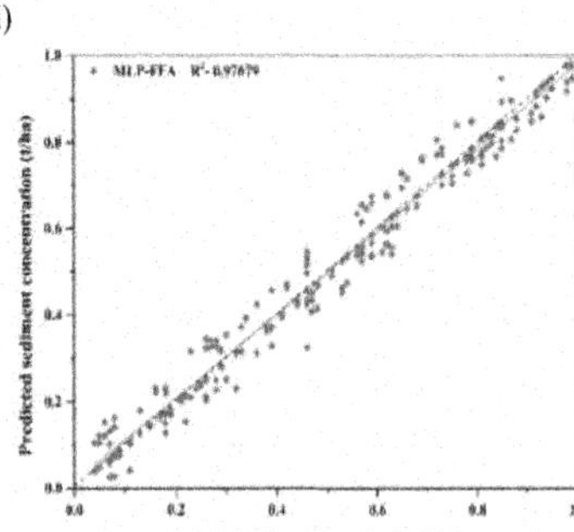

(c. iii)

FIGURE 12.5 (Continued)

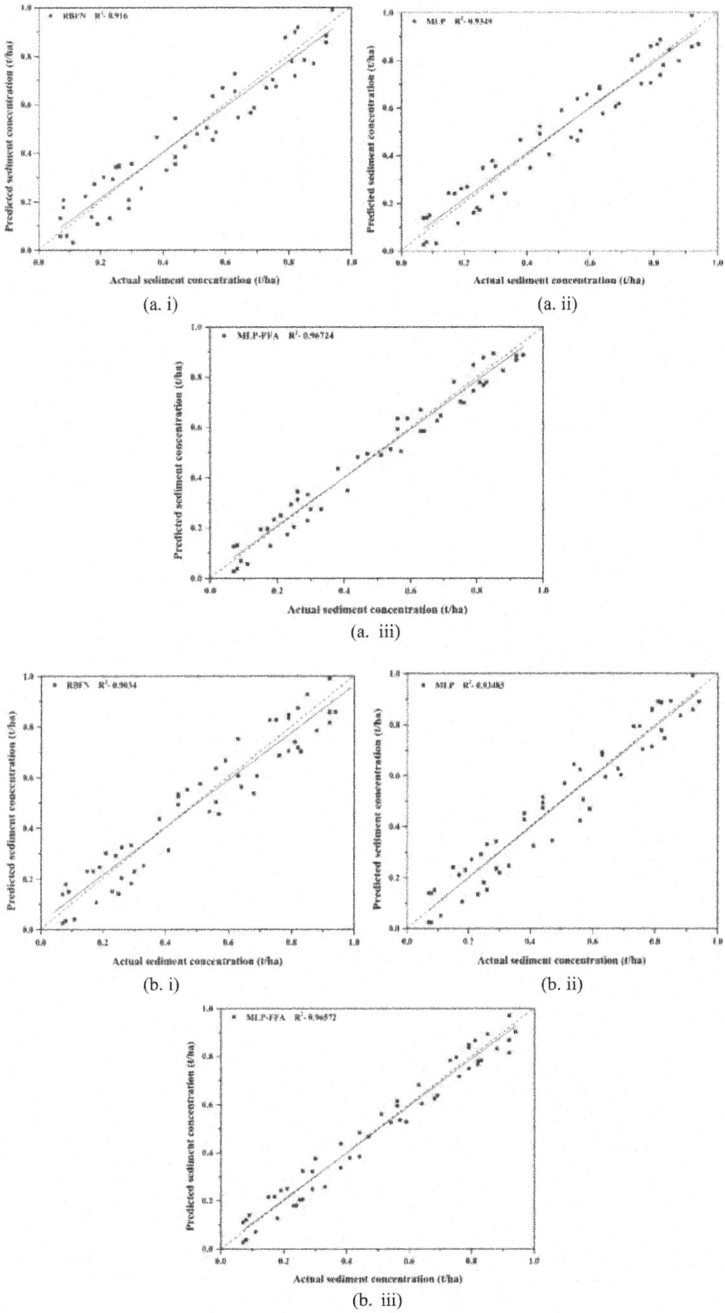

FIGURE 12.6 Scatter plot for actual and predict sediment using (i) RBFN, (ii) MLP, and (iii) MLP-FFA for (a) Sundergarh, (b) Salebhata, and (c) Tikarpada gauge stations during testing phase.

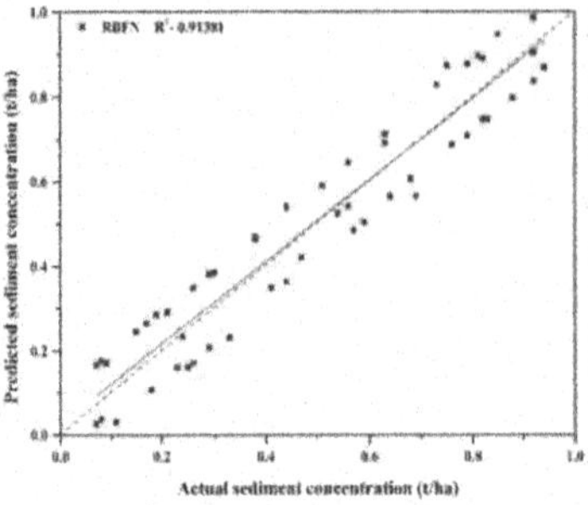

(c. i)

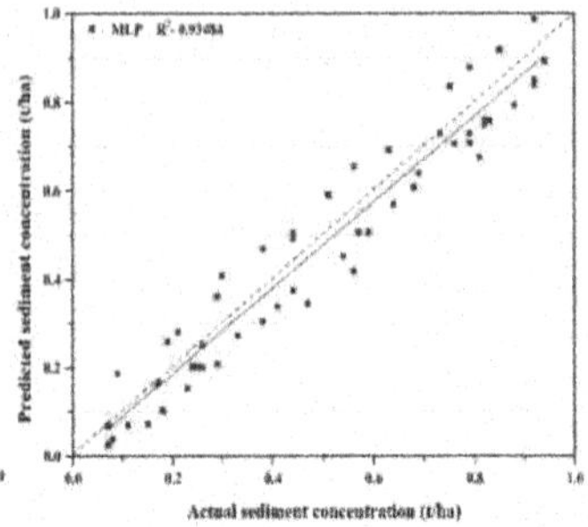

(c. ii)

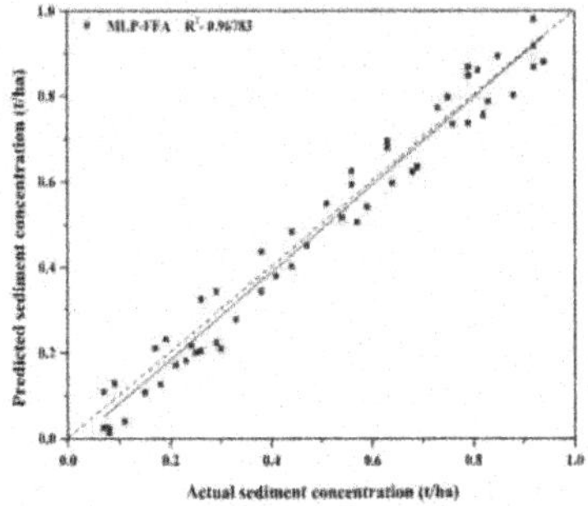

(c. iii)

FIGURE 12.6 (Continued)

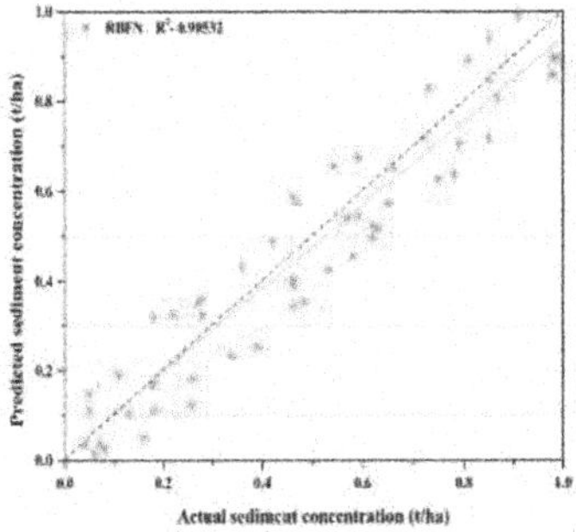

(a. i)

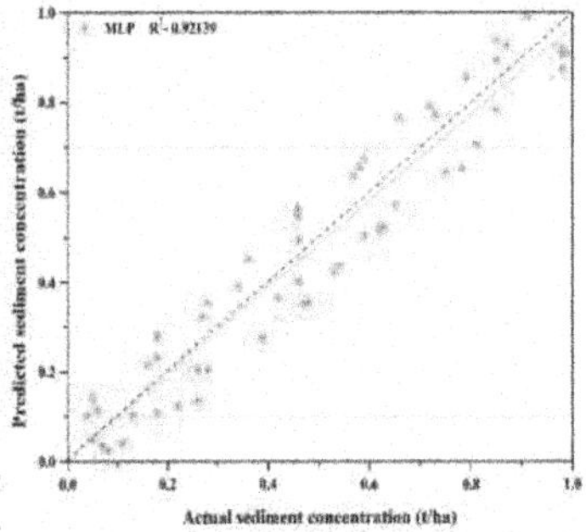

(a. ii)

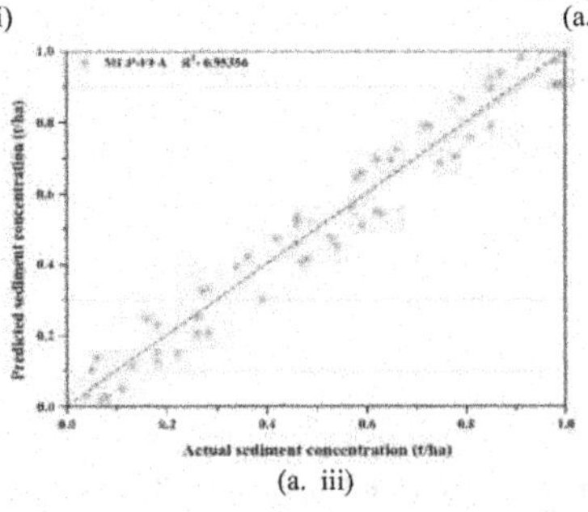

(a. iii)

FIGURE 12.7 Scatter plot for actual and predict sediment using (i) RBFN, (ii) MLP, and (iii) MLP-FFA for (a) Sundergarh, (b) Salebhata, and (c) Tikarpada gauge stations during validation phase.

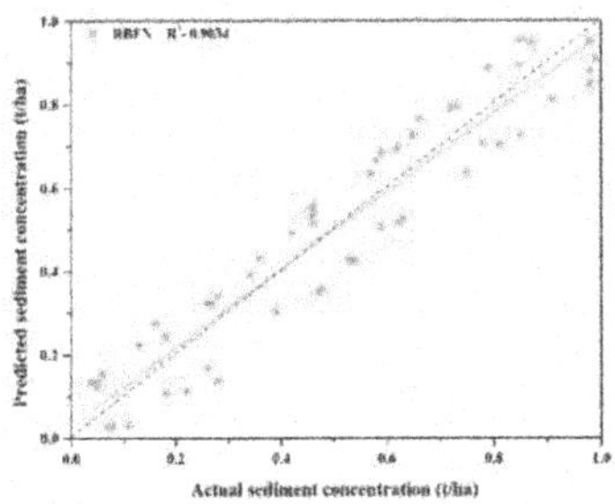

(b. i)

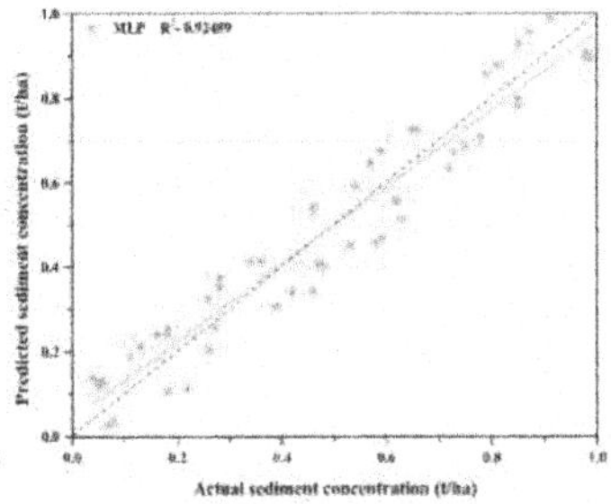

(b. ii)

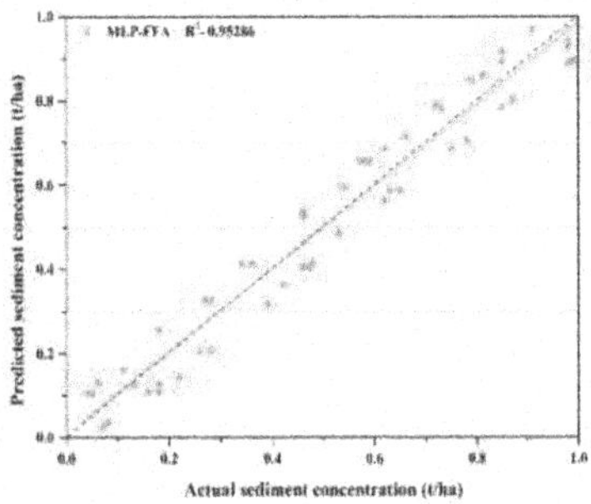

(b. iii)

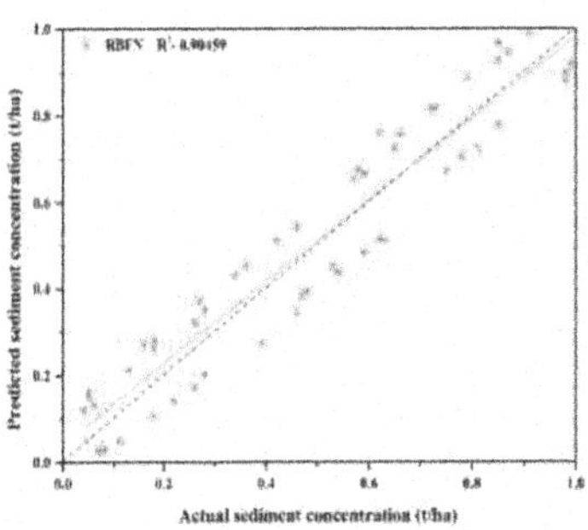

(c. i)

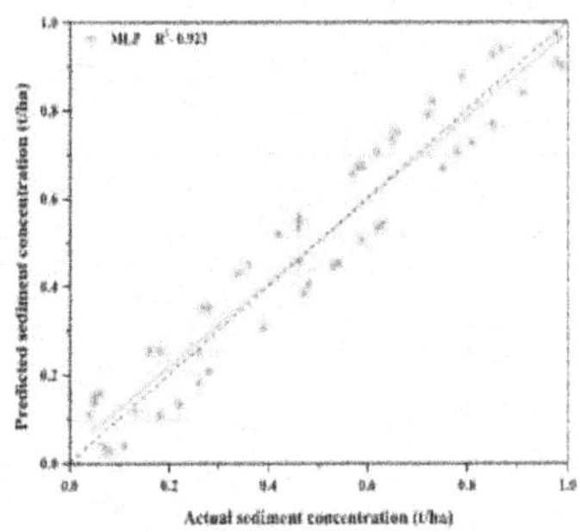

(c. ii)

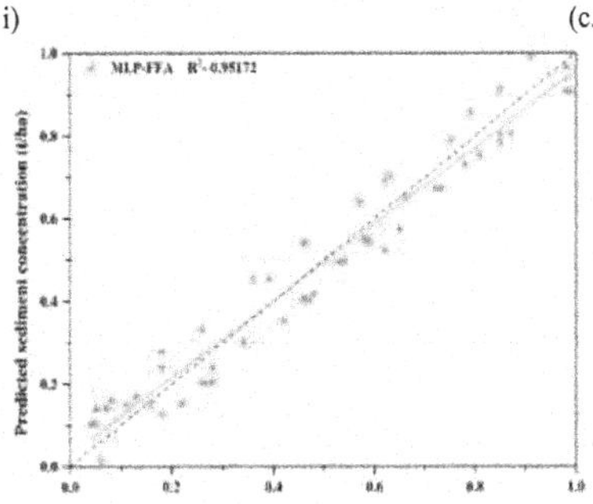

(c. iii)

FIGURE 12.7 (Continued)

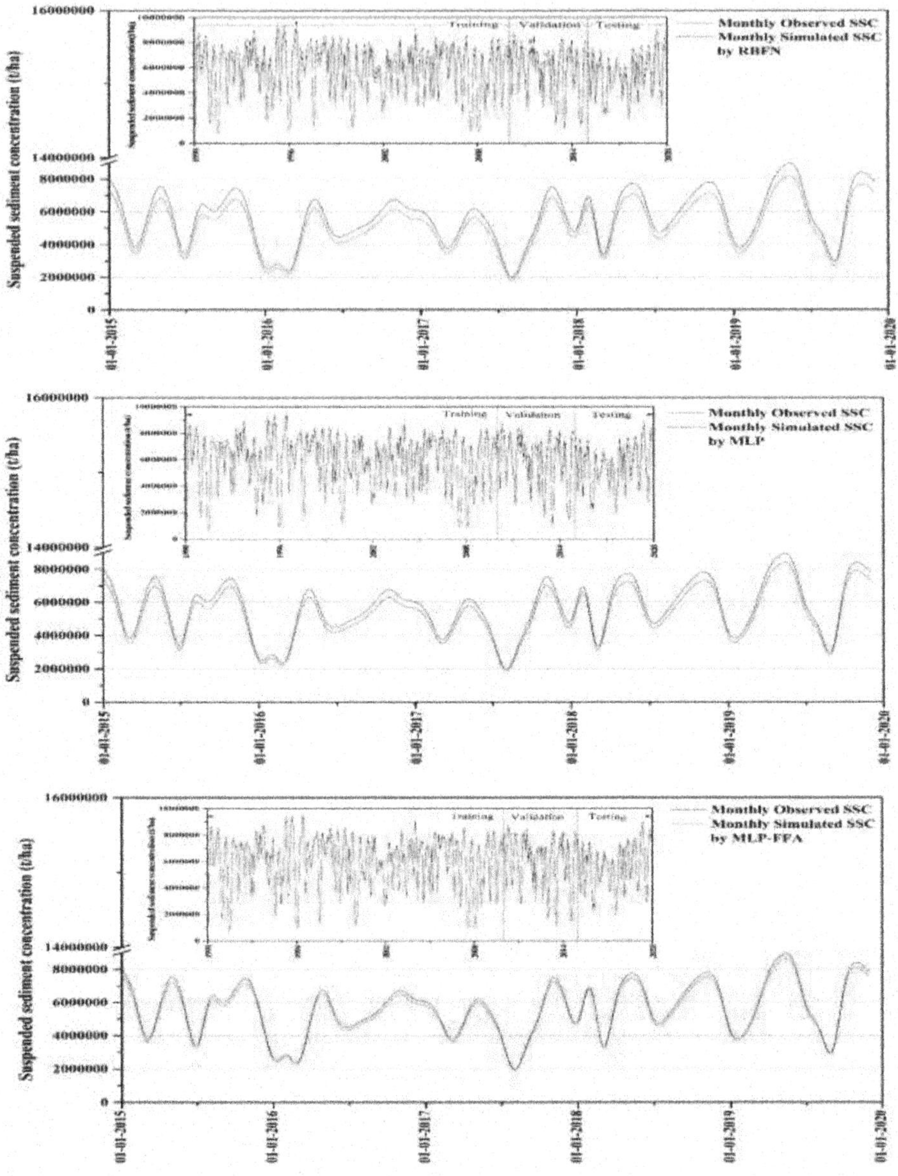

(a)

FIGURE 12.8 Deviance of actual versus predicted SSC at (a) Sundargarh, (b) Salabheta, and (c) Tikarpada gauging sites.

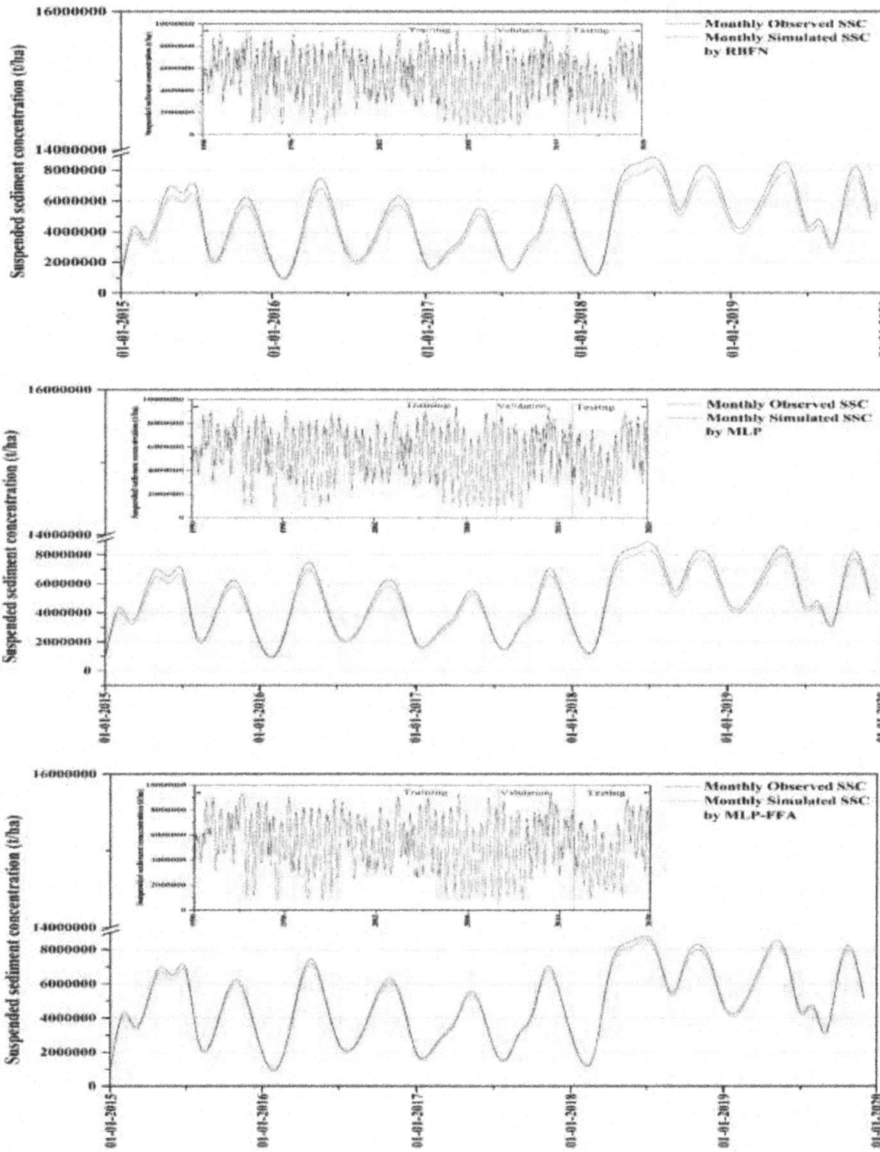

(b)

FIGURE 12.8 (Continued)

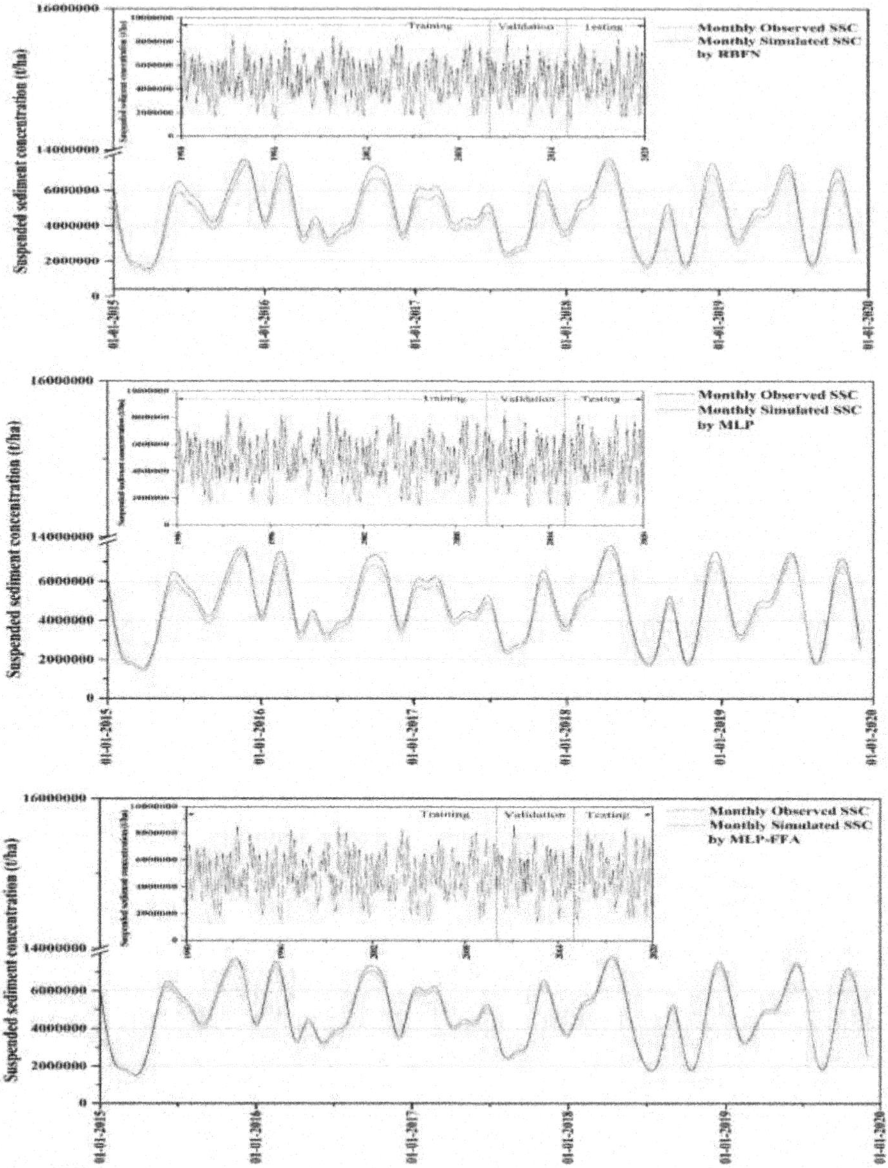

(c)

FIGURE 12.8 (Continued)

TABLE 12.4
Performance Index of Different Approaches for Sediment Load Prediction

Stations	RBFN			MLP			MLP-FFA		
	MAE	RMSE	R^2	MAE	RMSE	R^2	MAE	RMSE	R^2
Sundergarh	0.00556	0.06828	0.95397	0.00431	0.05851	0.97357	0.00334	0.05035	0.98104
Salabheta	0.00553	0.06822	0.95461	0.00427	0.05819	0.97406	0.00329	0.05007	0.98132
Tikarpada	0.00551	0.06819	0.95483	0.00426	0.05741	0.97499	0.00326	0.04997	0.98145

9,607,589 μg/L for RBFN, MLP, and MLP-FFA, respectively, adjacent to actual peak 9,869,121 μg/L for Salabheta site. For Tikarpada gauge site peak, SSC is 9,807,982 μg/L aligned with predicted SSC 9,177,329 μg/L, 9,384,277 μg/L, and 9,576,514 μg/L for RBFN, MLP, and MLP-FFA, respectively.

12.4.1 Comparison of Pre-eminent Outcomes

RBFN, MLP, and MLP-FFA models are utilized for performance evaluation based on RMSE, R^2, and MAE indictors for three gauging sites. Assessment results of performance measures are quantified in Table 12.4, illustrating effectiveness of each model. Estimating sediment load is valuable and thus approaches utilized here are important to demonstrate information about SSC. Hence, calculation of RMSE, R^2, and MAE values are vital to predict SSC. It is apparent that MLP-FFA model performed well in comparison to MLP and RBFN.

12.5 CONCLUSION

This research focused on developing an MLP-based forecasting model combined with FFA as an optimization algorithm for estimating suspended sediment concentration. This case study is conducted on Mahanadi River located in Odisha, India. Two ANN models– MLP and RBFN – are also employed in this study for comparing accuracy of proposed hybrid MLP-FFA model in estimating SSC at selected study area. Results obtained by MLP, RBFN, and MLP-FFA models were evaluated considering RMSE, R^2, and MAE, which specified that proposed hybrid model performed superior to stand-alone MLP and RBFN models. Hence, a conclusion can be made that NN-based models can enhance prediction ability of sediment concentration in a river basin. Also, it is observed that even though NN model is performing pre-eminent to predict sediment load in Mahanadi River, certain inconsistency still prevails in datasets that has not been apprehended by projected models. Overall, this research study will be useful for modelling hydrological processes where estimations regarding suspended sediment values are unavailable.

REFERENCES

Achite, Mohamed and Sylvain Ouillon. 2007. "Suspended sediment transport in a semiarid watershed, Wadi Abd, Algeria (1973–1995)." *Journal of Hydrology*, 343 (3–4), 187–202.

Afan, Haitham Abdulmohsin, Ahmed El-Shafie, Zaher Mundher Yaseen, Mohammed Majeed Hameed, Wan Hanna Melini Wan Mohtar, and Aini Hussain. 2014. "ANN based sediment prediction model utilizing different input scenarios." *Water Resources Management*, 29 (4), 1231–1245.

Alp, Murat and H. Kerem Cigizoglu. 2007. "Suspended sediment load simulation by two artificial neural network methods using hydrometeorological data." *Environmental Modelling and Software*, 22 (1), 2–13.

Bouzeria, Housseyn, Abderrahmane N. Ghenim, and Kamel Khanchoul. 2017. "Using artificial neural network (ANN) for prediction of sediment loads, application to the Mellah catchment, Northeast Algeria." *Journal of Water and Land Development*, 33 (1), 47–55.

Broomhead, David S. and David Lowe. 1988. "Radial basis functions, multi-variable functional interpolation and adaptive networks." Royal Signals and Radar Establishment Malvern (United Kingdom), RSRE Memorandum No. 4148.

Buyukyildiz, Meral and Serife Yurdagul Kumcu. 2017. "An estimation of the suspended sediment load using adaptive network based fuzzy inference system, support vector machine and artificial neural network models." *Water Resources Management*, 31 (4), 1343–1359.

Ch, Sudheer, S. K. Sohani, Deepak Kumar, Anushree Malik, B. R. Chahar, A. K. Nema, Bijaya K. Panigrahi, and R. C. Dhiman. 2014. "A support vector machine-firefly algorithm based forecasting model to determine malaria transmission." *Neurocomputing*, 129, 279–288.

Choubin, Bahram, Hamid Darabi, Omid Rahmati, Farzaneh Sajedi-Hosseini, and Bjørn Kløve. 2018. "River suspended sediment modelling using the CART model: a comparative study of machine learning techniques." *Science of the Total Environment*, 615, 272–281.

Cigizoglu, H. Kerem. 2004. "Estimation and forecasting of daily suspended sediment data by multi-layer perceptrons." *Advances in Water Resources*, 27 (2), 185–195.

Cobaner, M., B. Unal, and O. Kisi. 2009. "Suspended sediment concentration estimation by an adaptive neuro-fuzzy and neural network approaches using hydro-meteorological data." *Journal of Hydrology*, 367 (1–2), 52–61.

Cybenko, G., 1989. "Approximation by super positions of a sigmoidal function." Mathematics of control, signals and systems, *2*(4), pp.303–314.

Cybenko, George. 1992. "Approximation by superpositions of a sigmoidal function." Mathematics of Control, Signals and Systems, 5 (4), 455.

Emamgholizadeh, Samad and Razieh Karimi Demneh. 2019. "A comparison of artificial intelligence models for the estimation of daily suspended sediment load: a case study on the Telar and Kasilian Rivers in Iran." *Water Science and Technology: Water Supply*, 19 (1), 165–178.

Evsukoff, Alexandre Gonçalves, Beatriz S. L. P. De Lima, and Nelson F. F. Ebecken. 2011. "Long-term runoff modeling using rainfall forecasts with application to the Iguaçu River basin." *Water Resources Management*, 25 (3), 963–185.

Fernando, D. Achela and Asaad Y. Shamseldin. 2009. "Investigation of internal functioning of the radial-basis-function neural network river flow forecasting models." *Journal of Hydrologic Engineering*, 14 (3), 286–292.

Feyzolahpour, Mehdi. 2012. "Estimating suspended sediment concentration using neural differential evolution (NDE), multilayer perceptron (MLP) and radial basis function (RBF) models." *International Journal of the Physical Sciences*, 7 (29), 5106–5117.

Ghorbani, Mohammad Ali, Ravinesh C. Deo, Vahid Karimi, Zaher Mundher Yaseen, and Ozlem Terzi. 2018. "Implementation of a hybrid MLP-FFA model for water level prediction of lake Egirdir, Turkey." *Stochastic Environmental Research and Risk Assessment*, 32 (6), 1683–1697.

Ghose, Dillip K. and Sandeep Samantaray. 2018. "Modelling sediment concentration using back propagation neural network and regression coupled with genetic algorithm." *Procedia Computer Science*, 125, 85–92.

Ghose, Dillip K. and Sandeep Samantaray. 2019. "Sedimentation process and its assessment through integrated sensor networks and machine learning process." In Computational Intelligence in Sensor Networks. 473–488, Springer.

Guven, Aytac and Özgür Kişi. 2011. "Estimation of suspended sediment yield in natural rivers using machine-coded linear genetic programming." *Water Resources Management*, 25 (2), 691–704.

Hornik, K., Stinchcombe, M. and White, H., 1989. "Multilayer feedforward networks are universal approximators." *Neural Networks*, *2*(5), pp.359–366.

Jimmy, S. R, A. Sahoo, Sandeep Samantaray, and Dillip K. Ghose. 2020. "Prophecy of runoff in a river basin using various neural networks." In Communication Software and Networks. 709–718, Springer.

Ju, Qin, Zhongbo Yu, Zhenchun Hao, Gengxin Ou, Jian Zhao, and Dedong Liu. 2009. "Division-based rainfall-runoff simulations with BP neural networks and xinanjiang model." *Neurocomputing*, 72 (13–15), 2873–2883.

Kayarvizhy, N., S. Kanmani, and R. V. Uthariaraj. 2014. "ANN models optimized using swarm intelligence algorithms." WSEAS Transactions on Computers, 13 (45), 501–519.

Khan, Mohd Yawar Ali, Faisal Hasan, and Fuqiang Tian. 2019. "Estimation of suspended sediment load using three neural network algorithms in Ramganga river catchment of Ganga basin, India." *Sustainable Water Resources Management*, 5 (3), 1115–1131.

Kişi, Özgür. 2004. "Multi-layer perceptrons with Levenberg-Marquardt training algorithm for suspended sediment concentration prediction and estimation." *Hydrological Sciences Journal*, 49 (6), 1025–1040.

Kisi, Ozgur, Tefaruk Haktanir, Mehmet Ardiclioglu, Ozgur Ozturk, Ekrem Yalcin, and Salih Uludag. (2009). "Adaptive neuro-fuzzy computing technique for suspended sediment estimation." *Advances in Engineering Software*, 40 (6), 438–444.

Kisi, Ozgur, Alireza Moghaddam Nia, Mohsen Ghafari Gosheh, Mohammad Reza Jamalizadeh Tajabadi, and Azadeh Ahmadi. 2012. "Intermittent streamflow forecasting by using several data driven techniques." *Water Resources Management*, 26 (2), 457–474.

Kumar, Dheeraj, Ashish Pandey, Nayan Sharma, and Wolfgang Albert Flügel. 2016. "Daily suspended sediment simulation using machine learning approach." CATENA, 138, 77–90.

Łukasik, Szymon and Sławomir Żak. 2009. "Firefly algorithm for continuous constrained optimization tasks." In International Conference on Computational Collective Intelligence, 97–106.

Malik, Anurag, Anil Kumar, Ozgur Kisi, and Jalal Shiri. 2019. "Evaluating the performance of four different heuristic approaches with gamma test for daily suspended sediment concentration modeling." *Environmental Science and Pollution Research*, 26 (22), 22670–22687.

Malik, Anurag, Anil Kumar, and Jamshid Piri. 2017. "Daily suspended sediment concentration simulation using hydrological data of Pranhita River basin, India." *Computers and Electronics in Agriculture*, 138, 20–28.

Melesse, A. M., S. Ahmad, M. E. McClain, X. Wang, and Y. H. Lim. 2011. "Suspended sediment load prediction of river systems: an artificial neural network approach." *Agricultural Water Management*, 98 (5), 855–866.

Memarian, Hadi and Siva Kumar Balasundram. 2012. "Comparison between multi-layer perceptron and radial basis function networks for sediment load estimation in a tropical watershed." *Journal of Water Resource and Protection*, 4 (10), 870–876.

Mingzhou, Qin, Richard H. Jackson, Yuan Zhongjin, Mark W. Jackson, and Sun Bo. 2007. "The effects of sediment-laden waters on irrigated lands along the lower Yellow River in China." *Journal of Environmental Management*, 85 (4), 858–865.

Mohamadi, S., M. Ehteram, and A. El-Shafie. 2020. "Accuracy enhancement for monthly evaporation predicting model utilizing evolutionary machine learning methods." *International Journal of Environmental Science and Technology*, 17, 3373–3396.

Mohanta, N. R., P. Biswal, S. S. Kumari, S. Samantaray, and Abinash Sahoo. 2020. "Estimation of sediment load using adaptive neuro-fuzzy inference system at Indus River basin, India." In Intelligent Data Engineering and Analytics. 427–434, Springer, Singapore.

Mohanta, N. R., P. Biswal, S. S. Kumari, S. Samantaray, and Abinash Sahoo. 2020b. "Estimation of sediment load using adaptive neuro-fuzzy inference system at Indus River basin, India." In Intelligent Data Engineering and Analytics. 427–434, Springer, Singapore.

Mustafa, M. R. and M. H. Isa. 2014. "Comparative study of MLP and RBF neural networks for estimation of suspended sediments in Pari River, Perak." *Research Journal of Applied Sciences, Engineering and Technology*, 7 (18), 3837–3841.

Mustafa, M. R., M. H. Isa, and R. B. Rezaur. 2011. "A comparison of artificial neural networks for prediction of suspended sediment discharge in river: a case study in Malaysia." *World Academy of Science, Engineering and Technology*, 81, 372–376.

Mustafa, M. R., R. B. Rezaur, S. Saiedi, and M. H. Isa. 2012. "River suspended sediment prediction using various multilayer perceptron neural network training algorithms: a case study in Malaysia." *Water Resources Management*, 26 (7), 1879–1897.

Nourani, Vahid, Mohammad T. Alami, and Mohammad H. Aminfar. 2009. "A combined neural-wavelet model for prediction of Ligvanchai watershed precipitation." *Engineering Applications of Artificial Intelligence*, 22 (3), 466–472.

Olyaie, Ehsan, Hamid Zare Abyaneh, and Ali Danandeh Mehr. 2017. "A comparative analysis among computational intelligence techniques for dissolved oxygen prediction in Delaware river." *Geoscience Frontiers*, 8 (3), 517–527.

Rezapour, Omolbani Mohamad, Lee Teang Shui, and Desa Bin Ahmad. 2010. "Review of artificial neural network model for suspended sediment estimation." *Australian Journal of Basic and Applied Sciences*, 4 (8), 3347–3353.

Sahoo, A., S. Samantaray, and D. K. Ghose. 2019. "Stream flow forecasting in Mahanadi river basin using artificial neural networks." *Procedia Computer Science*, 157, 168–174.

Sahoo, A., U. K. Singh, M. H. Kumar, and Sandeep Samantaray. 2020. "Estimation of flood in a river basin through neural networks: a case study." In Communication Software and Networks. 755–763, Springer, Singapore.

Samantaray, Sandeep and Dillip K. Ghose. 2018. "Evaluation of suspended sediment concentration using descent neural networks." *Procedia Computer Science*, 132, 1824–1831.

Samantaray, S. and Ghose, D. K. 2019. "Sediment assessment for a watershed in arid region via neural networks." *Sādhanā*, 44 (10), 219.

Samantaray, S. and Ghose, D. K. 2020. "Assessment of suspended sediment load with neural networks in arid watershed." *Journal of the Institution of Engineers (India): Series A*, 101, 371–380.

Samantaray, S., A. Sahoo, and D. K. Ghose. 2019a. "Assessment of runoff via precipitation using neural networks: watershed modelling for developing environment in arid region." *Pertanika Journal of Science and Technology*, 27 (4), 2245–2263.

Samantaray, S., A. Sahoo, and D. K. Ghose. 2019b "Assessment of groundwater potential using neural network: a case study." In International Conference on Intelligent Computing and Communication. 655–664, Springer.

Samantaray, S. and A. Sahoo. 2020a. "Assessment of sediment concentration through RBNN and SVM-FFA in arid watershed, India." Smart Intelligent Computing and Applications, 701–709.

Samantaray, S. and A. Sahoo. 2020b. "Appraisal of runoff through BPNN, RNN, and RBFN in Tentulikhunti watershed: a case study." In Frontiers in Intelligent Computing. 258–267, Springer.

Samantaray, S. and A. Sahoo. 2020c. "Prediction of runoff using BPNN, FFBPNN, CFBPNN algorithm in arid watershed: a case study." *International Journal of Knowledge-based and Intelligent Engineering Systems*, 24 (3), 243–251.

Samantaray, S., O. Tripathy, A. Sahoo, and D. K. Ghose. 2020a. "Rainfall forecasting through ANN and SVM in Bolangir watershed, India." Smart Intelligent Computing and Applications, 767–774.

Samantaray, Sandeep, Abinash Sahoo, and Dillip K. Ghose. 2020b. "Assessment of sediment load concentration using SVM, SVM-FFA and PSR-SVM-FFA in arid watershed, India: a case study." *KSCE Journal of Civil Engineering*. 24, 1944–1957.

Samantaray, Sandeep, Abinash Sahoo, and Dillip K. Ghose. 2020c. "Prediction of sedimentation in an arid watershed using BPNN and ANFIS." ICT Analysis and Applications, 295–302.

Samet, Komeil, Khosrow Hoseini, Hojat Karami, and Mirali Mohammadi. 2019. "Comparison between soft computing methods for prediction of sediment load in rivers: Maku Dam case study." *Iranian Journal of Science and Technology*, 43 (1), 93–103.

Senthil Kumar, A. R., C. S. P. Ojha, Manish Kumar Goyal, R. D. Singh, and P. K. Swamee. 2012. "Modeling of suspended sediment concentration at Kasol in India using ANN, fuzzy logic, and decision tree algorithms." *Journal of Hydrologic Engineering*, 17 (3), 394–404.

Soleymani, Seyed Ahmad, Shidrokh Goudarzi, Mohammad Hossein Anisi, Wan Haslina Hassan, Mohd Yamani Idna Idris, Shahaboddin Shamshirband, Noorzaily Mohamed Noor, and Ismail Ahmedy. 2016. "A novel method to water level prediction using RBF and FFA." *Water Resources Management*, 30 (9), 3265–3283.

Tang, Zaiyong, Chrys De Almeida, and Paul A Fishwick. 1991. "Time series forecasting using neural networks vs. Box-Jenkins methodology." *Simulation*, 57 (5), 303–310.

Tfwala, Samkele S. and Yu Min Wang. 2016. "Estimating sediment discharge using sediment rating curves and artificial neural networks in the Shiwen River, Taiwan." *Water (Switzerland)*, 8 (2), 53.

Tou, J. T. and R. C. Gonzalez. 1974. Pattern Recognition Principles, Addison-Wesley, Reading, MA, 377.

Yadav, Arvind, Snehamoy Chatterjee, and S. K. Md. Equeenuddin. 2018. "Prediction of suspended sediment yield by artificial neural network and traditional mathematical model in Mahanadi River basin, India." *Sustainable Water Resources Management*, 4 (4), 745–59.

Yang, Xin-She. 2010a. "Firefly algorithm, stochastic test functions and design optimisation." *International Journal of Bio-Inspired Computation*, 2 (2), 78–84.

Yang, Xin-She. 2010b. "Nature-inspired metaheuristic algorithms," Luniver Press, United Kingdom.

13 Scheming of Runoff Using Hybrid ANFIS for a Watershed: Western Odisha, India

13.1 INTRODUCTION

Hydrological modelling is an influential method for investigating hydrologic systems essential for practising water resource engineers and research hydrologists working on design and improvement of combined procedure for water resources management. Runoff prediction is one of the most valuable processes involved in hydrological systems. Precise runoff estimation by utilizing evaporation, rainfall, and other hydrologic parameters is a significant problem in water resources engineering. Rainfall-runoff procedure is a complicated non-linear result of several hydrological constraints such as intensity of rainfall, evaporation, and geographical condition of watershed, penetration of water into soil, and depression storage along with interaction between surface water and groundwater flows. Hence, model of rainfall-runoff relationship cannot be developed by simple models. Hydrological study and its influences in a watershed depend on the prime weather parameters, maximum and minimum temperature, infiltration capacity, precipitation, and evapotranspiration loss.

Rainfall-runoff model was developed with the help of artificial neural network (ANN) models (Lin and Chen, 2004; Chen and Adams, 2006; Sridharam et al., 2020). Infiltration rate with consideration of slope was measured (Joshi and Tambe, 2010). Luo et al. (2011) employed regularized BPNN to model rainfall-runoff relationship of Xitiaoxi catchment and used Broyden Fletcher Goldfarb Shanno (BFGS) algorithm for comparing prediction performance of proposed models. Outcomes showed that regularized BPNN enhanced generalization capability and avoided overfitting efficiently outperforming BFGS. Lee et al. (2010) applied RBFN and BPNN models for predicting regional runoff and utilizing them in an effective way. Results revealed that RBFN model provided better stability and reliability in runoff prediction. Truatmoraka et al. (2016) applied BPNN to develop a model to predict water levels considering capacity of water discharge, height of basin at gauge sites, average rainfall-runoff, water level, and maximum capacity of water discharge at gauge sites. Result of prediction model showed high accurateness. Remesan et al. (2008) explored the capability of NN autoregressive with exogenous input (NARX) and ANFIS, for efficiently modelling rainfall-runoff relationship from antecedent information regarding rainfall and runoff. Findings from study revealed that both

NARX and ANFIS worked proficiently to model rainfall-runoff and provided high accuracy and consistency in predicting runoff values. Jothiprakash et al. (2009) used ANFIS for modelling rainfall-runoff relationship with lumped data of an intermittent river system called River Kanand in Maharashtra, India. El-shafie et al. (2011) applied an ANN model for predicting rainfall-runoff relation in a watershed situated in Tanakami area of Japan. Results revealed that applied ANN described behaviour of rainfall-runoff relationship more precisely compared to conventional regression model. Data-driven approaches for estimating runoff under different climatic conditions was assessed by Sarzaeim et al. (2017). Talei et al. (2010) investigated selection of inputs to be utilized by ANFIS for developing an application for runoff forecasting. The influence of algorithm of back propagation on rainfall runoff model was studied (Kumar et al., 2016). Rainfall-runoff model using ANN on a watershed of Pakistan was established (Ghumman et al., 2012). Zounemat-kermani et al. (2013) studied impact of flow records collected from upstream stations of Cahaba River, Alabama, on performance of Levenberg-Marquardt (LM) learning algorithm and RBFN models to predict daily runoff from a watershed and compared with that of a regression model. Results revealed that ANN models were more accurate in predicting runoff dynamics. Dastorani et al. (2010) explored the potential of ANN and ANFIS models for reconstructing missing flow data from gauging sites located at different climatic regions of Iran. Obtained results were compared with those of results from traditional methods and found that applied data-driven techniques are efficient and reliable for finding missing data using data from adjacent stations. The applicability and capability of different ANFIS techniques to predict daily and monthly runoff at multiple gauge sites are investigated (Nourani et al., 2009; Nourani and Komasi, 2013). Fuzzy logic (FL) is utilized to show flexibility and relationship amid input and output variables for rainfall-runoff modelling (Noury et al., 2008). Discharge in southern Taiwan using HEC-HMS and ANN model was predicted (Young and Liu, 2015). ANN-QPSO for forecasting daily runoff was used (Cheng et al., 2015). Kisi et al. (2013) used ANN, ANFIS, and Gene Expression Programming (GEP) for modelling process involved in rainfall and runoff relation for a small watershed in Turkey, utilizing 4 years of rainfall as well as runoff values. Runoff model which is influenced by density of grassed waterway was established (Singh, 2007). Dorado et al. (2003) used GP and ANN for modelling the impact of rain on the runoff in an urban basin. ANNs are promising tools for modelling of complex processes and to understand the procedure under analysis for assessment of proposed models (Chakravarti et al., 2015). The performance of MLP and RBFN models are comprehensively evaluated in terms of generalization properties, with hydrograph characteristics to predict uncertainty. Various neural networks (NNs) are employed to predict runoff at various gauge stations in India (Jimmy et al., 2020; Ghose and Samantaray, 2019; Samantaray et al., 2019a; Samantaray and Ghose, 2020; Samantaray and Sahoo, 2020a, c). Dehghani et al. (2019) investigated the application of integrated ANFIS-GWO to forecast multistep ahead influence rate of flow of a wastewater treatment plant situated in Isfahan town, Iran, and obtained results were compared with conventional ANFIS technique. Findings from study showed that hybrid ANFIS-GWO considerably improved prediction accuracy and performed more proficiently than ANFIS in nearly all of prediction prospects. Panahi et al.

(2020) used hybrid ANFIS-GWO, SVM-GWO, combination of bee algorithm (Bee) with both ANFIS and SVM, i.e. ANFIS-Bee and SVM-Bee, for developing landslide susceptibility maps (LSM) for Icheon Town located in South Korea and compared their accuracy in prediction. Results revealed that SVM-Bee and SVM-GWO models were found to be very effective techniques to evaluate landslide vulnerability mapping than other proposed models. Paryani et al. (2020) applied combination of ANFIS and GWO (ANFIS-GWO) models and also ANFIS and particle swarm optimization (ANFIS-PSO) models to generate LSM. Accuracy of proposed hybrid models were assessed using statistical evaluation measures. Results demonstrated that ANFIS-PSO model performed better than ANFIS-GWO in developing LSM in Karun watershed, Iran. Pourghasemi et al. (2019a) used a new collaborative model combining stepwise weight assessment ratio analysis (SWARA) with ANFIS-GWO for developing a multiple hazard map for earthquakes, landslides, and floods for managing hazard-susceptible zones of Lorestan Region, Iran. Performance of proposed model was satisfactory in creating combined multiple hazard map of selected study area. Hadavandi et al. (2018) proposed a new GWO-based ANN simulation model named GWNN for predicting tensile strength of siro-spun yarn in spinning mills. GWO model was utilized as a global search for determining optimal weights of MLP. Prediction accurateness of GWNN was assessed with MLP trained by BP algorithm and multiple linear regression (MLR) model and was found that developed GWNN model showed higher precision than other models. Tikhamarine et al. (2020) proposed effective hybrid models by combining GWO algorithm with SVM, MLP, and autoregression (AR) models for forecasting monthly stream flow of River Nile, Egypt. Potential of proposed models were validated based on various applied statistical measures. Results showed that proposed integrated models outperformed conventional methods and provided superior forecasts throughout training and testing periods for monthly inflow for all input conditions. Barman and Choudhury (2020) used a novel integrated model combining SVM with GWO for forecasting power system load on days of regional special events. Efficiency of proposed model was compared with SVM, ANN, SVM-PSO, and SVM combined with genetic algorithm (SVM-GA). Results demonstrated that forecasting accuracy of proposed hybrid SVM-GWO model outperformed all other models used in this study for Assam, India.

The objective of this study is to assess the impact of infiltration capacity of soil for predicting runoff using BPNN, ANFIS, and ANFIS-GWO algorithm.

13.2 STUDY AREA

Cuttack is the second largest city of Odisha, India, and one of the oldest. It is situated within geographic coordinates of 20°03″–20°40″N latitude and 84°58″–86°20″E longitude. It is bordered by River Mahanadi on northern side and River Kathajodi on southern side of the city. This district covers 3932 km^2 area and is densely inhabited. Cuttack experience tropical climatic conditions, with winter being cold and summer hot. Maximum temperature is well above 40°C in summer months and minimum is around 10°C recorded in winter months. Generally, summer continues from March till June and winter from month of October till February. Precipitation is usually

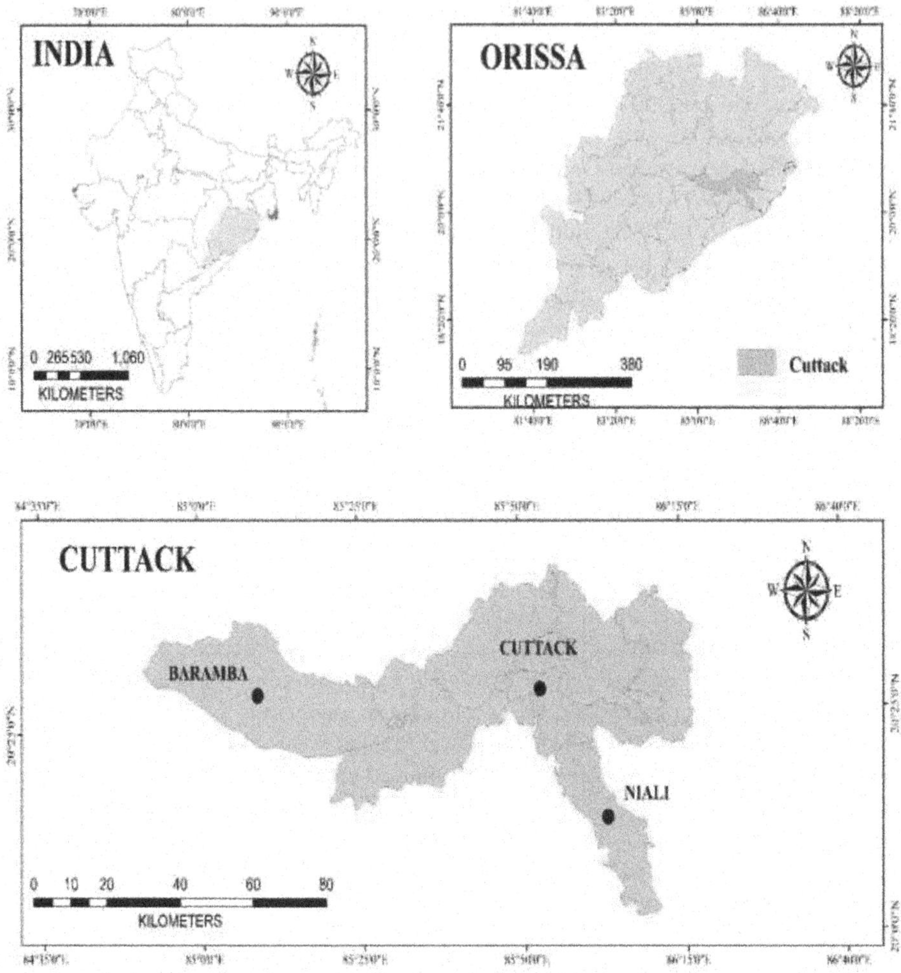

FIGURE 13.1 Study area showing Cuttack watershed.

heavy in monsoon occurring in months of July and August. This district receives average precipitation of about 1892.55 mm. Southwest monsoon is predominantly accountable for rain. Because of inadequate industrial facilities, agriculture is the key source of living, with around 75% of population depending on it. In this study, Baramba, Cuttack, and Niali watersheds are considered for runoff prediction, as shown in Figure 13.1.

The data on daily basis for rainfall, maximum temperature, minimum temperature, runoff, and infiltration loss are accumulated from India Meteorological Department (IMD), Bhubaneswar, for the period of the monsoon months (May to October) from 1990 to 2019. The data from 1990 to 2009 are used for training and data from 2010 to 2019 are used for testing the network.

13.3 METHODOLOGY

13.3.1 Artificial Neural Network

ANN is used for many applications to describe neural processing operations. Studies show the response of neuron randomly by averaging many observations to obtain predictable results. Transmission of signals through connection links associated with weights for some input represented in terms of output in any architecture. The signal of the output is achieved by activating some input through processor. BPNN and ANFIS are studied in individual sections.

13.3.2 BPNN

Back-propagation network helps in performing the error correction with the help of adjustable weights. In hidden layer, non-linear activation functions trigger the multiplicity of functions to activate the neurons and encompass the progress of activation towards output layer. During training, error propagation activation occurs from input to output through the transition of hidden layer. BPNN is performed through training, testing, and validation. Figures 13.2 and 13.3 show the architecture of two scenarios for developing BPNN model. This model consists of an input layer, one or more hidden, and one output layer in a feed-forward multiple layer NN (Samantaray and Ghose, 2018, 2019a, b). Input layer comprises I nodes; hidden layer comprises J nodes and output layer comprises k nodes. Hence, output z_k can be represented by the following equation:

$$z_k = f\left(b_{0k} + \sum_{j=1}^{J} b_{jk} . f\left(a_{0j} + \sum_{i=1}^{I} a_{ij} . x_i\right)\right) \tag{13.1}$$

where function f is the activation function, x_i is the quantity of input, a_{ij} and b_{jk} ($i = 1, 2,\ldots, I$; $j = 1, 2,\ldots, J$; $k = 1, 2,\ldots, K$) are weighted values, and a_{0j} and b_{0k} are deviations.

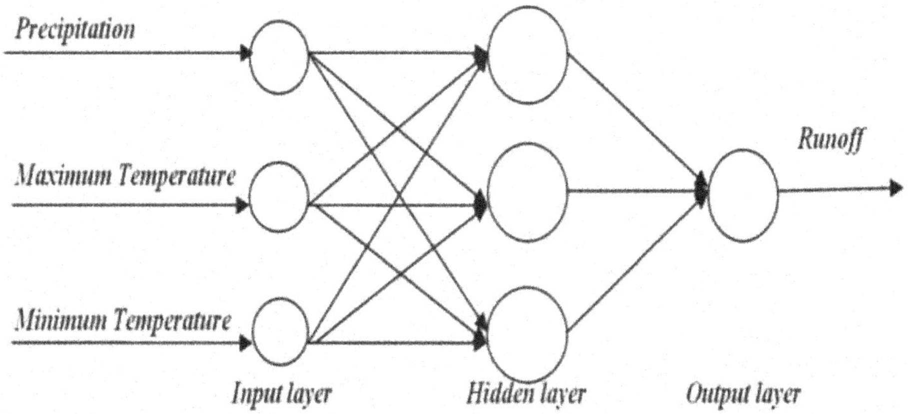

FIGURE 13.2 Architecture of BPNN for scenario 1.

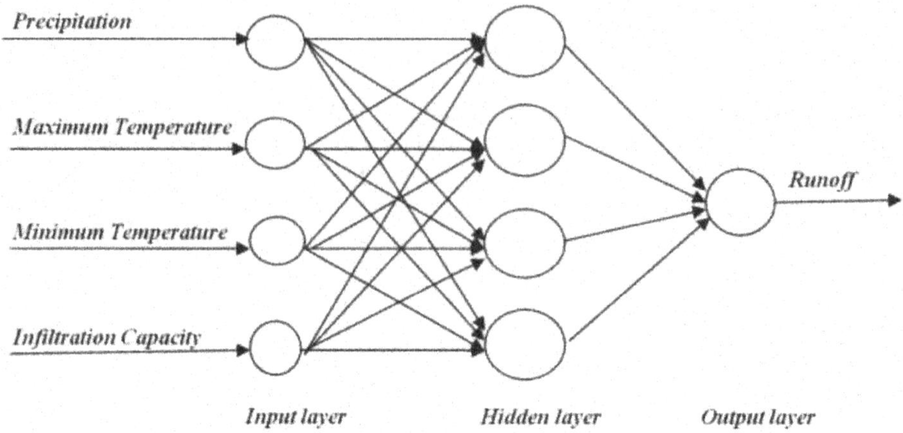

FIGURE 13.3 Architecture of BPNN for scenario 2.

Selecting transfer function of BPNN contains sufficient options; however, in BP unit, subsequent common selection principles can be utilized: monotonous, non-decreasing continuous function, and differentiable. Present research selects most common binary logistic sigmoid function. The following equation defines the sigmoid function:

$$f(x) = \frac{1}{1+e^{-x}} \tag{13.2}$$

where value ranges between (0, 1). In addition, if a linear function is chosen for this transfer function, for example $f(x) = x$, complete ANN architecture will be linearly influenced from input to output layer (Ghose and Samantaray, 2018; Sahoo et al., 2020b; Samantaray et al., 2019b).

13.3.3 ANFIS

Because of predictive ability in numerous field of study, FL is a very well accepted machine learning algorithm (Mathur et al., 2016; Dehnavi et al., 2015). Jang (1993) introduced this hybridization modelling technique. ANFIS gives an authorization for selecting various parameters and membership functions (Aghdam et al., 2016). FL provides exceptional results in accordance to fuzzy inference system (Chen et al., 2017). Proposed model could resolve model overfitting and non-linear complications (Pourghasemi et al., 2019b). ANFIS proposed here Takagi-Sugeno fuzzy inference system. ANFIS is modification over neural network intermingled through FL principles. The FL principle corresponds to a set of rules: a set of fuzzy if-then rules to learn capability of approximating non-linear functions. It is a hybrid network called neuro fuzzy networks consisting properties of both neural networks and FL principles (Samantaray et al., 2020a, b; Sahoo et al., 2020a, b). Model architecture of proposed ANFIS model is shown in Figure 13.4. The neural network easily learns from the data, but the interpretation of knowledge is difficult in case of neural network. So

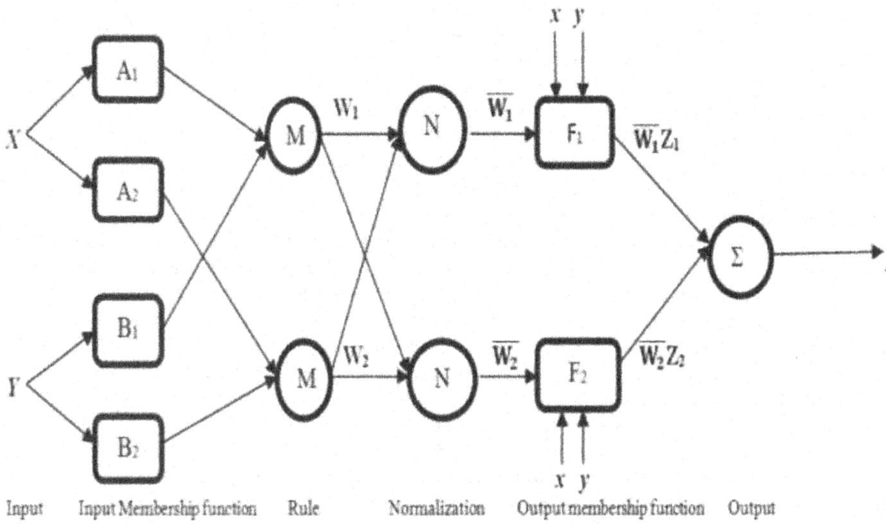

FIGURE 13.4 Architecture of ANFIS.

interpretation of knowledge acquired by neural network is advanced through complex comprehension by modification of weight and architecture. Since FL itself is not capable of learning data, it is hybridized through the algorithm of neural network and achievement conceived is called ANFIS.

The Takagi and Sugeno system is based on two if-then rules, which are as follows:

$$\text{Rule 1: If } x = A_1 \text{ and } y = B_1, \text{ then } Z_1 = p_1x + q_1y + r_1 \tag{13.3}$$

$$\text{Rule 2: If } x = A_2 \text{ and } y = B_2, \text{ then } Z_2 = p_2x + q_2y + r_2 \tag{13.4}$$

where $x(A_1, A_2)$ and y (B_1, B_2) are inputs; A_1, A_2, B_1, and B_2 are fuzzy sets found during training process; and p_{ij}, q_{ij} and r_{ij} $(i, j = 1, 2)$ are parameters attained in training period (Zhang et al., 2010).

Applying previously used trial and error technique to achieve best values of aforementioned parameters consumes a lot of time, which is a major issue in ANFIS. Several research hydrologists applied various robust methodologies for optimizing ANFIS model parameters (Aghdam et al., 2017; Khosravi et al., 2018; Ahmadlou et al., 2018; Tien Bui et al., 2012; Chen et al., 2019;). For solving this issue, a new optimizer called GWO was integrated with ANFIS in this research to find optimal parameters and finally predict runoff.

13.3.4 GWO

Mirjalili et al. (2014) proposed GWO algorithm mimicking social behaviour and hierarchy of grey wolves. GWO is a novel metaheuristic optimization algorithm. In general, the wolves pack is distributed into four categories: alpha (α), beta (β), delta (δ) and omega (ω). Alpha (α) wolf is the most dominant wolf and is the leader

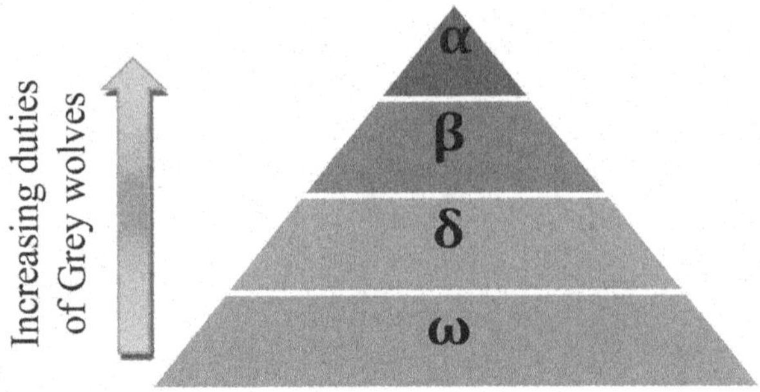

FIGURE 13.5 Social hierarchy of grey wolves.

of wolves pack. The level of domination continues, decreasing from α to ω, as presented in Figure 13.5. Mechanism involved in GWO is carried out by splitting a solution set into four groups for a specified optimization problem. α, β, and δ wolves are first three solutions, whereas residual solutions fall in group of ω wolves. For implementing this mechanism, hierarchical way is updated in each iteration on the basis of three optimal solutions. The significant approach involved in GWO is to search, encircle, hunt, and finally attack the prey.

Previous to the hunting procedure, grey wolves encircle the victim. Encircling conduct of grey wolves is represented in subsequent equation:

$$\vec{X}(t+1) = \overrightarrow{X_P}(t) - \vec{A}.\vec{D} \tag{13.5}$$

where $\vec{X}(t+1)$ is the next position of any wolf, $\overrightarrow{X_P}(t)$ is the grey wolf position vector, t is the present iteration, $\vec{A}$ is the coefficient of matrix, and $\vec{D}$ is the distance that separates wolf and victim. This is estimated using the following equation:

$$\vec{D} = \left|\vec{C}.\overrightarrow{X_P}(t) - \vec{X}(t)\right| \tag{13.6}$$

$$\vec{A} = 2\vec{a}.\vec{r_1} - \vec{a} \tag{13.7}$$

$$\vec{C} = 2.\vec{r_2} \tag{13.8}$$

where $\vec{r_1}$ and $\vec{r_2}$ are arbitrarily produced from (0–1).

Preceding equations allow a solution for relocating around prey in a two-dimensional search space. Nevertheless, this is insufficient for simulating societal intellect of grey wolves. For simulating the prey, pre-eminent solution achieved until now considered as α wolf is nearer to position of prey; however, global optimum solution is not known. Hence, it is anticipated that the topmost three results have a better awareness about their position, so remaining wolves must be obliged for updating their positions utilizing resulting equations:

$$\vec{X}(t+1) = \frac{\overrightarrow{X_1} + \overrightarrow{X_2} + \overrightarrow{X_3}}{3} \tag{13.9}$$

where $\overrightarrow{X_1}$, $\overrightarrow{X_2}$ and $\overrightarrow{X_3}$ are calculated utilizing subsequent equations:

$$\overrightarrow{X_1} = \overrightarrow{X_\alpha}(t) - \overrightarrow{A_1} * \overrightarrow{D_\alpha} \quad (13.10)$$

$$\overrightarrow{X_2} = \overrightarrow{X_\beta}(t) - \overrightarrow{A_2} * \overrightarrow{D_\beta} \quad (13.11)$$

$$\overrightarrow{X_3} = \overrightarrow{X_\delta}(t) - \overrightarrow{A_3} * \overrightarrow{D_\delta} \quad (13.12)$$

where are given by: D_α, D_β, and D_δ

$$\overrightarrow{D_\alpha} = \left|\overrightarrow{C_1}.\overrightarrow{X_\alpha}(t) - \vec{X}\right| \quad (13.13)$$

$$\overrightarrow{D_\beta} = \left|\overrightarrow{C_2}.\overrightarrow{X_\beta}(t) - \vec{X}\right| \quad (13.14)$$

$$\overrightarrow{D_\delta} = \left|\overrightarrow{C_1}.\overrightarrow{X_\delta}(t) - \vec{X}\right| \quad (13.15)$$

The encircling and attacking of prey repeatedly continues till an optimal solution is achieved or it reaches maximum iterations.

13.3.5 Model Evaluation

In scenario 1 monthly precipitation, minimum and maximum temperatures are taken as input. Monthly rainfall, minimum temperature, maximum temperature, and infiltration capacity are taken as input to predict runoff as output for scenario 2. A dataset of 30 years is taken into consideration for model development.

The following evaluation indices are used to measure the performance of the model:

$$\text{Coefficient of determination } \left(R^2\right) = 1 - \frac{\left(\sum_{i=1}^{N} x_{comp}^{i} - \bar{x}_{comp}\right)^2}{\left(\sum_{i=1}^{N} x_{obs}^{i} - \bar{x}_{obs}\right)^2} \quad (13.16)$$

$$\text{Mean squared error MSE} = \frac{1}{n}\sum_{j=1}^{n}\left(x_{comp}^{i} - x_{obs}^{i}\right)^2 \quad (13.17)$$

$$\text{Root mean squared error RMSE} = \frac{\sum_{i=1}^{N}\left(x_{comp}^{i} - \bar{x}_{comp}\right)\left(x_{obs}^{i} - \bar{x}_{obs}\right)}{\sqrt{\sum_{i=1}^{N}\left(x_{comp}^{i} - \bar{x}_{comp}\right)^2\left(x_{obs}^{i} - \bar{x}_{obs}\right)^2}} \quad (13.18)$$

where

x_{comp}^{i} = predicted data
x_{obs}^{i} = observed data
$\bar{x}_{comp}$ = mean predicted data
$\bar{x}_{obs}$ = mean observed data

The RMSE is applied for measuring forecasting accuracy, which produce a positive value by squaring errors. RMSE increase from zero to large positive value as differences amid forecasting and observation become progressively high. High value for R^2 (up to 1) and small value of RMSE illustrate high model efficiency and vice versa.

13.4 RESULTS AND DISCUSSIONS

13.4.1 BPNN

To evaluate various architectures and their efficiencies, various transfer functions – tangential sigmoidal, logarithmic sigmoidal, and purelin – are applied. Evaluation of model architecture MSE (training and testing), RMSE (training and testing), and R^2 values is presented in Table 13.1. The result found that Tan-sig function provides prominent value of performance for all three stations.

In Table 13.1 for Tan-sig, Log-sig, and Purelin function with various hidden layers 1, 2, 3, 4, …, 9 are considered for performance evaluation. At Baramba station, Tan-sig function found the best model (3-9-1) with MSE training and testing value 0.00624 and 0.00947, RMSE training and testing value 0.07695 and 0.08112, and R^2 for training and testing are 0.93195 and 0.90657, respectively. Similarly, at Cuttack and Niali, Tan-sig function found prominent R^2 value as 0.93201 and 0.93219 during training period, respectively. For Log-sig function, best model architecture is observed to be 3-7-1 with MSE training and testing values 0.00648 and 0.00958, RMSE training and testing values 0.07885 and 0.08138, and R^2 training and testing values 0.93179 and 0.90579, respectively, at Niali station. In case of Purelin function, best model architecture is 3-5-1 possessing MSE training and testing values 0.00637 and 0.00951, RMSE training and testing values 0.07729 and 0.08126, and R^2 training and testing values 0.93187 and 0.90613, respectively, at Niali gauge station.

From Table 13.2, it can be found that Tan-sig function (4-4-1) shows best values of model performance which possess MSE training and testing value 0.00523 and 0.00873, RMSE training and testing values 0.06566 and 0.07975, and R^2

TABLE 13.1
Comparison of Best Results of BPNN for Scenario 1

	Transfer		MSE		RMSE		R^2	
Station	**Function**	**Architecture**	**Training**	**Testing**	**Training**	**Testing**	**Training**	**Testing**
Baramba	Tan-sig	3-9-1	**0.00624**	**0.00947**	**0.07695**	**0.08112**	**0.93195**	**0.90657**
	Log-sig	3-4-1	0.00686	0.00997	0.08263	0.08194	0.93116	0.9044
	Purelin	3-6-1	0.00672	0.00983	0.08092	0.08188	0.93147	0.90475
Cuttack	Tan-sig	3-4-1	**0.00616**	**0.00938**	**0.06918**	**0.08094**	**0.93201**	**0.90691**
	Log-sig	3-6-1	0.00667	0.00976	0.08058	0.08176	0.93148	0.90502
	Purelin	3-3-1	0.00653	0.00964	0.07919	0.08153	0.93163	0.90531
Niali	Tan-sig	3-2-1	**0.00601**	**0.0093**	**0.06891**	**0.08087**	**0.93219**	**0.90724**
	Log-sig	3-7-1	0.00648	0.00958	0.07885	0.08138	0.93179	0.90579
	Purelin	3-5-1	0.00637	0.00951	0.07729	0.08126	0.93187	0.90613

TABLE 13.2
Comparison of Best Results of BPNN for Scenario 2

Station	Transfer Function	Architecture	MSE		RMSE		R^2	
			Training	Testing	Training	Testing	Training	Testing
Baramba	Tan-sig	4-5-1	**0.00539**	**0.00883**	**0.06597**	**0.07993**	**0.93418**	**0.91051**
	Log-sig	4-8-1	0.00595	0.00924	0.06857	0.08075	0.93226	0.90776
	Purelin	4-2-1	0.00588	0.00916	0.06687	0.08052	0.93233	0.9081
Cuttack	Tan-sig	4-7-1	**0.00531**	**0.00879**	**0.06579**	**0.07986**	**0.9369**	**0.91612**
	Log-sig	4-3-1	0.00572	0.00909	0.06686	0.08037	0.93253	0.90845
	Purelin	4-9-1	0.00563	0.00897	0.06677	0.08025	0.93258	0.90887
Niali	Tan-sig	4-4-1	**0.00523**	**0.00873**	**0.06566**	**0.07975**	**0.93727**	**0.91784**
	Log-sig	4-8-1	0.00557	0.00891	0.06652	0.08014	0.93278	0.90929
	Purelin	4-6-1	0.00545	0.00888	0.06611	0.08005	0.93528	0.90973

training and testing values 0.93727 and 0.91784 at Niali station. Similarly, at Cuttack and Baramba station, Tan-sig function produces prominent value of performance with R^2 values 0.9369 and 0.93418, respectively. On overall basis, it is found that scenario 2 performs better than the scenario 1. The graphs with best values for runoff from maximum temperature, precipitation, minimum temperature, and infiltration loss utilizing (BPNN) with Tan-sig transfer function are shown in Figure 13.6.

13.4.2 ANFIS

Membership function like Tri, Trap, Gbell, Gauss, Gauss2, and Pi are developed to find the performance of various architectures in scenario 1 and scenario 2. For each membership function, evaluation of model architecture are done with the help of MSE training, testing, RMSE training, testing, and coefficient of determination.

From Table 13.3, we found that infiltration capacity shows best results as compared to the exclusion of infiltration. The performance of best result for scenario 2 with Gbell membership function shows MSE training, testing are 0.00332 and 0.0027 with R^2 values as 0.95737 and 0.9276 for training and testing, respectively, at Niali gauge station. It is found that for all three stations, Gbell function shows best value of performance for both the scenarios. Best results for runoff from maximum temperature, precipitation, minimum temperature, infiltration capacity using ANFIS with Gbell membership function are represented in Figure 13.7.

13.4.3 ANFIS-GWO

The outcomes of model performance utilizing training dataset (Figure 13.8) presented that ANFIS-GWO model (R^2 = 0.97282) for scenario 2 performed better than the scenario 1 (R^2 = 0.96595). In testing phase, ANFIS-GWO model once again attained highest accuracy with MSE and R^2 values of 0.00182 and 0.95074,

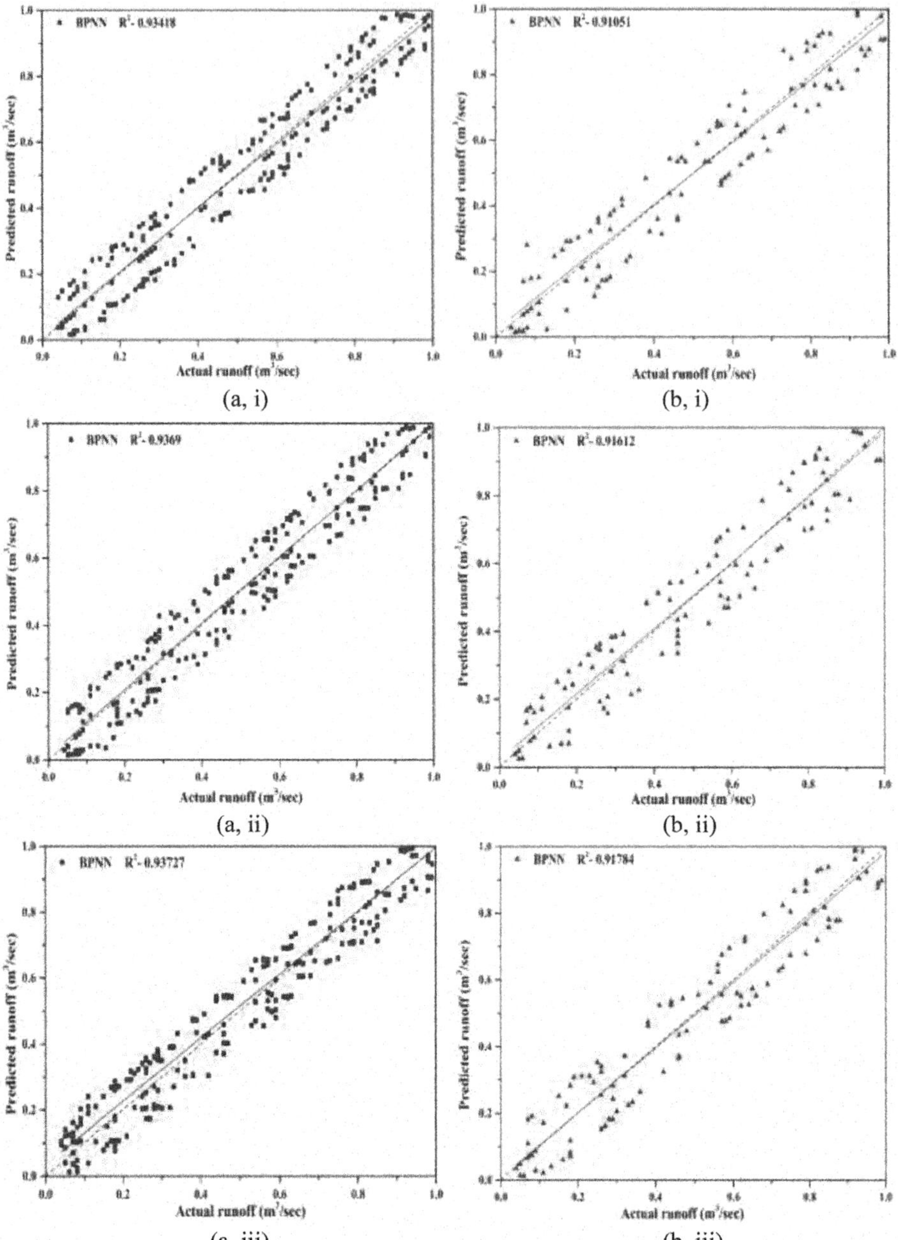

FIGURE 13.6 Observed versus predicted runoff model using BPNN (Log-sig) for (a) training and (b) testing phases at (i) Baramba, (ii) Cuttack, and (iii) Niali station.

TABLE 13.3
Comparison of Best Results of ANFIS for Scenario 1 and Scenario 2

Scenario	Station	Membership Function	MSE		RMSE		Coefficient of Determination (R^2)	
			Training	Testing	Training	Testing	Training	Testing
I	Baramba	Tri	0.00458	0.02962	0.0646	0.07634	0.93899	0.91868
		Trap	0.00451	0.00734	0.06455	0.06776	0.93942	0.91884
		Gbell	**0.00366**	**0.00363**	**0.05922**	**0.02079**	**0.94489**	**0.9208**
		Gauss	0.00375	0.00385	0.06211	0.02874	0.94191	0.91978
		Gauss2	0.00374	0.00376	0.06202	0.02227	0.94231	0.91994
		Pi	0.00516	0.00867	0.0655	0.07968	0.93769	0.91808
	Cuttack	Tri	0.00424	0.00529	0.06446	0.06087	0.93975	0.91899
		Trap	0.00379	0.00492	0.0639	0.05802	0.94022	0.91919
		Gbell	**0.00365**	**0.00362**	**0.05894**	**0.02077**	**0.94527**	**0.92107**
		Gauss	0.00372	0.00369	0.06168	0.02085	0.94273	0.92013
		Gauss2	0.00371	0.00367	0.0608	0.02084	0.94338	0.92034
		Pi	0.00499	0.00834	0.06509	0.07917	0.93831	0.91826
	Niali	Tri	0.00377	0.00406	0.06291	0.05712	0.94064	0.91936
		Trap	0.00376	0.00392	0.06257	0.03546	0.94147	0.91955
		Gbell	**0.00364**	**0.0036**	**0.05891**	**0.02071**	**0.94578**	**0.92116**
		Gauss	0.00368	0.00365	0.05977	0.02082	0.94385	0.92056
		Gauss2	0.00367	0.00364	0.05975	0.0208	0.94441	0.92072
		Pi	0.00483	0.00802	0.06489	0.07875	0.93867	0.91843
II	Baramba	Tri	0.00357	0.00347	0.05802	0.02055	0.94796	0.92179
		Trap	0.00355	0.00345	0.05779	0.0205	0.94836	0.92188
		Gbell	**0.00335**	**0.00285**	**0.0554**	**0.01838**	**0.95473**	**0.92415**
		Gauss	0.00343	0.00325	0.05648	0.01967	0.95077	0.92277
		Gauss2	0.00341	0.00322	0.05635	0.01946	0.95136	0.92294
		Pi	0.00363	0.00352	0.05866	0.02066	0.94639	0.9213
	Cuttack	Tri	0.00352	0.00342	0.05757	0.02048	0.94878	0.92204
		Trap	0.0035	0.0034	0.05724	0.02045	0.94902	0.92218
		Gbell	**0.00334**	**0.00274**	**0.05522**	**0.01819**	**0.95735**	**0.92693**
		Gauss	0.0034	0.00315	0.05617	0.01924	0.95173	0.92306
		Gauss2	0.00339	0.00314	0.05589	0.01899	0.95259	0.92341
		Pi	0.00362	0.00351	0.05834	0.02062	0.94687	0.92143
	Niali	Tri	0.00348	0.00334	0.05701	0.02041	0.94954	0.92235
		Trap	0.00346	0.00328	0.05672	0.01979	0.95016	0.92258
		Gbell	**0.00332**	**0.0027**	**0.05496**	**0.01794**	**0.95737**	**0.9276**
		Gauss	0.00338	0.00309	0.05574	0.0186	0.95297	0.92364
		Gauss2	0.00337	0.00306	0.05557	0.01855	0.95325	0.92389
		Pi	0.00359	0.0035	0.0582	0.0206	0.94758	0.92156

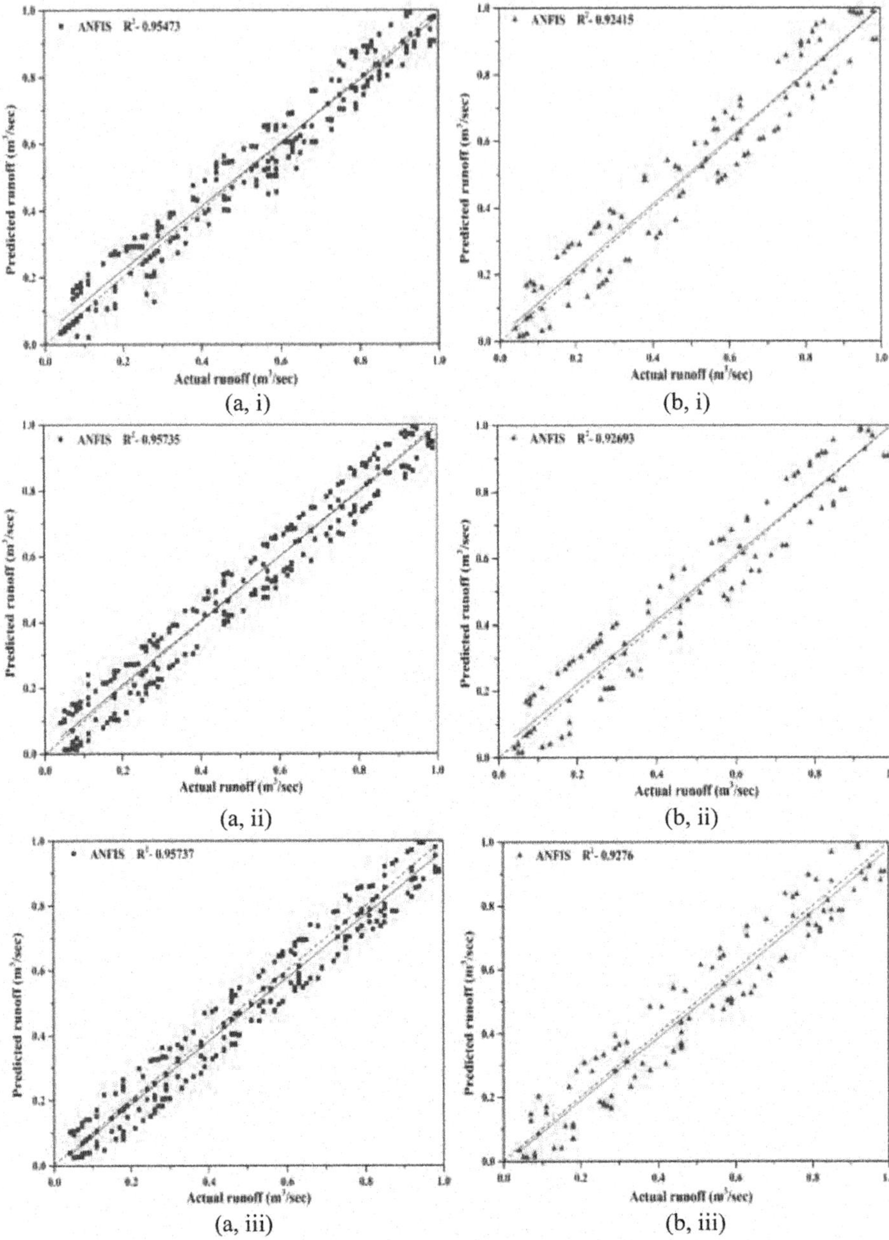

FIGURE 13.7 Observed versus predicted runoff model using ANFIS (Gbell) for (a) training and (b) testing phases at (i) Baramba, (ii) Cuttack, and (iii) Niali station.

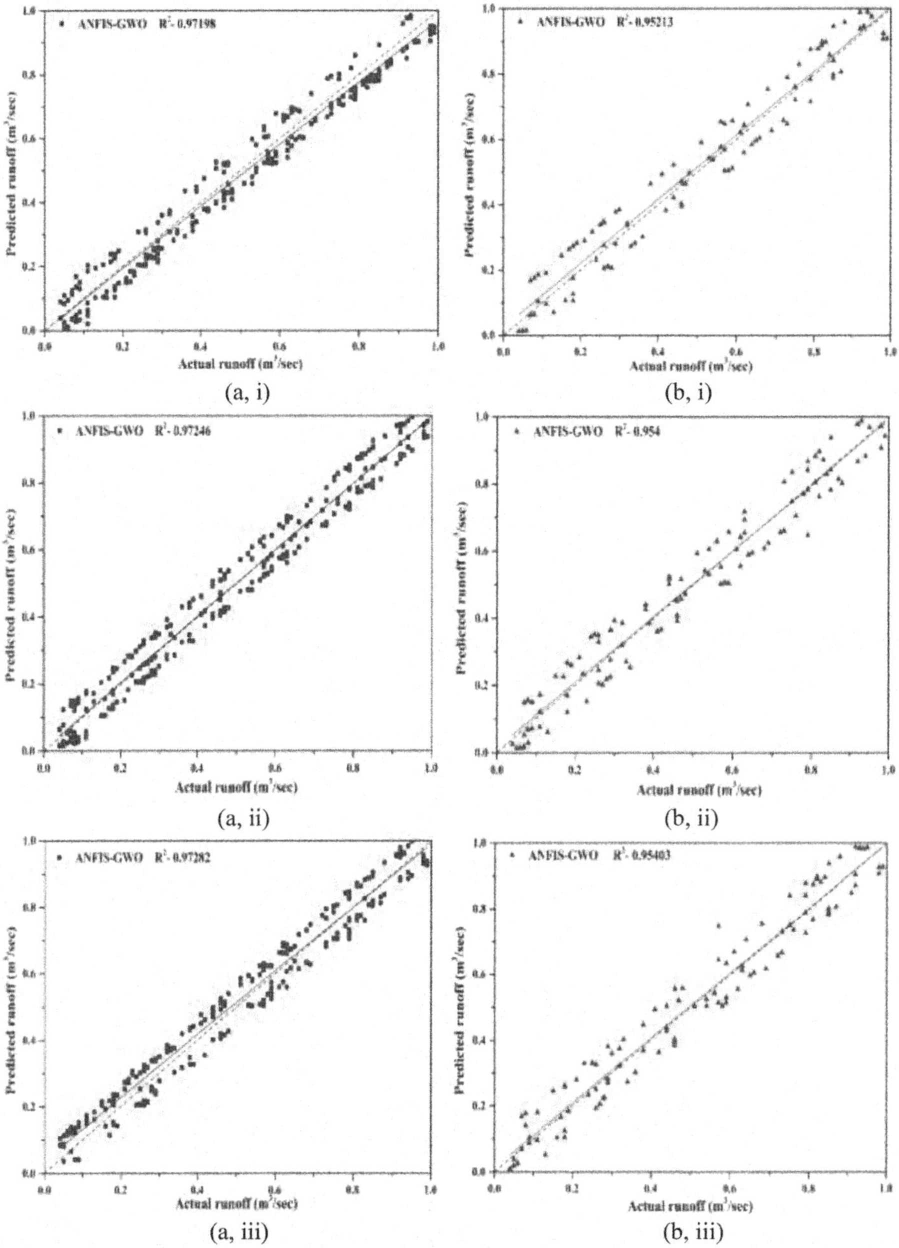

FIGURE 13.8 Observed versus predicted runoff model using ANFIS-GWO (Gbell) for (a) training and (b) testing phases at (i) Baramba, (ii) Cuttack, and (iii) Niali station.

respectively, for all three gauge stations. Here also scenario 2 gives best value of performance than scenario 1 for all three stations. A detailed comparison of both scenarios for ANFIS-GWO is shown in Table 13.4.

Linear scale plot of actual versus predicted runoff for developed models of proposed sites are presented in Figure 13.9. Results demonstrate that estimated peak

TABLE 13.4
Comparison of Best Results of ANFIS-GWO for Scenario 1 and Scenario 2

		Membership	MSE		RMSE		Coefficient of Determination (R^2)	
Scenario	Station	Function	Training	Testing	Training	Testing	Training	Testing
I	Baramba	Trap	0.00319	0.00254	0.05384	0.01674	0.96135	0.93004
		Tri	0.00323	0.00257	0.05416	0.01707	0.96077	0.92987
		Gbell	0.00296	0.00226	0.0511	0.01398	0.96518	0.93761
		Pi	0.0033	0.00264	0.05479	0.01769	0.95823	0.92818
		Gauss	0.00304	0.0024	0.05279	0.01555	0.96325	0.93331
		Gauss2	0.00303	0.00238	0.05245	0.01508	0.96366	0.93464
	Cuttack	Trap	0.00312	0.00248	0.05362	0.01624	0.96223	0.93141
		Tri	0.00318	0.00251	0.05375	0.01658	0.96184	0.93059
		Gbell	0.00294	0.00224	0.05091	0.0137	0.96553	0.93855
		Pi	0.00329	0.00262	0.05457	0.0175	0.95869	0.92869
		Gauss	0.00302	0.00237	0.05221	0.0148	0.96397	0.93599
		Gauss2	0.003	0.00235	0.05202	0.01459	0.96418	0.93628
	Niali	Trap	0.00307	0.00243	0.05304	0.01582	0.96291	0.93237
		Tri	0.00309	0.00245	0.0534	0.01603	0.96267	0.93183
		Gbell	**0.00293**	**0.00222**	**0.05065**	**0.01345**	**0.96595**	**0.93899**
		Pi	0.00326	0.00259	0.05441	0.01725	0.96008	0.92948
		Gauss	0.00299	0.00233	0.05172	0.0144	0.96443	0.93671
		Gauss2	0.00297	0.00229	0.05137	0.01416	0.96484	0.93706
II	Baramba	Trap	0.00282	0.00208	0.0498	0.01228	0.96749	0.94283
		Tri	0.00286	0.00211	0.0499	0.01253	0.96713	0.94191
		Gbell	0.00213	0.0018	0.03063	0.00934	0.97198	0.95213
		Pi	0.00291	0.0022	0.05057	0.01324	0.96609	0.93998
		Gauss	0.00262	0.00193	0.04859	0.01062	0.96966	0.94834
		Gauss2	0.00267	0.00195	0.04872	0.01098	0.96927	0.94749
	Cuttack	Trap	0.00277	0.00202	0.04931	0.01167	0.96815	0.94461
		Tri	0.00278	0.00205	0.04938	0.01196	0.96791	0.94354
		Gbell	0.00184	0.00177	0.02971	0.00917	0.97246	0.954
		Pi	0.0029	0.00217	0.05044	0.01298	0.96647	0.94016
		Gauss	0.00256	0.00188	0.04694	0.01014	0.97009	0.95002
		Gauss2	0.00259	0.0019	0.04704	0.01039	0.96997	0.94966
	Niali	Trap	0.0027	0.00198	0.04893	0.01116	0.96889	0.94658
		Tri	0.00275	0.002	0.04909	0.01149	0.96844	0.94545
		Gbell	**0.00115**	**0.00174**	**0.02692**	**0.00908**	**0.97282**	**0.95403**
		Pi	0.00289	0.00215	0.05018	0.01272	0.96686	0.94074
		Gauss	0.00224	0.00182	0.03292	0.00966	0.97052	0.95074
		Gauss2	0.00254	0.00184	0.03395	0.00983	0.97026	0.95041

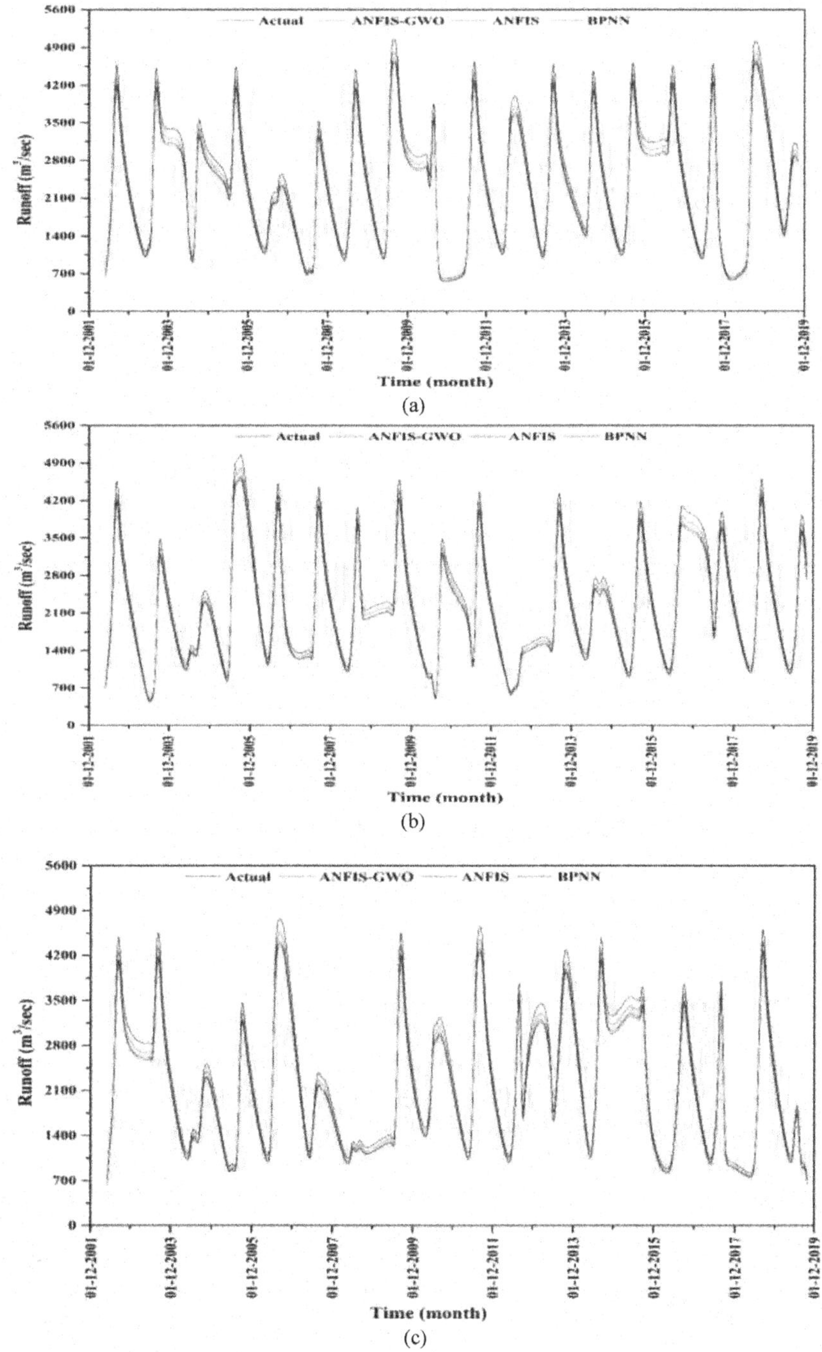

FIGURE 13.9 Deviation of actual and predicted runoff at (a) Baramba, (b) Cuttack, and (c) Niali gauge stations during testing phase.

TABLE 13.5
Performance Index of Different Approaches for Sediment Load Prediction

	BPNN			ANFIS			ANFIS-GWO		
Stations	**MSE**	**RMSE**	R^2	**MSE**	**RMSE**	R^2	**MSE**	**RMSE**	R^2
Baramba	0.00539	0.06597	0.93418	0.00335	0.0554	0.95473	0.00213	0.03063	0.97198
Cuttack	0.00531	0.06579	0.9369	0.00334	0.05522	0.95735	0.00184	0.02971	0.97246
Niali	0.00523	0.06566	0.93727	0.00332	0.05496	0.95737	0.00115	0.02692	0.97282

runoff is 4711.142 m^3/s, 4766.598 m^3/s, 4925.262 m^3/s for BPNN, ANFIS, and ANFIS-GWO against actual peak 5134.76 m^3/s for station Niali. Approximated peak runoff is 4656.016 m^3/s, 4707.08 m^3/s, and 4863.846 m^3/s for BPNN, ANFIS, and ANFIS-GWO, respectively, adjacent to actual peak 5106.4 m^3/s for Cuttack station. For Baramba gauging station, tangible runoff is 5054.42 m^3/s aligned with predicted runoff 4637.43 m^3/s, 4692.018 m^3/s, 4848.2 m^3/s for BPNN, ANFIS, and ANFIS-GWO, respectively. From present research, it is observed that Niali station among the proposed gauging stations indicate best value of model performance for all three techniques.

13.4.4 Comparison of Pre-eminent Outcomes

MSE, RMSE, and R^2 are applied for evaluating performance of BPNN, ANFIS, and ANFIS-GWO models for three gauging stations. Assessment of performance indices are indicated in Table 13.5 illustrating effectiveness of each model. Assessment of runoff is very significant and therefore techniques utilized in this study are important to demonstrate runoff information. Thus, computation of RMSE, R^2, and MSE values are vital to predict runoff. It is observed that ANFIS-GWO model performed well compared to BPNN and ANFIS.

13.5 CONCLUSIONS

Two scenarios have been considered for predicting runoff in different gestures with specific impacts. When infiltration capacity is added to scenario 1, the performance of model architecture improves, which indicates the influence of infiltration capacity of soil in the proximity of study area. From the above results and discussion, it is found that infiltration loss plays an important role for developing the model to predict runoff. Scenario 2 performs best results for all three networks. The best performance is found with Gbell membership function having coefficient of determination in training and testing as 0.97282 and 0.95403, respectively, through the ANFIS-GWO algorithm. The influence of infiltration capacity improves the model prediction for the watershed. This work is useful for estimating the runoff and their abstracts for controlling the soil erosion, planning, design, and managing of the watershed. However, the integration of advanced technique required to be investigated for improving results through management of watershed.

REFERENCES

Aghdam, I. N., M. H. M. Varzandeh, and B. Pradhan. 2016. "Landslide susceptibility mapping using an ensemble statistical index (Wi) and adaptive neuro-fuzzy inference system (ANFIS) model at Alborz Mountains (Iran)." *Environmental Earth Sciences*, 75 (7), 553.

Aghdam, I. N., B. Pradhan, and M. Panahi. 2017. "Landslide susceptibility assessment using a novel hybrid model of statistical bivariate methods (FR and WOE) and adaptive neuro-fuzzy inference system (ANFIS) at southern Zagros Mountains in Iran." *Environmental Earth Sciences*, 76 (6), 237.

Ahmadlou, M., M. Karimi, S. Alizadeh, A. Shirzadi, D. Parvinnejhad, H. Shahabi, and M. Panahi. 2018. "Flood susceptibility assessment using integration of adaptive network-based fuzzy inference system (ANFIS) and biogeography-based optimization (BBO) and BAT algorithms (BA)." *Geocarto International*, 34, 1–21.

Barman, Mayur and Nalin Behari Dev Choudhury. 2020. "A similarity based hybrid GWO-SVM method of power system load forecasting for regional special event days in anomalous load situations in Assam, India." *Sustainable Cities and Society*, 61, 102311.

Chakravarti, Ankit, Nitin Joshi, and Himanshu Panjiar. 2015. "Rainfall runoff analysis using the artificial neural network." *Indian Journal of Science and Technology*, 8 (14), 1–7.

Chen, Jieyun and Barry J. Adams. 2006. "A framework for urban storm water modeling and control analysis with analytical models." *Water Resources Research*, 42 (6), W06419, 1–13.

Chen, W., Pourghasemi, H.R., Panahi, M., Kornejady, A., Wang, J., Xie, X. and Cao, S., 2017. "Spatial prediction of landslide susceptibility using an adaptive neuro-fuzzy inference system combined with frequency ratio, generalized additive model, and support vector machine techniques." *Geomorphology*, 297, 69–85.

Chen, W., Panahi, M., Tsangaratos, P., Shahabi, H., Ilia, I., Panahi, S., and Ahmad, B. B. 2019. "Applying population-based evolutionary algorithms and a neuro-fuzzy system for modeling landslide susceptibility." *Catena*, 172, 212–231.

Cheng, Chun-tian, Wen-jing Niu, Zhong-kai Feng, Jian-jian Shen, and Kwok-wing Chau. 2015. "Daily reservoir runoff forecasting method using artificial neural network based on quantum-behaved particle swarm optimization." *Water*, 7 (8), 4232–46.

Dastorani, Mohammad T., Alireza Moghadamnia, Jamshid Piri, and Miguel Rico-Ramirez. 2010. "Application of ANN and ANFIS models for reconstructing missing flow data." *Environmental Monitoring and Assessment*, 166 (1–4), 421–34.

Dehghani, Majid, Akram Seifi, and Hossien Riahi-Madvar. 2019. "Novel forecasting models for immediate-short-term to long-term influent flow prediction by combining ANFIS and grey wolf optimization." *Journal of Hydrology*, 576, 698–725.

Dehnavi, A., Aghdam, I. N., Pradhan, B., and Varzandeh, M. H. M. 2015. "A new hybrid model using step-wise weight assessment ratio analysis (SWARA) technique and adaptive neuro-fuzzy inference system (ANFIS) for regional landslide hazard assessment in Iran." *Catena*, 135, 122–148.

Dorado, Julian, Juan R. Rabuñal, Alejandro Pazos, Daniel Rivero, Antonino Santos, and JerÓNIMO Puertas. 2003. "Prediction and modeling of the rainfall-runoff transformation of a typical urban basin using ANN and GP." *Applied Artificial Intelligence*, 17 (4), 329–43.

El-Shafie, A., Muhammad Mukhlisin, Ali A. Najah, and Mohd Raihan Taha. 2011. "Performance of artificial neural network and regression techniques for rainfall-runoff prediction." International Journal of Physical Sciences, 6 (8), 1997–2003.

Ghose, D.K. and Samantaray, S., 2018. "Modelling sediment concentration using back propagation neural network and regression coupled with genetic algorithm." *Procedia Computer Science*, 125, 85–92.

Ghose, D. K. and Samantaray, S. 2019. "Estimating runoff using feed-forward neural networks in scarce rainfall region." In Smart Intelligent Computing and Applications. 53–64, Springer, Singapore.

Ghumman, A. R., Yousry M. Ghazaw, A. R. Sohail, and K. Watanabe. 2011. "Runoff forecasting by artificial neural network and conventional model." *Alexandria Engineering Journal*, 50 (4), 345–50.

Hadavandi, Esmaeil, Sobhan Mostafayi, and Parham Soltani. 2018. "A grey wolf optimizer-based neural network coupled with response surface method for modeling the strength of siro-spun yarn in spinning mills." *Applied Soft Computing*, 72, 1–13.

Jang, J.-S. 1993. "ANFIS: adaptive-network-based fuzzy inference system." *IEEE Transactions on Systems, Man, and Cybernetics Systems*, 23, 665–685.

Jimmy, S. R., Sahoo, A., Samantaray, Sandeep, and Dillip K. Ghose. 2020. "Prophecy of runoff in a river basin using various neural networks." In Communication Software and Networks. 709–718, Springer.

Joshi, Veena U. and Devidas T. Tambe. 2010. "Estimation of infiltration rate, run-off and sediment yield under simulated rainfall experiments in upper Pravara basin, India: effect of slope angle and grass-cover." *Journal of Earth System Science*, 119 (6), 763.

Jothiprakash, V, R. B. Magar, and Sunil Kalkutki. 2009. "Rainfall-runoff models using adaptive neuro-fuzzy inference system (ANFIS) for an intermittent river." *International Journal of Artificial Intelligence*, 3, 1–23.

Khosravi, K., M. Panahi, and Dieu Tien Bu. 2018. "Spatial prediction of groundwater spring potential mapping based on an adaptive neuro-fuzzy inference system and metaheuristic optimization." *Hydrology and Earth System Sciences*, 22, 4771–4792.

Kisi, Ozgur, Jalal Shiri, and Mustafa Tombul. 2013. "Modeling rainfall-runoff process using soft computing techniques." Computers & Geosciences, 51, 108–17.

Kumar, P. Sundara, T. V. Praveen, and M. Anjanaya Prasad. 2016. "Artificial neural network model for rainfall-runoff-a case study." *International Journal of Hybrid Information Technology*, 9 (3), 263–72.

Lee, S. C., Hsien-Te Lin, and T. Y. Yang. 2010. "Artificial neural network analysis for reliability prediction of regional runoff utilization." *Environmental Monitoring and Assessment*, 161 (1–4), 315–26.

Lin, Gwo-Fong and Lu-Hsien Chen. 2004. "A non-linear rainfall-runoff model using radial basis function network." *Journal of Hydrology*, 289 (1–4), 1–8.

Luo, Xian, You-Peng Xu, and Jin-Tao Xu. 2011. "Regularized back-propagation neural network for rainfall-runoff modeling." In 2011 International Conference on Network Computing and Information Security. 2:85–88, IEEE.

Mathur, N., I. Glesk, and Buis, A. 2016. "Comparison of adaptive neuro-fuzzy inference system (ANFIS) and Gaussian processes for machine learning (GPML) algorithms for the prediction of skin temperature in lower limb prostheses." *Medical Engineering & Physics*, 38 (10), 1083–1089.

Mirjalili, S., S. M. Mirjalili and A. Lewis. 2014. "Grey wolf optimizer." *Advances in Engineering Software*, 69, 46–61.

Nourani, Vahid, M. Keynezhad, and L. Makani. 2009. "Using adaptive neuro-fuzzy inference system rainfall-runoff modeling." *Journal of Civil and Environmental Engineering*, 39, 75–81.

Nourani, Vahid and Mehdi Komasi. 2013. "A geomorphology-based ANFIS model for multi-station modeling of rainfall–runoff process." *Journal of Hydrology*, 490, 41–55.

Noury, M., H. R. Khatami, M. T. Moeti, and G. Barani. 2008. "Rainfall-runoff modeling using fuzzy methodology." *Journal of Applied Sciences*, 8 (16), 2851–58.

Panahi, Mahdi, Amiya Gayen, Hamid Reza Pourghasemi, Fatemeh Rezaie, and Saro Lee. 2020. "Spatial prediction of landslide susceptibility using hybrid support vector regression (SVR) and the adaptive neuro-fuzzy inference system (ANFIS) with various metaheuristic algorithms." *Science of the Total Environment*, 741, 139937.

Paryani, Sina, Aminreza Neshat, Saman Javadi, and Biswajeet Pradhan. 2020. "Comparative performance of new hybrid ANFIS models in landslide susceptibility mapping." *Natural Hazards*, 103, 1961–1988.

Pourghasemi, Hamid Reza, Amiya Gayen, Mahdi Panahi, Fatemeh Rezaie, and Thomas Blaschke. 2019a. "Multi-hazard probability assessment and mapping in Iran." *Science of the Total Environment*, 692, 556–71.

Pourghasemi, H. R., Kornejady, A., Kerle, N., and Shabani, F., 2019b. "Investigating the effects of different landslide positioning techniques, landslide partitioning approaches, and presence-absence balances on landslide susceptibility mapping." *Catena*, 104364.

Remesan, Renji, Muhammad Ali Shamim, Dawei Han, and Jimson Mathew. 2008. "ANFIS and NNARX based rainfall-runoff modeling." In 2008 IEEE International Conference on Systems, Man and Cybernetics. 1454–59, IEEE.

Sahoo, A., Samantaray, S., Bankuru, S. and Ghose, D. K. 2020a. Prediction of flood using adaptive neuro-fuzzy inference systems: a case study. In Smart Intelligent Computing and Applications. 733–739, Springer, Singapore.

Sahoo, A., U. K. Singh, M. H. Kumar, and Samantaray, Sandeep. 2020b. "Estimation of flood in a river basin through neural networks: a case study." In Communication Software and Networks, 755–763, Springer, Singapore.

Samantaray, S. and Ghose, D. K. 2018. "Evaluation of suspended sediment concentration using descent neural networks." *Procedia Computer Science*, 132, 1824–1831.

Samantaray, S. and Ghose, D. K. 2019a. "Sediment assessment for a watershed in arid region via neural networks." *Sādhanā*, 44 (10), 219.

Samantaray, S. and Ghose, D. K. 2019b. "Dynamic modelling of runoff in a watershed using artificial neural network." In Smart Intelligent Computing and Applications. 561–568, Springer, Singapore.

Samantaray, S. and Ghose, D. K. 2020. "Assessment of suspended sediment load with neural networks in arid watershed." *Journal of the Institution of Engineers (India): Series A*, 101, 371–380.

Samantaray, S., Sahoo, A. and Ghose, D. K. 2019b, June. "Assessment of groundwater potential using neural network: a case study." In International Conference on Intelligent Computing and Communication. 655–664, Springer, Singapore.

Samantaray, S., Sahoo, A. and Ghose, D. K., 2020a. "Prediction of sedimentation in an arid watershed using BPNN and ANFIS." In ICT Analysis and Applications. 295–302, Springer, Singapore.

Samantaray, S., Tripathy, O., Sahoo, A. and Ghose, D. K. 2020b. "Rainfall forecasting through ANN and SVM in Bolangir watershed, India." In Smart Intelligent Computing and Applications, 767–774.

Samantaray, S., Sahoo, A. and Ghose, D. K. 2019a. "Assessment of runoff via precipitation using neural networks: watershed modelling for developing environment in arid region." *Pertanika Journal of Science & Technology*, 27 (4), 2245–2263.

Samantaray, S. and Sahoo, A. 2020a. "Appraisal of runoff through BPNN, RNN, and RBFN in Tentulikhunti watershed: a case study." In Frontiers in Intelligent Computing: Theory and Applications, 258–267, Springer, Singapore.

Samantaray, S. and Sahoo, A. 2020b. "Estimation of runoff through BPNN and SVM in Agalpur watershed." In Frontiers in Intelligent Computing: Theory and Applications, 268–275, Springer, Singapore.

Samantaray, S. and Sahoo, A. 2020c. "Prediction of runoff using BPNN, FFBPNN, CFBPNN algorithm in arid watershed: a case study." *International Journal of Knowledge-based and Intelligent Engineering Systems*, 24 (3), 243–251.

Samantaray, S. and Ghose, D. K. (2020). "Modelling runoff in a river basin, India: an integration for developing un-gauged catchment." *International Journal of Hydrology Science and Technology*, 10 (3), 248–266.

Sarzaeim, Parisa, Omid Bozorg-Haddad, Atiyeh Bozorgi, and Hugo A Loáiciga. 2017. "Runoff projection under climate change conditions with data-mining methods." *Journal of Irrigation and Drainage Engineering*, 143 (8), 4017026.

Singh, Sushil K. 2007. "Use of gamma distribution/Nash model further simplified for runoff modeling." *Journal of Hydrologic Engineering*, 12 (2), 222–24.

Sridharam, S., Sahoo, A., Samantaray, S. and Ghose, D. K. 2020. "Assessment of flow discharge in a river basin through CFBPNN, LRNN and CANFIS." In Communication Software and Networks. 765–773, Springer, Singapore.

Talei, Amin, Lloyd Hock Chye Chua, and Tommy S. W. Wong. 2010. "Evaluation of rainfall and discharge inputs used by adaptive network-based fuzzy inference systems (ANFIS) in rainfall–runoff modeling." *Journal of Hydrology*, 391 (3–4), 248–62.

Tien Bui, D., Pradhan, B., Lofman, O., Revhaug, I., and Dick, O. B. 2012. "Landslide susceptibility mapping at HoaBinh province (Vietnam) using an adaptive neuro-fuzzy inference system and GIS." *Computers & Geosciences*, 45, 199–211.

Tikhamarine, Yazid, Doudja Souag-Gamane, Ali Najah Ahmed, Ozgur Kisi, and Ahmed El-Shafie. 2020. "Improving artificial intelligence models accuracy for monthly streamflow forecasting using grey wolf optimization (GWO) algorithm." *Journal of Hydrology*, 582, 124435.

Truatmoraka, Panjaporn, Narongrit Waraporn, and Dhanasite Suphachotiwatana. 2016. "Water level prediction model using back propagation neural network: case study: the lower of Chao Phraya Basin." In 2016 4th International Symposium on Computational and Business Intelligence (ISCBI), 200–205, IEEE.

Young, Chih-Chieh and Wen-Cheng Liu. 2015. "Prediction and modelling of rainfall–runoff during typhoon events using a physically-based and artificial neural network hybrid model." *Hydrological Sciences Journal*, 60 (12), 2102–16.

Zhang, L, Xiong, G., Liu, H., Zou, H., Guo, W. 2010. "Bearing fault diagnosis using multiscale entropy and adaptive neuro-fuzzy inference." *Expert Syst Appl*, 37, 6077–6085.

Zounemat-Kermani, Mohammad, Ozgur Kisi, and Taher Rajaee. 2013. "Performance of radial basis and LM-feed forward artificial neural networks for predicting daily watershed runoff." *Applied Soft Computing*, 13 (12), 4633–44.

14 Application of Hybrid Neural Network Techniques for Drought Forecasting

14.1 INTRODUCTION

Among many natural disasters, drought is one of the major disasters causing severe havocs in regions around the world. It is one of the most complicated natural hazards and has broadly adverse effect on water resources, economy, tourism, agriculture, and ecosystem (Wambua et al., 2016; Maca and Pech, 2015). It is a natural portion of climate which happens in both low and high precipitation regions and virtually all climatic systems (Wilhite and Buchanan, 2005, Wilhite, 2009). Temperature and precipitation are very significant natural elements or variables; and several studies revealed that they are important aspects affecting drought intensity (Sun and Ma, 2015; Easterling et al., 2007). Prediction of upcoming dry events in an area is very essential to find viable answers regarding assessment of risk related to drought occurrences and management of water (Bordi and Sutera, 2007). Moreover, forecasting drought conditions play a vital part in mitigating impact of drought on water management (Kim and Valdes, 2003). Occurrence of aridity is a perpetual climatic factor, whereas drought is a temporary irregularity (Zhang and Lin, 2016; Ndehedehe et al., 2016).

An objective drought condition assessment in a specific region is preliminary step for water resources planning for preventing and mitigating adverse effects of future happenings. Temporal and spatial severity and extent of drought can be found out with support of these measures (McKee et al., 1993; Palmer, 1995; Guttmann, 1998; Edwards and Mckee, 1997; Hayes, 2000). McKee et al. (1993) developed SPI, which is an efficient drought index having numerous benefits compared to others. SPI calculation is very easy than more complicated drought index, for example Palmer Drought Severity Index (PDSI; Palmer, 1965), as SPI necessitates rainfall data only, while PDSI utilizes many parameters. SPI classifies different kinds of drought as environmental, agricultural, or hydrological and has been broadly utilized for analysis of drought events occurring at many parts around the world.

Some specific applications of ANN in water resources involve forecasting river flows (Dibike and Solomatine, 2001; Imrie et al., 2000; Mohanta et al., 2020a), model evapotranspiration (Trajkovic et al., 2003; Sudheer et al., 2002), modelling sediment yield and runoff (Agarwal et al., 2006; Mohanta et al., 2020b; Samantaray and Sahoo,

2020a, d; Samantaray et al., 2019a; Samantaray et al., 2020a), water quality modelling (Milot et al., 2002; Schmid and Koskiaho, 2006; Keskin et al., 2015), groundwater level prediction (Maiti and Tiwari, 2014; Daliakopoulos et al., 2005; Samantaray et al., 2019b; Samantaray et al., 2020b; Sridharam et al., 2020a), estimating sediment concentration (Nagy et al., 2002; Samantaray and Sahoo, 2020b), and rainfall-runoff process (Jeong and Kim, 2005; Samantaray and Sahoo, 2020c, Jimmy et al., 2020). Mishra et al. (2007) urbanized an amalgam model, conjoining a linear stochastic model and a non-linear ANN model for forecasting drought conditions in River Kansabati, India, and compared its performance with individual stochastic and ANN models using SPI. Observation of findings reveals that proposed hybrid model produced drought forecasts with superior accuracy. Ali et al. (2017) investigated application of multilayer perceptron (MLP) algorithm for drought forecasting on the basis of SPI at Northern Area and KPK, Pakistan. On the basis of different evaluation criteria, results demonstrated that MLP has potential capability for SPI drought forecasting. Moghari and Araghinejad (2015) applied direct multi-step MLP, recursive multi-step MLP, direct multi-step RBF, recursive multi-step RBF, direct multi-step generalized regression neural network (DMSGRNN), recursive multi-step GRNN (RMSGRNN), and SPI time series approach for providing drought forecasting of Gorganroud basin situated in northern Iran. Outcomes based on performance indicators revealed that RBF and GRNN performed pre-eminent in drought index forecasting and drought class. Santos et al. (2009) applied ANN models to forecast drought of three areas in River San Francisco, Brazil. Results revealed that employed technique was proficient to forecast SPI. Barua et al. (2010) developed aggregated drought index (ADI), RMSNN, and DMSNN to present a drought forecasting approach of River Yarra in Victoria (Australia) utilizing monthly time step. Findings from study indicated that DMSNN produced slightly better results compared to RMSNN. Djerbouai and Souag-Gamane (2016) explored use of ANN model along with combined W-ANN to forecast drought events in Algerois River, Algeria, and compared with traditional stochastic models. Outcomes showed that W-ANN model gave best performance for all SPI time series than simple ANN model. Morid et al. (2007) scrutinized usability of ANN method for both likelihood and severity of drought forecasting with different timescales. Barua et al. (2012) developed DMSNN and RMSNN to classify drought condition of Yarra River in Victoria, Australia, based on non-linear ADI and compared with traditional ARIMA model. Results revealed that both the neural network models performed superiorly than ARIMA model. Rezaeianzadeh et al. (2016) established an ANN model on the basis of hydro-climatic parameters for forecasting inflow volume on monthly basis of succeeding month and evaluating Markov chain outcomes to forecast drought events of Doroodzan watershed in Fars Province, Iran. Findings from study revealed that both models predicted drought events accurately at proposed study area. Borji et al. (2016) assessed efficacy of ANN and SVM models to predict streamflow drought index (SDI) of Latian catchment situated in Iran. SVM approach provided better efficiency and reliability compared to ANN in forecasting long-term droughts. Belayneh and Adamowski (2012) compared and examined efficiency of ANN, SVM, and W-ANN to forecast drought events based on SPI in Awash Basin of Ethiopia. Forecasting results indicated that W-ANN gave best SPI forecasting values over multiple time series at desired study area.

In past decades, ANN was a successive tool commonly applied to the prediction of various hydrologic parameters in different watersheds (Mohanta, 2020a; Sahoo et al. 2019, 2020a, b; Samantaray and Ghose, 2019; Sridharam et al., 2020b). Mishra and Desai (2006) calculated SPI for multiple timescales and compared RMSNN and DMSNN with ARIMA for drought forecasting of River Kansabati located in Purulia district of West Bengal, India. Findings from the study revealed that RMSNN produced better results for a short lead time, while DMSNN model outperformed RMSNN and ARIMA models for a longer lead time to forecast drought conditions. Bacanli et al. (2009) explored applicability of adaptive neuro-fuzzy inference system (ANFIS) to forecast drought events and quantitative value of SPI at Central Anatolia, Turkey. Results demonstrated that ANFIS provided more precise and consistent results in forecasting drought events and can be effectively utilized. Shirmohammadi et al. (2013) developed and assessed potential of W-ANFIS and W-ANN models for drought forecasting in Ajabshir Plain, Iran, based on SPI and compared with simple ANN and ANFIS models. Comparative findings showed that W-ANFIS gave most accurate and reliable results followed by W-ANN, ANFIS, and ANN. Mokhtarzad et al. (2017) investigated usability of SVM, ANFIS, and ANN techniques for finding a suitable drought forecasting model in Bojnourd city of Khorasan Province, Tehran, based on SPI. Results showed that SVM model produced more precise and consistent values for drought forecasting. Rezaeian-Zadeh and Tabari (2012) explored the ability of MLP model to forecast SPI values in different timescales of drought at five stations in Iran. Outcomes revealed that MLP4 gave better prediction results with superior efficacy compared to other MLPs. Hosseini-Moghari et al. (2017) applied recursive MLP and recursive SVM for multistep ahead drought forecasting in Gorganrood, Iran, on the basis of monthly time series of SPI and compared the accuracy of obtained results with ARIMA. Findings from study in accordance to performance indices suggested that ANN models outperformed traditional ARIMA model. Bari Abarghouei et al. (2013) applied ANN model to predict drought conditions in Ardakan region of Yazd province computing different time series of SPI. Results revealed that ANN proved to be a potential model in drought prediction with greater accuracy and reliability. Jalalkamali et al. (2015) employed MLP, ANFIS, SVM, and ARIMA models to develop a suitable drought forecasting model using SPI for Yazd Province, Iran. Results demonstrated that ARIMA model produced SPI values and drought forecasting outcomes with better precision compared to other ANN models. Kousari et al. (2017) utilized ANN model and SPI for developing a regional drought forecasting model for Fars Province of Iran.

Belayneh et al. (2014) compared efficiency of ARIMA, ANN, SVM, W-ANN, and W-SVM to forecast long-term drought events in River Awash located in Ethiopia. Results indicated that combined ANN models performed better than simple ANN and ARIMA models for long-term drought forecasting. Komasi et al. (2018) proposed W-SVM for drought forecasting utilizing SPI Urmia Lake located in Iran and also evaluated the effectiveness of cuckoo search based SVM model to model and forecast SPI time series. Obtained results indicated that W-SVM and CS-SVM models performed better compared to simple SVM model in SPI time series forecasting. Zahraie and Nasseri (2011) utilized SVM model to develop seasonal SPI forecasting models which acts as an indicator to severity of drought events for Tehran city in Iran. Results showed that SVM model predicted SPI values with good accuracy and reliability that can be utilized for long-term drought forecasting. Zhang et al. (2020) investigated and compared forecasting capabilities of WNN, ARIMA, and SVM models to forecast drought conditions in Sanjiang Plain, China, based on SPI. Comparative results based on different performance evaluation criteria revealed that ARIMA model performed better than proposed data-driven models for desired study area. Masinde (2014) proposed hybridization of ANN and effective drought index (EDI) approach for short- and long-term drought forecasting with severity of drought conditions in Kenya. Proposed hybrid model was found to be an effective and consistent model for drought forecasting with enhanced results.

Forecasting drought events and their severity plays a significant part in drought mitigation and its impact on water resources management. Since SPI is most extensively utilized techniques linked to drought among many other applied approaches, precise and consistent SPI estimation is very much vital.

Major objective of present research is to investigate potential of various data-driven methods for drought forecasting. For solving aforementioned purpose, this study aims at developing W-ANN and W-SVM models for forecasting SPI for different prediction time steps. Also, W-ANN and W-SVM models were compared with simple ANN and SVM models in the Bharuch, Porbandar, Surendranagar, India. Performance of developed models was assessed and compared utilizing standard statistical performance measures.

14.2 STUDY AREA

Gujarat is a state on the western coast of India with a coastline of 1600 km bordered by Madhya Pradesh to the east, the Arabian Sea and the Pakistani province of Sindh to the west, Rajasthan to the northeast, Dadra and Nagar Haveli and Maharashtra to the southeast, and Daman and Diu to the south. It lies between 20°06′–24°42′N latitudes and 68°10′74°28′E longitudes. Plains of Gujarat are very dry and hot in cold and summer and in winter it is dry. Summer is milder in the hilly regions and the coast. In winter, temperature averages between 12°C and 27 °C and in summer between 25°C and 43°C. Monsoon season starts from June and lasts till September. Most of Gujarat receives scanty rainfall. Although typically dry, it is desertic in northwest, and wet in southern districts because of a heavy monsoon season. Climate of Gujarat involves diverse conditions, showing a wide variability ranging from arid, through semi-arid, to sub-humid tropical monsoon type. The state has a highly erratic rainfall pattern, making it subject to widespread droughts. Drought-prone area of Porbandar, Surendranagar, and Bharuch regions in Gujarat state, India, is taken up as case study (Figure 14.1).

14.3 METHODOLOGY

14.3.1 SPI

SPI was first introduced by McKee et al. (1993) with the aim to monitor and define drought. SPI is obtained by dividing the difference between precipitation and mean to standard deviation in a specific duration. It is a dimensionless index taking positive values for wet periods and negative values in drought periods. SPI entails rainfall

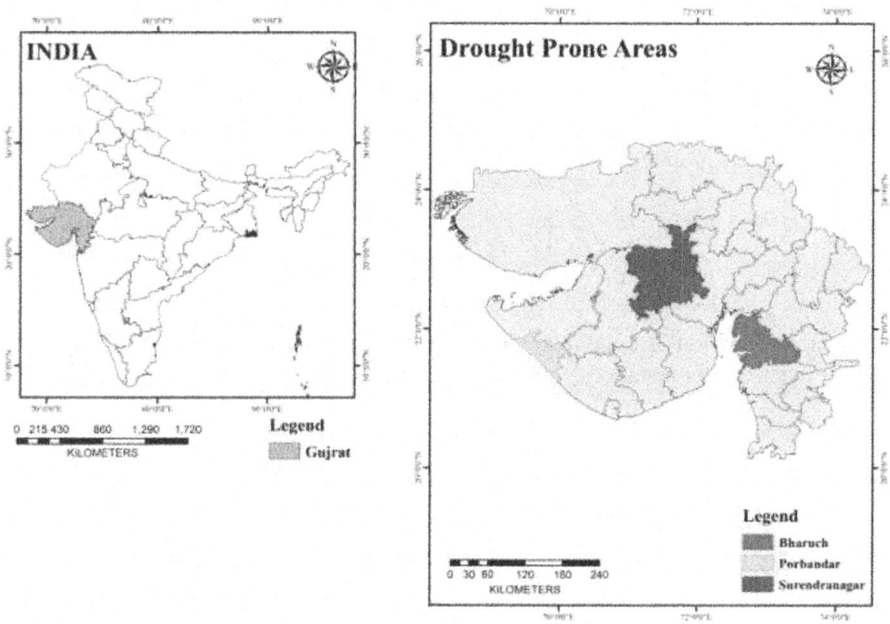

FIGURE 14.1 Study area.

data only and probability density function is fitted on this data to compute SPI at various timescales. In maximum circumstances, gamma distribution models optimal observational precipitation values. Assuming that precipitation values fit well on gamma distribution in an area and x indicates amount of precipitation, then density probability function is obtained using the following expression:

$$f(x)=\frac{1}{\beta^{\alpha}\Gamma(\alpha)}x^{\alpha-1}e^{\frac{-x}{\beta}}\quad x>0 \tag{14.1}$$

where α specifies shape factor, β is the distribution scale factor, and $\Gamma(\alpha)$ is the gamma function that is described by the equation:

$$\Gamma(\alpha)=\int_{0}^{\infty}y^{\alpha-1}e^{-y}dy \tag{14.2}$$

Additionally, α and β optimized coefficients are computed by equation given below (Thom, 1958):

$$\hat{a}=\frac{1}{4A}\left[1+\sqrt{1+\frac{4A}{3}}\right]\quad A=\ln(\overline{x})-\frac{\sum\ln(x)}{n} \tag{14.3}$$

$$\hat{\beta}=\frac{\overline{x}}{\hat{a}} \tag{14.4}$$

In above equation connected to computation of A, n signifies precipitation observation number. After coefficients α and β are estimated, integration of probability function $f(x)$ density is conducted with respect to x. Subsequently, an equation for cumulative probability $F(x)$ is determined, showing definite amount of rainfall that has been perceived in a specified month for a precise time period:

$$F(x)=\int_0^x f(x)dx=\frac{1}{\hat{\beta}^{\hat{\alpha}}}\int_0^x x^{\hat{\alpha}-1}e^{-x/\hat{\beta}}dx \quad (14.5)$$

As gamma function is undefined for $x = 0$, and rainfall data may comprise zero values, in such circumstances cumulative probability is computed by Equation 14.6:

$$H(x)=q+(1-q)F(x) \quad (14.6)$$

where q illustrates probability of no precipitation.

SPI was chosen for forecasting drought in this investigation because of its simplicity as it is based only on precipitation amount. Secondly, it can be computed on any timescale making it potential for dealing with several types of drought. Thirdly, because it is based on normal distribution, occurrences of severe and extreme drought categorizations for any position and any timescale are reliable.

14.3.2 ANN

ANN is a computational model for processing information motivated from human brain (Wambua et al., 2016; Maier et al., 2010). Among several ANN methods, MLP is used in this study. It can solve mathematical problems which require non-linear expressions by describing correct weights (Scarselli and Tsoi, 1998; Samantaray et al., 2020c). A distinctive MLP comprises three layers: the first layer is the input; the second layer is called hidden layer, and the last layer is known as output layer (Zhang et al., 2003). Architecture of MLP is presented in Figure 14.2.

Log-sigmoid activation function is utilized amid input and hidden layers, and linear activation function is utilized amid hidden and output layers (Adam et al., 2016). Both applied functions are specified below:

$$Y=f(x)=1/\left(1+e^{-x}\right) \quad (14.7)$$

$$Y=f(x)=x$$

Between various training approaches, Levenberg-Marquardt (LM) is an efficient NN training algorithm, which is utilized for training network with maximum effectiveness. Projected technique is the fastest technique and delivers a numerical solution for obtaining MSE (Kayri, 2016).

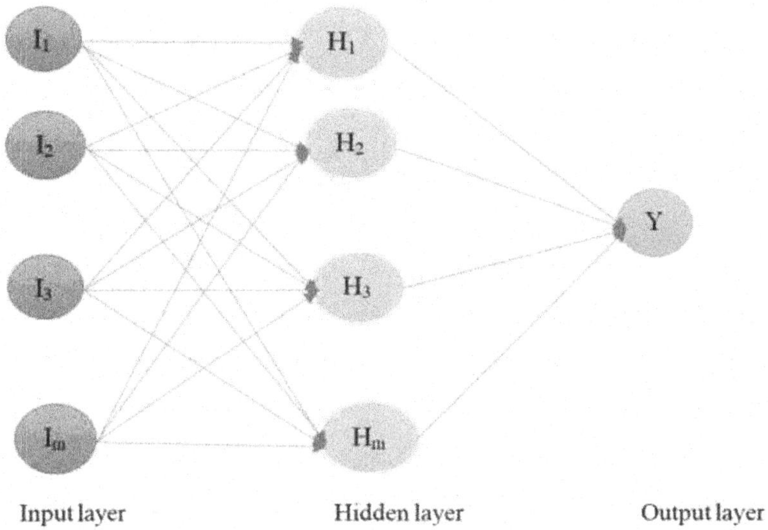

FIGURE 14.2 Architecture of ANN.

14.3.3 SVM

SVM is a type of classifier resulting from statistical learning concept on the basis of structural peril minimization. Hypothetically, it helps in minimization of a probable machine learning error and diminishes overfitting issue. Like ANNs, SVMs can be utilized for classifying and reversing numerous complications. As SVM minimizes investigational errors and intricacy concurrently, it can also advance its simplification for predicting real-world complexities (Yoon et al., 2011, Sahoo et al., 2020c; Samantaray et al., 2020d).

The elementary notion of SVM is mapping input vector x to high-dimension feature space $\phi(x)$ (kernel function) in a non-linear method where, hypothetically, a modest linear regression can handle itself with multifaceted non-linear regression of input space. A simple architecture of SVM is presented in Figure 14.3.

Let (x, y) be N number of sample set, where $x \in R^m$ is an input vector of m constituents, and y is the resultant output value. A SVM estimator (f) on regression is statistically characterized by Equation 14.8:

$$f(x) = w.\varnothing(x) + b \tag{14.8}$$

where w is the weight vector and b is the prejudice.

The problem of optimization can be written in terms of a convex optimizing problem with ε – insensitivity loss function for obtaining explanation to Equation 14.8 (Yoon et al., 2011). Hence, transformed objective function is given by Equation 14.9:

$$\frac{1}{2}w^T w + C\sum_{i=1}^{N}\xi_i + C\sum_{i=1}^{N}\xi_i^* \tag{14.9}$$

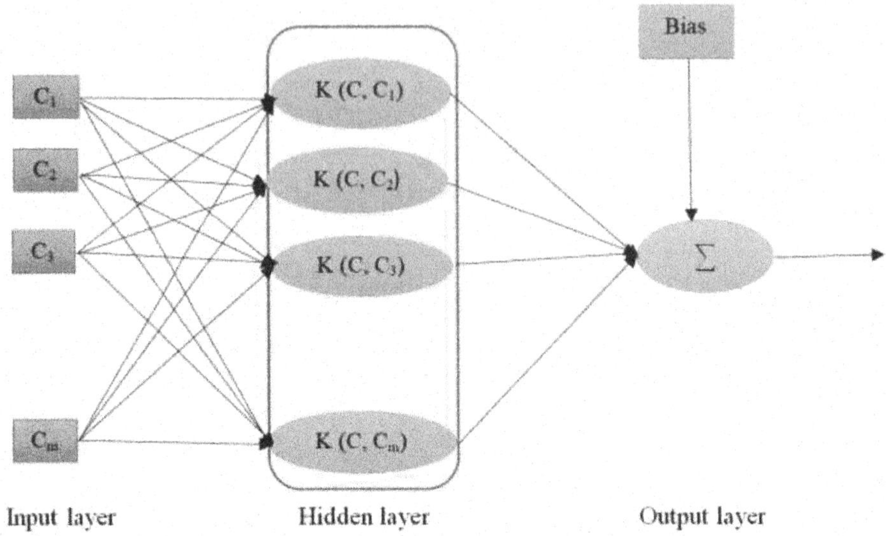

FIGURE 14.3 Architecture of SVM.

have to be minimalized bound by subsequent limitations:

$$\begin{cases} w^T\varnothing(x_i)+b-y_i \le \varepsilon+\xi_i \\ y_i-w^T\varnothing(x_i)-b \le \varepsilon+\xi_i \\ \xi_i,\xi_i^* \ge 0,\ i=1,\ldots,N \end{cases}$$

where C refers to positive adjustment constraint which regulates extent of observed error in optimization problems and ξ_i and ξ_i^* are slack variables which reprimand training errors by loss function above error lenience ε. This problem is generally resolved in a double procedure utilizing Lagrange multipliers.

14.3.4 Wavelet Transform (WT)

WT is a logical application which provides a representation of time and frequency for a signal in time field. Timescale of a continuous time series, $x(t)$, is described by Equation 14.10 (Mallat, 1999):

$$T(a,b)=\frac{1}{\sqrt{a}}\int_{-\infty}^{+\infty} g^*\frac{(t-b)}{a}\times t \cdot dt \tag{14.10}$$

where * denotes compound conjugate and g is known as mother wavelet or wavelet function. Variable a is presented as scale variable, which indicates duration and range of preferred time series. In the meantime, parameter b is communication parameter which determines location of wavelet on time axis. As a distinctive instance of present research, concept of WT is proposed for decomposing initial raw signal of SPI time series into several subseries utilizing mother wavelets stretched by parameters a and b.

14.3.5 Wavelet-ANN and Wavelet-SVM

For building hybrid W-ANN and W-SVM models, elements of subseries that are obtained from utilization of discrete WT (DWT) on original time series data were utilized as inputs for ANN models. Architecture of W-SVM is shown in Figure 14.4. Every subseries component plays a distinctive part in original time series, and performance of every subseries is unique. Firstly, original SPI data were disintegrated into a detailed series utilizing DWT. Decomposition procedure was repeated with consecutive estimation signals being disintegrated in sequence, so that original time series was fragmented down to several lesser resolution constituents (Adamowski and Chan, 2011). All the stated variables were disintegrated to 1, 2, 3, and 4 stages by six different types of wavelets: Haar, Db2, Sym3, Coif1, Mexican hat, and Morlet.

14.3.5.1 Preparation of Dataset

For this study, three rain gauging sites were taken into consideration and monthly rainfall data was acquired for a period of 1990–2019. For training purposes 70%

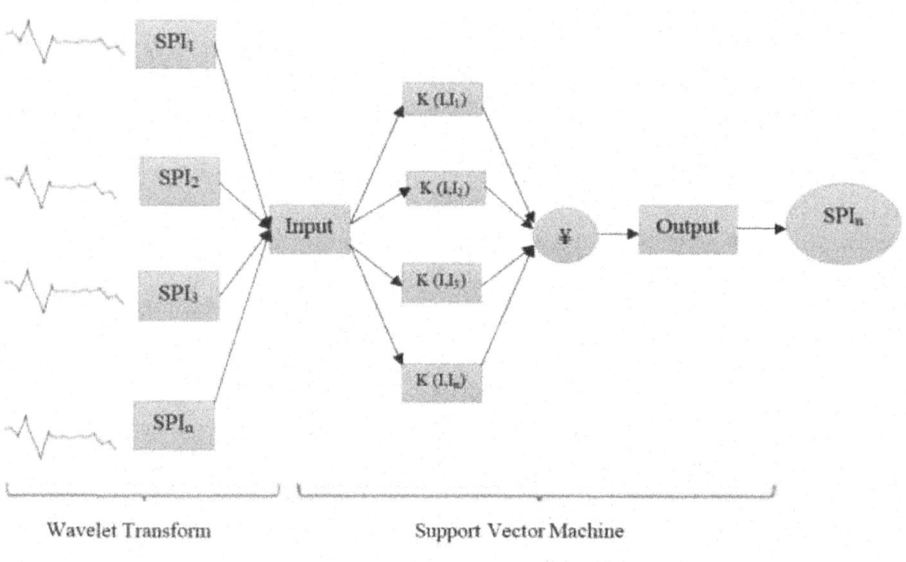

FIGURE 14.4 Architecture of W-SVM.

data (1990–2010) and for testing purposes 30% data (2011-2019) are considered. SPI values were computed based on monthly precipitation values:

$$\text{Scenario 1: SPI}(t+1) = f\{\text{SPIt, Pt}\}$$

$$\text{Scenario 2: SPI}(t+1) = f\{\text{SPIt, SPIt} - 1\text{, Pt, Pt} - 1\}$$

$$\text{Scenario 3: SPI}(t+1) = f\{\text{SPIt, SPIt} - 1\text{, SPIt} - 2\text{, Pt, Pt} - 1\text{, Pt} - 2\}$$

$$\text{Scenario 4: SPI}(t+1) = f\{\text{SPIt, SPIt} - 1\text{, SPIt} - 2\text{, SPIt} - 3\text{, Pt, Pt} - 1\text{, Pt} - 2\text{, Pt} - 3\}$$

The performance of each of these input vectors is estimated utilizing RMSE and WI. Before application of the algorithm, the data is normalized to [0.05, 0.95] utilizing the following transformation function (Rezaeian-Zadeh et al., 2010):

$$X_n = 0.05 + 0.9\frac{X_r - X_{\min}}{X_{\max} - X_{\min}} \tag{14.11}$$

where X_n and X_r are the normalized and original inputs, respectively.

$X_{\min}$ and $X_{\max}$ are the minimum and maximum of input data, respectively.

14.4 RESULTS AND DISCUSSION

The performance of different models was assessed utilizing WI, RMSE, and MSE for different lead times for all SPI series. Performance measures for SPI 3 are presented in Table 14.1. It is observed that for Bharuch station, the best forecasting result WI = 0.97586 was obtained for W-SVM in comparison to the W-ANN model having WI = 0.95567. When simple ANN models were considered, the neural network model performed lesser than other three approaches. When lead time was taken beyond two months, there was an increase in performance indices for all models. Hence, it can be observed that there is an increase in performance measures of all models with the increase in the lead time. When the performance was compared over a three-month lag time (scenario 4), the results for the ANN (WI = 0.92328) and SVM (WI = 0.94295) models were much poorer in comparison to W-ANN and W-SVM. The performance of W-SVM was better than the W-ANN, SVM, and ANN for all four scenarios. This shows that the performance improves in case of hybrid models.

The performance measures for SPI 6 are given in Table 14.2. The forecast measures for SPI 6 show the best forecasting results with W-SVM having WI = 0.97738

TABLE 14.1
Performance Measures for Comparison of Observed and Predicted Data for SPI 3

Station	Input		ANN		SVM		W-ANN		W-SVM	
			Training	Testing	Training	Testing	Training	Testing	Training	Testing
Bharuch	Scenario: 1	MSE	0.00929	0.00871	0.00635	0.00614	0.00526	0.00495	0.00401	0.00358
		RMSE	0.07967	0.05187	0.07111	0.02767	0.06878	0.0227	0.06724	0.02114
		WI	0.91938	0.88599	0.94087	0.91716	0.95429	0.93682	0.97406	0.94771
	Scenario: 2	MSE	0.00918	0.00858	0.00628	0.00605	0.00522	0.00489	0.00396	0.00353
		RMSE	0.07896	0.04647	0.07096	0.02752	0.06873	0.02254	0.06689	0.02109
		WI	0.92075	0.88714	0.94136	0.91801	0.95483	0.93733	0.97442	0.94709
	Scenario: 3	MSE	0.00905	0.00847	0.00622	0.00597	0.00519	0.00481	0.0039	0.00397
		RMSE	0.07787	0.03917	0.07088	0.02741	0.06862	0.02244	0.06616	0.021
		WI	0.92187	0.88835	0.94215	0.91876	0.95525	0.93765	0.97499	0.94796
	Scenario: 4	MSE	0.00893	0.00837	0.00617	0.0059	0.00516	0.00475	0.00387	0.00354
		RMSE	0.07732	0.03834	0.07075	0.02718	0.06852	0.02229	0.06546	0.02089
		WI	**0.92328**	**0.88953**	**0.94295**	**0.91936**	**0.95567**	**0.93812**	**0.97586**	**0.94858**
Porbandar	Scenario: 1	MSE	0.00925	0.00867	0.00634	0.00611	0.00524	0.00493	0.00399	0.00356
		RMSE	0.07934	0.05023	0.07105	0.02765	0.06876	0.02266	0.06723	0.02113
		WI	0.91993	0.88637	0.94109	0.91745	0.95451	0.93709	0.9741	0.94687
	Scenario: 2	MSE	0.00914	0.00854	0.00624	0.00601	0.00521	0.00484	0.00394	0.00351
		RMSE	0.07848	0.04186	0.07095	0.02747	0.0687	0.02252	0.06648	0.02108
		WI	0.92102	0.88758	0.94178	0.91832	0.95493	0.93738	0.97467	0.94738
	Scenario: 3	MSE	0.00901	0.00845	0.00621	0.00594	0.00518	0.00479	0.00389	0.00383
		RMSE	0.07768	0.03899	0.07081	0.02731	0.06857	0.0224	0.06603	0.02095
		WI	0.92243	0.88886	0.94257	0.91899	0.95542	0.93789	0.97528	0.94811
	Scenario: 4	MSE	0.00888	0.00834	0.00616	0.00588	0.00515	0.00473	0.00386	0.00343
		RMSE	0.07626	0.03805	0.0707	0.02713	0.06848	0.02226	0.06497	0.02084
		WI	**0.92371**	**0.88975**	**0.94327**	**0.91957**	**0.95574**	**0.9383**	**0.97589**	**0.94889**

(Continued)

TABLE 14.1
(Continued)

Surendranagar	Scenario: 1	MSE	0.00921	0.00862	0.00633	0.00608	0.00523	0.00491	0.00398	0.00355
		RMSE	0.07902	0.04998	0.07098	0.02758	0.06875	0.02265	0.06714	0.02111
		WI	0.92049	0.88665	0.94115	0.91779	0.95461	0.93711	0.97422	0.94695
	Scenario: 2	MSE	0.0091	0.00851	0.00623	0.00598	0.0052	0.00483	0.00393	0.00306
		RMSE	0.07805	0.03989	0.0709	0.02744	0.06867	0.02247	0.06634	0.02106
		WI	0.92146	0.88791	0.94183	0.91858	0.95516	0.93756	0.97474	0.94773
	Scenario: 3	MSE	0.00897	0.00841	0.00618	0.00592	0.00517	0.00478	0.00388	0.00376
		RMSE	0.07749	0.03886	0.07079	0.02728	0.06855	0.02237	0.06587	0.02091
		WI	0.92285	0.88902	0.94288	0.91912	0.95548	0.93801	0.97557	0.94837
	Scenario: 4	MSE	0.00885	0.00826	0.00615	0.00583	0.00514	0.00471	0.00385	0.00342
		RMSE	0.07611	0.03754	0.07067	0.02708	0.06845	0.02225	0.06483	0.02079
		WI	**0.92392**	**0.89012**	**0.94331**	**0.91974**	**0.95599**	**0.93847**	**0.97605**	**0.94901**

TABLE 14.2
Performance Measures for Comparison of Observed and Predicted Data for SPI 6

Station	Input		ANN		SVM		W-ANN		W-SVM	
			Training	Testing	Training	Testing	Training	Testing	Training	Testing
Bharuch	Scenario: 1	MSE	0.00882	0.00823	0.00612	0.00581	0.00513	0.0047	0.00384	0.00341
		RMSE	0.07509	0.03715	0.07063	0.02695	0.06844	0.02218	0.06427	0.02077
		WI	0.92435	0.89073	0.94363	0.91997	0.956	0.93857	0.97621	0.94924
	Scenario: 2	MSE	0.00871	0.00811	0.00608	0.00573	0.00509	0.00466	0.00377	0.00255
		RMSE	0.07465	0.03593	0.07047	0.02676	0.06837	0.02209	0.06014	0.02008
		WI	0.92536	0.89188	0.94536	0.92047	0.95663	0.93903	0.97666	0.95146
	Scenario: 3	MSE	0.00858	0.00798	0.00598	0.00568	0.00504	0.00461	0.0037	0.0025
		RMSE	0.07425	0.03448	0.07038	0.02656	0.06831	0.02201	0.05908	0.01953
		WI	0.92653	0.89281	0.94554	0.92098	0.95695	0.9394	0.97698	0.95239
	Scenario: 4	MSE	0.00842	0.00789	0.0059	0.00565	0.005	0.00458	0.00362	0.00319
		RMSE	0.07364	0.03358	0.07025	0.02622	0.06825	0.0219	0.05837	0.01867
		WI	**0.92762**	**0.89389**	**0.94615**	**0.92145**	**0.95746**	**0.93996**	**0.97738**	**0.95317**
Porbandar	Scenario: 1	MSE	0.00879	0.00819	0.00611	0.00578	0.00511	0.00468	0.00381	0.00338
		RMSE	0.07498	0.03684	0.0706	0.0269	0.06841	0.02215	0.06248	0.02074
		WI	0.92468	0.89124	0.94398	0.92009	0.95632	0.93886	0.97643	0.94962
	Scenario: 2	MSE	0.00867	0.00807	0.00605	0.00571	0.00508	0.00465	0.00374	0.00331
		RMSE	0.07438	0.03537	0.07045	0.02671	0.06835	0.02208	0.06012	0.01996
		WI	0.92575	0.89216	0.94579	0.92068	0.95668	0.93918	0.97675	0.95175
	Scenario: 3	MSE	0.00854	0.00795	0.00595	0.00567	0.00503	0.0046	0.00368	0.00325
		RMSE	0.07402	0.03401	0.07034	0.02644	0.06828	0.02195	0.05883	0.01928
		WI	0.92689	0.89305	0.94583	0.92106	0.95714	0.93947	0.97706	0.95262
	Scenario: 4	MSE	0.00839	0.00785	0.00588	0.00563	0.00498	0.00455	0.00361	0.00245
		RMSE	0.07357	0.03347	0.0702	0.02619	0.06822	0.02189	0.05819	0.01848
		WI	**0.92774**	**0.89426**	**0.94644**	**0.92163**	**0.95755**	**0.94**	**0.97766**	**0.95355**

(Continued)

TABLE 14.2
(Continued)

Surendranagar	Scenario: 1	MSE	0.00875	0.00814	0.00609	0.00575	0.0051	0.00467	0.00379	0.00336
		RMSE	0.07478	0.03636	0.07056	0.02689	0.06838	0.02212	0.06117	0.0207
		WI	0.92491	0.89159	0.94404	0.92028	0.95636	0.93891	0.97649	0.94995
	Scenario: 2	MSE	0.00863	0.00802	0.00601	0.00569	0.00507	0.00464	0.00372	0.00253
		RMSE	0.07428	0.03489	0.07039	0.02661	0.06833	0.02206	0.05931	0.01975
		WI	0.92618	0.89257	0.94527	0.92083	0.95691	0.93929	0.97682	0.95204
	Scenario: 3	MSE	0.00847	0.00791	0.00591	0.00566	0.00502	0.00459	0.00363	0.00248
		RMSE	0.07389	0.03367	0.07027	0.02632	0.06826	0.02192	0.05851	0.01889
		WI	0.92735	0.89348	0.94586	0.92124	0.95723	0.93971	0.97714	0.95288
	Scenario: 4	MSE	0.00836	0.00783	0.00586	0.00561	0.00496	0.00453	0.0036	0.00242
		RMSE	0.07332	0.03324	0.07018	0.02616	0.06821	0.02188	0.05796	0.01823
		WI	**0.92808**	**0.89475**	**0.94648**	**0.92187**	**0.9578**	**0.94027**	**0.97784**	**0.95379**

and W-ANN having WI = 0.95746 at Bharuch gauge station. The performance of W-SVM yielded a good result with WI = 0.97766 considering scenario 4 against ANN (WI = 0.92774), SVM (WI = 0.94644) and W-ANN (WI = 0.95755) at Porbandar watershed. Similarly, for Surendranagar gauge station, the prominent Wilmott index value for ANN, SVM, W-ANN, and W-SVM models during training phases are 0.92808, 0.94648, 0.9578, 0.97784, respectively. For all three stations, scenario 4 found best value performance.

For SPI 9, the performance measures are presented in Table 14.3. The forecast measures for SPI 9 show the best forecasting results for W-SVM (WI = 0.97924) and for W-ANN (WI = 0.95931) at Bharuch gauge station. Correspondingly, the performance of W-SVM produced a good result WI = 0.97935 at Porbandar gauge station. For all three stations, scenario 4 produced paramount value of performance. Also, for Surendranagar watershed, the paramount values of WI are 0.93215, 0.94985, 0.95955, and 0.97946 for ANN, SVM, W-ANN, and W-SVM, respectively, during training phase.

The performance measures for SPI 12 are given in Table 14.4. The forecast measures for SPI 12 shows the best forecasting results of WI = 0.98072 for W-SVM and WI = 0.96075 for W-ANN at Bharuch gauge station. The performance of W-SVM yielded a good result with WI = 0.98088 when the three-month lag time was considered against W-ANN having WI = 0.96078, SVM having WI = 0.95203, and ANN having WI = 0.93584 at Porbandar gauge station. Similarly, for Surendranagar gauge station, W-SVM produced best forecasting results considering scenario-4.

For SPI 24, the performance measures are revealed in Table 14.5. Forecast measures for SPI 24 for a three-month lag time provide best forecasting outcomes for W-SVM (WI = 0.98231) and W-ANN (WI = 0.97386) compared to simple SVM (0.93658) and ANN (0.94053) models. On the basis of all performance indices, it is found that hybrid models performed better than simple ANN models. It is observed from the figure that the error in the models reduced significantly over higher lead times. For all three stations, W-SVM performed best during both training and testing phases.

It can be stated that the model performances of W-SVM models for all stations are at an acceptable level for all SPI. Figure 14.5 shows the performances of ANN, SVM, W-ANN, and W-SVM models at Bharuch, Porbandar, and Surendranagar stations, respectively, for the data from 1 month to 12 months (SPI-1 to SPI-24). In this figure, the variations of MSE, WI, and RMSE criteria for SPI-1 to SPI-24 at Bharuch, Porbandar, and Surendranagar stations during the testing period are demonstrated. The pre-eminent values of WI for ANN, SVM, W-ANN, and W-SVM models are 0.883, 0.911, 0.956, and 0.974, respectively, for Bharuch station. Similarly, for Porbandar station, paramount values of WI are 0.891, 0.923, 0.961, and 0.978 for ANN, SVM, W-ANN, and W-SVM models, respectively. Correspondingly, Surendranagar position illustrates most excellent values of WI for ANN, SVM, W-ANN, and W-SVM models: 0.918, 0.951, and 0.969, respectively, during testing phase.

TABLE 14.3
Performance Measures for Comparison of Observed and Predicted Data for SPI 9

Station	Input		ANN		SVM		W-ANN		W-SVM	
			Training	Testing	Training	Testing	Training	Testing	Training	Testing
Bharuch	Scenario: 1	MSE	0.00832	0.00778	0.00584	0.00559	0.00495	0.00452	0.00359	0.00316
		RMSE	0.07317	0.03285	0.07016	0.02604	0.0682	0.02185	0.05774	0.01798
		WI	0.92853	0.89514	0.94676	0.92213	0.95787	0.94033	0.97812	0.95406
	Scenario: 2	MSE	0.00819	0.00767	0.00581	0.00556	0.00491	0.00448	0.00354	0.00276
		RMSE	0.07296	0.03214	0.07001	0.02565	0.06815	0.02178	0.05689	0.01736
		WI	0.92957	0.89647	0.94754	0.92288	0.95841	0.94077	0.97859	0.95495
	Scenario: 3	MSE	0.00796	0.00749	0.00575	0.00545	0.00485	0.00442	0.00345	0.00234
		RMSE	0.07277	0.03103	0.06997	0.02532	0.06811	0.02169	0.05634	0.01653
		WI	0.93048	0.90057	0.94837	0.92364	0.95884	0.94124	0.97892	0.95589
	Scenario: 4	MSE	0.00776	0.00702	0.00568	0.00542	0.0048	0.00437	0.00397	0.00261
		RMSE	0.07256	0.03098	0.06986	0.02475	0.06806	0.02163	0.05574	0.01584
		WI	**0.93155**	**0.91026**	**0.94916**	**0.92432**	**0.95931**	**0.94172**	**0.97924**	**0.95654**
Porbandar	Scenario: 1	MSE	0.00828	0.00776	0.00583	0.00558	0.00494	0.0045	0.00357	0.00239
		RMSE	0.07304	0.03255	0.07012	0.02587	0.06819	0.02182	0.05741	0.01779
		WI	0.92889	0.89559	0.94713	0.92248	0.95809	0.94054	0.97827	0.95447
	Scenario: 2	MSE	0.00817	0.00763	0.0058	0.00553	0.0049	0.00447	0.00352	0.00236
		RMSE	0.07289	0.03165	0.06999	0.02559	0.06814	0.02176	0.05665	0.01703
		WI	0.92989	0.90006	0.94786	0.92311	0.95847	0.94089	0.97874	0.95532
	Scenario: 3	MSE	0.00792	0.00734	0.00573	0.00544	0.00484	0.00441	0.00341	0.00213
		RMSE	0.07274	0.03102	0.06989	0.02522	0.06809	0.02168	0.05606	0.01632
		WI	0.93089	0.90089	0.94863	0.92389	0.95915	0.94141	0.97906	0.95601
	Scenario: 4	MSE	0.00775	0.00693	0.00566	0.00541	0.00479	0.00436	0.00365	0.00231
		RMSE	0.07248	0.03087	0.06985	0.02457	0.06804	0.02161	0.05557	0.01537
		WI	**0.93178**	**0.91047**	**0.94958**	**0.92466**	**0.95947**	**0.94176**	**0.97935**	**0.95678**

TABLE 14.3
(Continued)

Surendranagar	Scenario: 1	MSE	0.00823	0.00771	0.00582	0.00557	0.00492	0.00449	0.00356	0.00313
		RMSE	0.07304	0.03234	0.07003	0.02579	0.06817	0.0218	0.05718	0.01754
		WI	0.92928	0.89591	0.94749	0.92269	0.95812	0.94068	0.97837	0.95476
	Scenario: 2	MSE	0.00804	0.00752	0.00579	0.00549	0.00487	0.00444	0.00348	0.00305
		RMSE	0.07288	0.03112	0.06998	0.02547	0.06812	0.02173	0.05652	0.01687
		WI	0.93016	0.90023	0.94804	0.92336	0.95879	0.9412	0.97883	0.95557
	Scenario: 3	MSE	0.00788	0.00718	0.00571	0.00543	0.00482	0.00439	0.0034	0.00297
		RMSE	0.07269	0.03101	0.06987	0.02493	0.06808	0.02166	0.05591	0.01611
		WI	0.93116	0.91002	0.94897	0.92413	0.95916	0.94145	0.97914	0.95635
	Scenario: 4	MSE	0.00763	0.00685	0.00565	0.00539	0.00476	0.00433	0.00351	0.00229
		RMSE	0.07238	0.03058	0.06978	0.02436	0.06801	0.02157	0.05539	0.01509
		WI	**0.93215**	**0.91065**	**0.94985**	**0.92491**	**0.95955**	**0.94197**	**0.97946**	**0.95697**

TABLE 14.4
Performance Measures for Comparison of Observed and Predicted Data for SPI 12

Station	Input		ANN		SVM		W-ANN		W-SVM	
			Training	Testing	Training	Testing	Training	Testing	Training	Testing
Bharuch	Scenario: 1	MSE	0.00759	0.00672	0.00564	0.00538	0.00472	0.00429	0.00334	0.00228
		RMSE	0.07227	0.02998	0.06967	0.02433	0.068	0.02155	0.05513	0.01488
		WI	0.93247	0.91089	0.95001	0.92505	0.95963	0.94209	0.97967	0.95728
	Scenario: 2	MSE	0.00738	0.00657	0.00559	0.00534	0.00465	0.00422	0.00326	0.0022
		RMSE	0.07204	0.02937	0.06948	0.024	0.06792	0.02148	0.05458	0.01427
		WI	0.93379	0.91143	0.95086	0.92589	0.95994	0.94252	0.98007	0.95789
	Scenario: 3	MSE	0.00719	0.00647	0.00553	0.0053	0.00458	0.00399	0.00322	0.00211
		RMSE	0.07193	0.02905	0.06935	0.02376	0.06777	0.02141	0.05379	0.01353
		WI	0.93454	0.91206	0.95129	0.92667	0.96026	0.94289	0.98039	0.95846
	Scenario: 4	MSE	0.00703	0.0064	0.00545	0.00526	0.00439	0.00396	0.00318	0.00202
		RMSE	0.07168	0.02867	0.06923	0.02364	0.0677	0.02134	0.05296	0.01282
		WI	**0.93566**	**0.91265**	**0.95199**	**0.92766**	**0.96075**	**0.94332**	**0.98072**	**0.95889**
Porbandar	Scenario: 1	MSE	0.00747	0.00662	0.00563	0.00537	0.00471	0.00428	0.00331	0.00226
		RMSE	0.07216	0.02987	0.06956	0.02427	0.06798	0.02152	0.05496	0.01469
		WI	0.93283	0.91104	0.95033	0.92528	0.95969	0.94228	0.97992	0.95749
	Scenario: 2	MSE	0.00726	0.00651	0.00556	0.00532	0.00461	0.00402	0.00325	0.00217
		RMSE	0.07201	0.02926	0.06947	0.02396	0.06789	0.02146	0.05433	0.01399
		WI	0.93383	0.91168	0.95096	0.92613	0.96001	0.94265	0.98012	0.95803
	Scenario: 3	MSE	0.00718	0.00644	0.00551	0.00528	0.00454	0.00398	0.00321	0.00208
		RMSE	0.07189	0.02901	0.06934	0.02373	0.06777	0.02136	0.05357	0.01327
		WI	0.93479	0.91228	0.95161	0.92694	0.96046	0.94301	0.98056	0.95854
	Scenario: 4	MSE	0.00696	0.00638	0.00544	0.00525	0.00436	0.00393	0.00316	0.00199
		RMSE	0.07166	0.02865	0.0692	0.02358	0.06766	0.02132	0.05262	0.01257
		WI	**0.93584**	**0.91293**	**0.95203**	**0.92789**	**0.96078**	**0.94348**	**0.98088**	**0.95891**

TABLE 14.4
(Continued)

Surendranagar	Scenario: 1	MSE	0.00739	0.00659	0.0056	0.00536	0.00466	0.00423	0.00329	0.00224
		RMSE	0.07211	0.02979	0.06953	0.02421	0.06795	0.02149	0.05474	0.01445
		WI	0.93325	0.91125	0.95054	0.92557	0.95987	0.9423	0.98003	0.95764
	Scenario: 2	MSE	0.00722	0.00649	0.00554	0.00531	0.00459	0.00401	0.00324	0.00215
		RMSE	0.07196	0.02914	0.06938	0.02381	0.06784	0.02144	0.05401	0.01374
		WI	0.93426	0.91185	0.95128	0.92641	0.96021	0.94274	0.98024	0.95837
	Scenario: 3	MSE	0.00711	0.00642	0.00548	0.00527	0.00447	0.00397	0.00319	0.00206
		RMSE	0.07173	0.02894	0.06925	0.0237	0.06774	0.02135	0.05321	0.01301
		WI	0.93527	0.91241	0.95167	0.92735	0.96053	0.94316	0.98065	0.95872
	Scenario: 4	MSE	0.00684	0.00636	0.00541	0.00524	0.00431	0.00388	0.00315	0.00196
		RMSE	0.07157	0.02844	0.06918	0.02339	0.06763	0.02131	0.05238	0.01225
		WI	**0.93636**	**0.91307**	**0.95231**	**0.92804**	**0.96107**	**0.94361**	**0.981**	**0.95902**

TABLE 14.5
Performance Measures for Comparison of Observed and Predicted Data for SPI 24

Station	Input		ANN		SVM		W-ANN		W-SVM	
			Training	Testing	Training	Testing	Training	Testing	Training	Testing
Bharuch	Scenario: 1	MSE	0.00679	0.00635	0.00538	0.00523	0.00429	0.00386	0.00313	0.00193
		RMSE	0.07149	0.02844	0.06916	0.02332	0.06759	0.0213	0.05189	0.01196
		WI	0.93667	0.91339	0.95235	0.92839	0.96197	0.94376	0.98104	0.95908
	Scenario: 2	MSE	0.00669	0.00629	0.00535	0.00518	0.00424	0.00381	0.00304	0.00186
		RMSE	0.07138	0.02824	0.06908	0.02319	0.06751	0.02126	0.05082	0.01127
		WI	0.93775	0.91418	0.95286	0.92927	0.96357	0.94416	0.98136	0.95947
	Scenario: 3	MSE	0.00658	0.00622	0.00532	0.00511	0.00418	0.00375	0.00292	0.00179
		RMSE	0.07125	0.02808	0.06897	0.02301	0.06744	0.02121	0.05007	0.01043
		WI	0.93882	0.91513	0.95332	0.93012	0.9646	0.94469	0.98168	0.95989
	Scenario: 4	MSE	0.00647	0.00618	0.00529	0.00502	0.00409	0.00366	0.00281	0.00171
		RMSE	0.07117	0.02788	0.0689	0.02283	0.06733	0.02118	0.00278	0.00168
		WI	**0.93986**	**0.91631**	**0.95397**	**0.93107**	**0.97357**	**0.94501**	**0.98199**	**0.96559**
Porbandar	Scenario: 1	MSE	0.00675	0.00633	0.00537	0.00522	0.00427	0.00384	0.00311	0.00191
		RMSE	0.07145	0.02837	0.06913	0.02327	0.06757	0.02128	0.05154	0.01178
		WI	0.93691	0.91353	0.95254	0.92872	0.96229	0.94399	0.98113	0.95923
	Scenario: 2	MSE	0.00667	0.00625	0.00534	0.00517	0.00422	0.00379	0.003	0.00184
		RMSE	0.07133	0.02815	0.06904	0.02311	0.0675	0.02125	0.05061	0.01091
		WI	0.93821	0.91449	0.95295	0.92957	0.96371	0.94438	0.98145	0.95952
	Scenario: 3	MSE	0.00655	0.0062	0.00531	0.00507	0.00415	0.00372	0.00289	0.00175
		RMSE	0.07123	0.02801	0.06895	0.02297	0.06741	0.0212	0.04997	0.01012
		WI	0.93909	0.91554	0.95364	0.93046	0.96492	0.94481	0.98175	0.96008
	Scenario: 4	MSE	0.00642	0.00617	0.00528	0.00499	0.00405	0.00362	0.04948	0.00963
		RMSE	0.07116	0.02787	0.06887	0.02275	0.0673	0.02116	0.04926	0.00946
		WI	**0.94014**	**0.91638**	**0.95397**	**0.93315**	**0.97374**	**0.9466**	**0.98207**	**0.96648**

TABLE 14.5
(Continued)

Surendranagar	Scenario: 1	MSE	0.00671	0.0063	0.00536	0.0052	0.00426	0.00383	0.00308	0.00189
		RMSE	0.0714	0.02827	0.06909	0.02324	0.06753	0.02127	0.05127	0.01145
		WI	0.93732	0.91385	0.95263	0.92898	0.96325	0.94405	0.98132	0.95933
	Scenario: 2	MSE	0.00662	0.00623	0.00533	0.00516	0.00421	0.00378	0.00297	0.00181
		RMSE	0.07126	0.0281	0.06899	0.02305	0.06747	0.02123	0.05035	0.01068
		WI	0.93847	0.91484	0.95327	0.92983	0.96403	0.94453	0.98159	0.95964
	Scenario: 3	MSE	0.00651	0.00619	0.0053	0.00504	0.00411	0.00368	0.00284	0.00173
		RMSE	0.07118	0.02798	0.06892	0.02286	0.06737	0.02119	0.04955	0.00995
		WI	0.93953	0.91599	0.95365	0.93078	0.97325	0.94497	0.98191	0.96072
	Scenario: 4	MSE	0.00639	0.00615	0.00527	0.00497	0.00402	0.0036	0.00276	0.00165
		RMSE	0.07113	0.0277	0.06882	0.02271	0.06726	0.02115	0.04889	0.00937
		WI	**0.94053**	**0.9164**	**0.95422**	**0.93658**	**0.97386**	**0.94663**	**0.98231**	**0.9699**

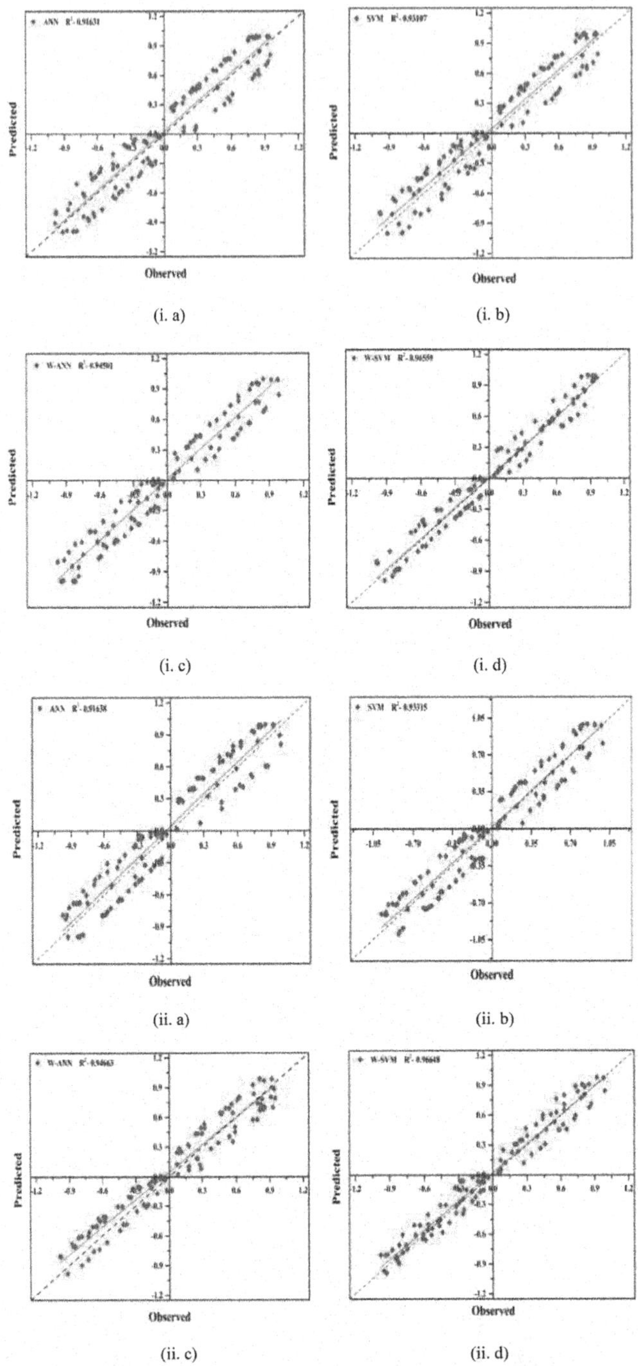

FIGURE 14.5 Actual versus predicted flood using (a) ANN, (b) SVM, (c) W-ANN, and (d) W-SVM at (i) Bharuch, (ii) Porbandar, and (iii) Surendranagar gauge stations during testing phase.

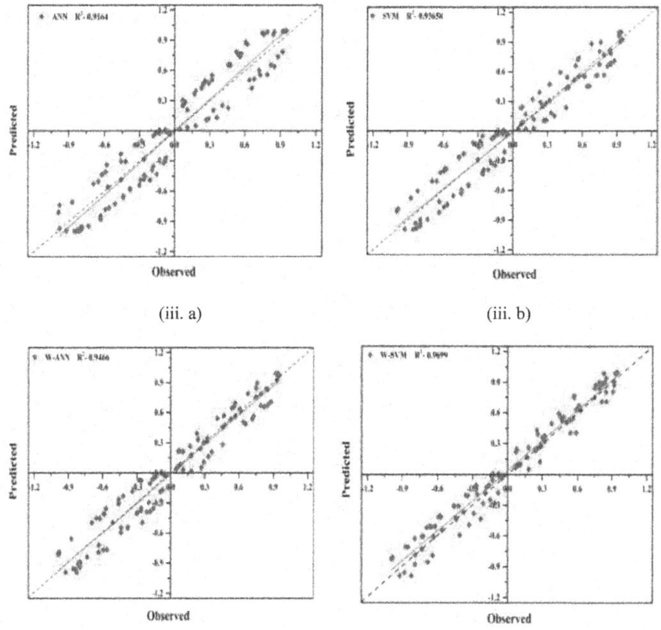

(iii. a) (iii. b) (iii. c) (iii. d)

FIGURE 14.5 (Continued)

14.5 COMPARISON OF MODEL PERFORMANCE

Generally, W-SVM and W-ANN models were found to give more accurate drought prediction results in comparison to ANN and SVM models. Similarly, it was observed that W-SVM model is capable of forecasting SPI more precisely than W-ANN. Since W-SVM models comprise neural network and wavelet transform, they could model SPI more precisely. The figures presented below illustrate the best ANN, SVM, Wavelet-ANN, and Wavelet-SVM models which could provide better SPI predictions in a good terms with observed values for all gauge stations. In accordance to their accuracy, the order of models is as follows: W-SVM, W-ANN, SVM, and ANN, respectively. Linear scale plot of actual versus predicted drought for developed models of proposed sites are presented in Figure 14.6.

Overall, all algorithms and models provided technical sustenance for regional drought forecasting, and forecasting outcomes from present research are significant for upcoming investigations and favour tactical improvement strategies for preventing drought conditions in study area. Moreover, forecasting drought events is a complicated procedure, and blindly making an attempt to forecast drought is not prudent. Advantages of various techniques are dependent on both basic principles of algorithms or models and features of data series. For selecting an appropriate approach, it may be more convincing and reasonable, depending on characteristics of data in different areas of study. Future research should give emphasis to improve forecasting accurateness by utilizing ensemble models or developed algorithms.

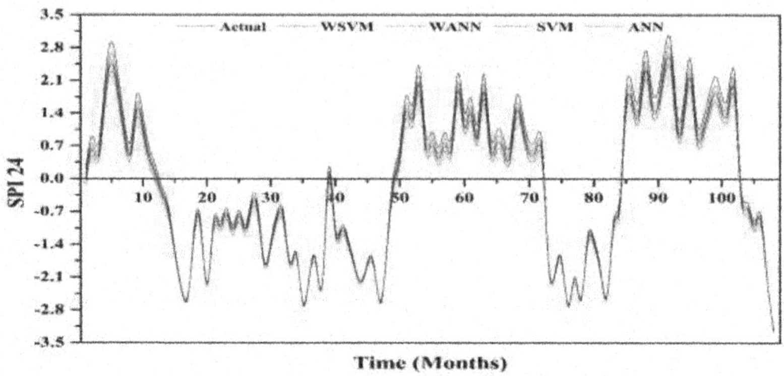

(a)

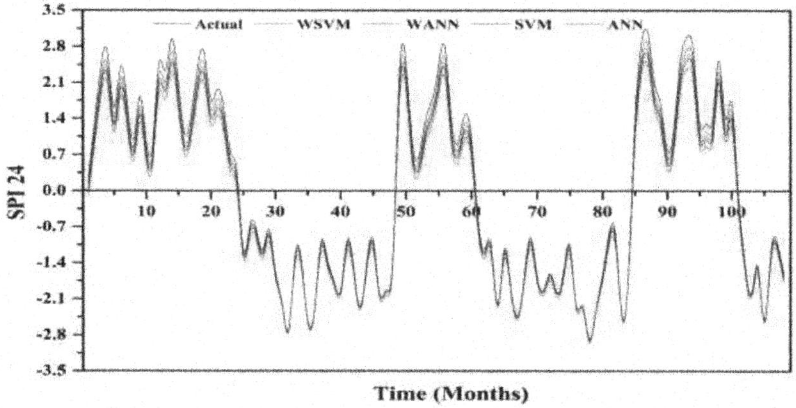

(b)

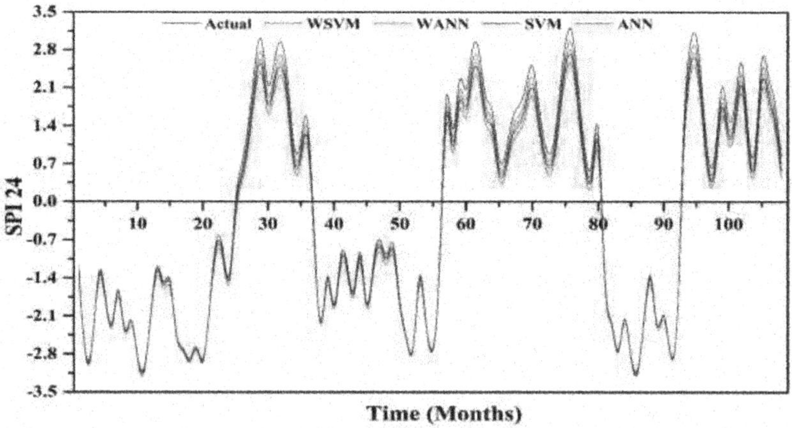

(c)

FIGURE 14.6 Deviation of actual and predicted drought at (a) Salabheta, (b) Suktel, and (c) Lant gauge stations during testing phase.

14.6 CONCLUSION

In this study, efficiency of four ANN models, namely ANN, SVM, W-ANN, and W-SVM for one, two and three months lag of forecasted SPI were evaluated to forecast drought conditions in Bharuch, Porbandar, and Surendranagar, India. WT was combined with ANN and SVM methods for developing two amalgam models for forecasting SPI for different time steps. The SPI was computed for 3, 6, 9, 12, and 24 months. Based on several performance evaluation criteria, namely WI, RMSE, and MSE, W-SVM model had substantial gain and performed best over W-ANN, ANN, and SVM models with accurate and precise drought forecasting results in proposed station. The results revealed that at short-term timescales of SPI, accurateness of techniques were weak; however, with increasing timescale, the accurateness was increased so that worst forecasts were associated with SPI 3 and best ones were related to SPI 24. The present study validated that preprocessing of data could increase SPI forecasting. Moreover, it was established that every modelling methodologies are proficient approximately to forecast SPI values reliably and precisely.

REFERENCES

Abarghouei, H. B., Kousari, M. R. and Zarch, M. A. A. 2013. "Prediction of drought in dry lands through feedforward artificial neural network abilities." *Arabian Journal of Geosciences*, 6 (5), 1417–1433.

Adam, S. P., Magoulas, G. D., Karras, D. A. and Vrahatis, M. N. 2016. "Bounding the search space for global optimization of neural networks learning error: an interval analysis approach." *The Journal of Machine Learning Research*, 17 (1), 5898–5937.

Adamowski, J. and Fung Chan, H. (2011). "A wavelet neural network conjunction model for SPI index forecasting." *Journal of Hydrology*, 407 (1–4), 28–40.

Agarwal, A., Mishra, S. K., Ram, S. and Singh, J. K. 2006. "Simulation of runoff and sediment yield using artificial neural networks." *Biosystems Engineering*, 94 (4), 597–613.

Ali, Z., Hussain, I., Faisal, M., Nazir, H. M., Hussain, T., Shad, M. Y., Mohamd Shoukry, A. and Hussain Gani, S. 2017. "Forecasting drought using multilayer perceptron artificial neural network model." *Advances in Meteorology*, 2017 (2).

Bacanli, U. G., Firat, M. and Dikbas, F. 2009. "Adaptive neuro-fuzzy inference system for drought forecasting." *Stochastic Environmental Research and Risk Assessment*, 23 (8), 1143–1154.

Barua, S., Perera, B. J. C., Ng, A. W. M. and Tran, D. 2010. "Drought forecasting using an aggregated drought index and artificial neural network." *Journal of Water and Climate Change*, 1 (3), 193–206.

Barua, S., Ng, A. W. M. and Perera, B. J. C. 2012. "Artificial neural network–based drought forecasting using a nonlinear aggregated drought index." *Journal of Hydrologic Engineering*, 17 (12), 1408–1413.

Belayneh, A. and Adamowski, J. 2012. "Standard precipitation index drought forecasting using neural networks, wavelet neural networks, and support vector regression." *Applied Computational Intelligence and Soft Computing*, 2012, 794061.

Belayneh, A., Adamowski, J., Khalil, B. and Ozga-Zielinski, B. 2014. "Long-term SPI drought forecasting in the Awash River Basin in Ethiopia using wavelet neural network and wavelet support vector regression models." *Journal of Hydrology*, 508, 418–429.

Bordi, I. and Sutera, A. 2007. "Drought monitoring and forecasting at large scale." In Methods and Tools for Drought Analysis and Management. 3–27, Springer, Dordrecht.

Borji, M., Malekian, A., Salajegheh, A. and Ghadimi, M. 2016. "Multi-time-scale analysis of hydrological drought forecasting using support vector regression (SVR) and artificial neural networks (ANN)." *Arabian Journal of Geosciences*, 9 (19), 725.

Daliakopoulos, I. N., Coulibaly, P., and Tsanis, I.K. (2005). "Groundwater level forecasting using artificial neural networks." Journal of Hydrology, 309 (1), 229–240.

Dibike, Y. B. and Solomatine, D. P. 2001. "River flow forecasting using artificial neural networks." Physics and Chemistry of the Earth, Part B: *Hydrology, Oceans and Atmosphere*, 26 (1), 1–7.

Djerbouai, S. and Souag-Gamane, D. 2016. "Drought forecasting using neural networks, wavelet neural networks, and stochastic models: case of the Algerois Basin in North Algeria." *Water Resources Management*, 30 (7), 2445–2464.

Easterling, D. R., Wallis, T. W. R., Lawrimore, J. H., and Heim, R. R., Jr. 2007. "Effects of temperature and precipitation trends on US drought." Geophysical Research Letters, 34 (20), 396.

Edwards, D. C. and McKee, T. B. (1997) Characteristics of 20th Century Droughts in the United States at Multiple Time Scales. Climatology Report, 97–2, Department of Atmospheric Sciences, Colorado State University, Fort Collins, CO, pp. 155.

Guttman, N. B. 1998. "Comparing the Palmer Drought Index and the Standardized Precipitation Index 1." *JAWRA Journal of the American Water Resources Association*, 34 (1), 113–121.

Hayes, M.J. (2000). Revisiting the SPI: Clarifying the Process. Drought Network News, A Newsletter of the International Drought Information Center and the National Drought Mitigation Center 12/1 (Winter 1999–Spring 2000), 13–15.

Hosseini-Moghari, S. M., Araghinejad, S. and Azarnivand, A. 2017. "Drought forecasting using data-driven methods and an evolutionary algorithm." *Modeling Earth Systems and Environment*, 3 (4), 1675–1689.

Hosseini-Moghari, S. M. and Araghinejad, S. 2015. "Monthly and seasonal drought forecasting using statistical neural networks." *Environmental Earth Sciences*, 74 (1), 397–412.

Imrie, C. E., Durucan, S., and Korre, A. (2000). "River flow prediction using artificial neural networks: generalisation beyond the calibration range." *Journal of Hydrology*, 233 (1), 138–153.

Jalalkamali, A., Moradi, M. and Moradi, N. 2015. "Application of several artificial intelligence models and ARIMAX model for forecasting drought using the Standardized Precipitation Index." *International Journal of Environmental Science and Technology*, 12 (4), 1201–1210.

Jeong, D., and Kim, Y. O. (2005) "Rainfall-runoff models using artificial neural networks for ensemble stream flow prediction." *Hydrological Processes*, 19, 3819–3835.

Jimmy, S. R, Sahoo, A., Samantaray, Sandeep, and Dillip K. Ghose. 2020. "Prophecy of runoff in a river basin using various neural networks." In Communication Software and Networks. 709–718, Springer.

Kayri, M. 2016. "Predictive abilities of Bayesian regularization and Levenberg–Marquardt algorithms in artificial neural networks: a comparative empirical study on social data." *Mathematical and Computational Applications*, 21 (2), 20.

Keskin, T. E., Düğenci, M. and Kaçaroğlu, F. 2015. "Prediction of water pollution sources using artificial neural networks in the study areas of Sivas, Karabük and Bartın (Turkey)." *Environmental Earth Sciences*, 73 (9), 5333–5347.

Kim, T. and Valdes, J.B. (2003). "Nonlinear model for drought forecasting based on a conjunction of wavelet transforms and neural networks." *Journal of Hydrologic Engineering*, 8, 319–328.

Komasi, M., Sharghi, S. and Safavi, H. R. 2018. "Wavelet and cuckoo search-support vector machine conjugation for drought forecasting using standardized precipitation index (case study: Urmia Lake, Iran)." *Journal of Hydroinformatics*, 20 (4), 975–988.

Kousari, M. R., Hosseini, M. E., Ahani, H. and Hakimelahi, H. 2017. "Introducing an operational method to forecast long-term regional drought based on the application of artificial intelligence capabilities." *Theoretical and Applied Climatology*, 127 (1–2), 361–380.

Maca, P. and Pech, P. 2016. "Forecasting SPEI and SPI drought indices using the integrated artificial neural networks." *Computational Intelligence and Neuroscience*, 2016, 3868519, 1-17.

Maier, A. R., Jain, A., Dandy, G. C., Sudheer, K. P. (2010). "Methods used for development of neural networks for the prediction of water resource variables in river systems: current status and future directions." *Environmental Modelling & Software*, 25 (8), 891–909.

Maiti, S., and Tiwari, R.K. (2014). "A comparative study of artificial neural networks, Bayesian neural networks and adaptive neuro-fuzzy inference system in groundwater level prediction." *Environmental Earth Sciences*, 71 (7), 3147–3160.

Mallat, S. 1999. A Wavelet Tour of Signal Processing, 2nd ed. Academic Publishers, San Diego, CA.

Masinde, M. 2014. "Artificial neural networks models for predicting effective drought index: factoring effects of rainfall variability." *Mitigation and Adaptation Strategies for Global Change*, 19 (8), 1139–1162.

McKee, T. B., Doesken, N. J. and Kleist, J. 1993. The relationship of drought frequency and duration to time scales. In *Proceedings of the 8th Conference on Applied Climatology*, 17 (22), 179–183.

Milot, J., Rodriguez, M. J., Se´rodes, J. B. (2002). "Contribution of neural networks for modeling trihalomethanes occurrence in drinking water." *Journal of Water Resources Planning and Management*, 128 (5), 370–376.

Mishra, A. K. and Desai, V. R. 2006. "Drought forecasting using feed-forward recursive neural network." *Ecological Modelling*, 198 (1–2), 127–138.

Mishra, A. K., Desai, V. R. and Singh, V. P. 2007. "Drought forecasting using a hybrid stochastic and neural network model." *Journal of Hydrologic Engineering*, 12 (6), 626–638.

Mohanta, N. R., N. Patel, K. Beck, S. Samantaray, and Abinash Sahoo. 2020a. "Efficiency of river flow prediction in river using Wavelet-CANFIS: a case study." In Intelligent Data Engineering and Analytics. 435–443, Springer, Singapore.

Mohanta, N. R., P. Biswal, S. S. Kumari, S. Samantaray, and Abinash Sahoo. 2020b. "Estimation of sediment load using adaptive neuro-fuzzy inference system at Indus River basin, India." In Intelligent Data Engineering and Analytics. 427–434, Springer, Singapore.

Mokhtarzad, M., Eskandari, F., Vanjani, N. J. and Arabasadi, A. 2017. "Drought forecasting by ANN, ANFIS, and SVM and comparison of the models." *Environmental Earth Sciences*, 76 (21), 729.

Morid S, Smakhtin V, Bagherzadeh K (2007) Drought forecasting using artificial neural networks and time series of drought indices. *International Journal of Climatology*, 27(15), 2103–2111.

Nagy, H.M., Watanabe K., and Hirano, M. (2002) "Prediction of sediment load concentration in rivers using artificial neural network model." *Journal of Hydraulic Engineering*, 128, 588–595.

Ndehedehe, C. E., Awange, J. L., Corner, R. J., Kuhn, M. and Okwuashi, O. 2016. "On the potentials of multiple climate variables in assessing the spatio-temporal characteristics of hydrological droughts over the Volta Basin." Science of the Total Environment, 557, 819–837.

Palmer, W. C. (1965). Meteorological Drought. Research Paper, U.S. Department of Commerce Weather Bureau, Washington, D.C, USA.

Rezaeianzadeh, M., Amin, S., Khalili, D. and Singh, V.P. 2010. "Daily outflow prediction by multi-layer perceptron with logistic sigmoid and tangent sigmoid activation functions." *Water Resources Management*, 24(11), 2673–2688.

Rezaeian-Zadeh, M. and Tabari, H. 2012. “MLP-based drought forecasting in different climatic regions.” *Theoretical and Applied Climatology*, 109 (3–4), 407–414.

Rezaeianzadeh, M., Stein, A. and Cox, J. P. 2016. “Drought forecasting using Markov chain model and artificial neural networks.” *Water Resources Management*, 30 (7), 2245–2259.

Sahoo, A., S. Samantaray, and D. K. Ghose. 2019. “Stream flow forecasting in Mahanadi river basin using artificial neural networks.” In Procedia Computer Science, 157. doi:10.1016/j.procs.2019.08.154.

Sahoo, A., S. Samantaray, S. Bankuru, and D. K. Ghose. 2020a. “Prediction of flood using adaptive neuro-fuzzy inference systems: a case study.” Smart Innovation, Systems and Technologies, 159. doi:10.1007/978-981-13-9282-5_70.

Sahoo, A., U. K. Singh, M. H. Kumar, and Samantaray, Sandeep. 2020b. “Estimation of flood in a river basin through neural networks: a case study.” In Communication Software and Networks. 755–763, Springer, Singapore.

Sahoo, A., A. Barik, S. Samantaray, and D. K. Ghose. 2020c. “Prediction of sedimentation in a watershed using RNN and SVM.” In Communication Software and Networks. 701–708, Springer, Singapore.

Santos, C. A. G., Morais, B. S. and Silva, G. B. 2009. “Drought forecast using an artificial neural network for three hydrological zones in San Francisco River basin, Brazil.” IAHS Publication, 333, 302.

Samantaray, Sandeep, and Dillip K. Ghose. 2019. “Sediment assessment for a watershed in arid region via neural networks.” *Sādhanā*, 44 (10), 219.

Samantaray, S., A. Sahoo, and D. K. Ghose. 2019a. “Assessment of runoff via precipitation using neural networks: watershed modelling for developing environment in arid region.” *Pertanika Journal of Science and Technology*, 27 (4).

Samantaray, Sandeep, A. Sahoo, and Dillip K. Ghose. 2019b. “Assessment of groundwater potential using neural network: a case study.” In International Conference on Intelligent Computing and Communication. 655–664, Springer.

Samantaray, S. and A. Sahoo. 2020a. “Estimation of runoff through BPNN and SVM in Agalpur Watershed.” Advances in Intelligent Systems and Computing, 1014. doi:10.1007/978-981-13-9920-6_27.

Samantaray, S. and A. Sahoo. 2020b. “Assessment of sediment concentration through RBNN and SVM-FFA in arid watershed, India.” Smart Innovation, Systems and Technologies, 159. doi:10.1007/978-981-13-9282-5_67.

Samantaray, S. and A. Sahoo. 2020c. “Appraisal of runoff through BPNN, RNN, and RBFN in Tentulikhunti watershed: a case study.” Advances in Intelligent Systems and Computing, 1014. doi:10.1007/978-981-13-9920-6_26.

Samantaray, S. and A. Sahoo. 2020d. “Prediction of runoff using BPNN, FFBPNN, CFBPNN algorithm in arid watershed: a case study.” *International Journal of Knowledge-based and Intelligent Engineering Systems*, 24 (3), 243–251.

Samantaray, S., A. Sahoo, and D. K. Ghose. 2020a. “Prediction of sedimentation in an arid watershed using BPNN and ANFIS.” Lecture Notes in Networks and Systems, 93. doi:10.1007/978-981-15-0630-7_29.

Samantaray, S., A. Sahoo, and D. K. Ghose. 2020b. “Infiltration loss affects toward groundwater fluctuation through CANFIS in arid watershed: a case study.” Smart Innovation, Systems and Technologies, 159. doi:10.1007/978-981-13-9282-5_76.

Samantaray, S., O. Tripathy, A. Sahoo, and D. K. Ghose. 2020c. “Forecasting through ANN and SVM in Bolangir watershed, India.” Smart Innovation, Systems and Technologies, 159. doi:10.1007/978-981-13-9282-5_74.

Samantaray, Sandeep, Abinash Sahoo, and Dillip K. Ghose. 2020d. “Assessment of sediment load concentration using SVM, SVM-FFA and PSR-SVM-FFA in arid watershed, India: a case study.” *KSCE Journal of Civil Engineering*, 24, 1944–1957.

Scarselli, F. and Tsoi, A. C. 1998. "Universal approximation using feedforward neural networks: a survey of some existing methods, and some new results." *Neural Networks*, 11 (1), 15–37.

Schmid, B. H. and Koskiaho, J. (2006). "Artificial neural network modeling of dissolved oxygen in a wetland pond: the case of Hovi, Finland." *Journal of Hydrologic Engineering*, 11 (2), 188–192.

Shirmohammadi, B., Moradi, H., Moosavi, V., Semiromi, M. T. and Zeinali, A. 2013. "Forecasting of meteorological drought using Wavelet-ANFIS hybrid model for different time steps (case study: southeastern part of east Azerbaijan province, Iran)." *Natural Hazards*, 69 (1), 389–402.

Sridharam, S., Sahoo, A., Sandeep Samantaray, and Ghose, Dillip K. 2020a. "Estimation of water table depth using wavelet-ANFIS: a case study." In Communication Software and Networks. 747–754, Springer, Singapore.

Sridharam, S., Sahoo, A., Samantaray, S. and Ghose, D. K. 2020b. "Assessment of flow discharge in a river basin through CFBPNN, LRNN and CANFIS." In Communication Software and Networks, 765–773, Springer, Singapore.

Sudheer, K. P., Gosain, A. K., Mohana Rangan, D. and Saheb, S. M. 2002. "Modelling evaporation using an artificial neural network algorithm." *Hydrological Processes*, 16 (16), 3189–3202.

Sun, C. and Ma, Y. (2015). "Effects of non-linear temperature and precipitation trends on Loess Plateau droughts." Quaternary International, 372, 175–179.

Thom, H.C.S. 1958. "A note on the gamma distribution." *Mon Weather Rev*, 86(4), 117–122.

Trajkovic, S., Todorovic, B. and Stankovic, M. 2003. "Forecasting of reference evapotranspiration by artificial neural networks." *Journal of Irrigation and Drainage Engineering*, 129 (6), 454–457.

Wambua, R. M., Mutua, B. M., and Raude, J. M. (2016). "Prediction of missing hydro-meteorological data series using artificial neural networks (ANN) for Upper Tana River Basin." *American Journal of Water Resources*, 4 (2), 35–43.

Wilhite, D. A. 2009. "The role of monitoring as a component of preparedness planning: delivery of information and decision support tools." Coping with Drought Risk in Agriculture and Water Supply Systems: Drought Management and Policy Development in the Mediterranean. Springer Publishers, Dordrecht, The Netherlands.

Wilhite, D. A. and Buchanan-Smith, M. 2005. "Drought as hazard: understanding the natural and social context." Drought and Water Crises: Science, Technology, and Management Issues, CRC Press, 3–29.

Yoon, H., Jun, S.C., Hyun, Y., Bae, G. O. and Lee, K. K. 2011. "A comparative study of artificial neural networks and support vector machines for predicting groundwater levels in a coastal aquifer." *Journal of Hydrology*, 396 (1–2), 128–138.

Zahraie, B. and Nasseri, M. 2011. Basin scale meteorological drought forecasting using support vector machine (SVM). In International Conference on Drought Management Strategies in Arid and Semi-arid Regions, Muscat, Oman, December.

Zhang, Q. J., Gupta, K. C. and Devabhaktuni, V. K. 2003. "Artificial neural networks for RF and microwave design-from theory to practice." *IEEE Transactions on Microwave Theory and Techniques*, 51 (4), 1339–1350.

Zhang, T. and Lin, X. 2016. "Assessing future drought impacts on yields based on historical irrigation reaction to drought for four major crops in Kansas." *Science of the Total Environment*, 550, 851–860.

Zhang, Y., Yang, H., Cui, H. and Chen, Q. 2019. "Comparison of the ability of ARIMA, WNN and SVM models for drought forecasting in the Sanjiang Plain, China." *Natural Resources Research*, 29, 1447–1464.

Zhang, Y., Yang, H., Cui, H. and Chen, Q. 2020. "Comparison of the ability of ARIMA, WNN and SVM models for drought forecasting in the Sanjiang Plain, China." *Natural Resources Research*, 29, 1447–1464.

Zadeh, M. R., Amin, S., Khalili, D. and Singh, V. P. 2010. "Daily outflow prediction by multi-layer perceptron with logistic sigmoid and tangent sigmoid activation functions." *Water Resources Management*, 24 (11), 2673–2688.

Index

I

K

L

M

P

R

S

T